Lecture Notes in Physics

For information about Vols. 1–151, please contact your bookseller or Springer-Verlag.

Vol. 152: Physics of Narrow Gap Semiconductors. Proceedings, 1981. Edited by E. Gornik, H. Heinrich and L. Palmetshofer. XIII, 485 pages. 1982.

Vol. 153: Mathematical Problems in Theoretical Physics. Proceedings, 1981. Edited by R. Schrader, R. Seiler, and D.A. Uhlenbrock. XII, 429 pages. 1982.

Vol. 154: Macroscopic Properties of Disordered Media. Proceedings, 1981. Edited by R. Burridge, S. Childress, and G. Papanicolaou. VII, 307 pages. 1982.

Vol. 155: Quantum Optics. Proceedings, 1981. Edited by C.A. Engelbrecht. VIII, 329 pages. 1982.

Vol. 156: Resonances in Heavy Ion Reactions. Proceedings, 1981. Edited by K.A. Eberhard. XII, 448 pages. 1982.

Vol. 157: P. Niyogi, Integral Equation Method in Transonic Flow. XI, 189 pages. 1982.

Vol. 158: Dynamics of Nuclear Fission and Related Collective Phenomena. Proceedings, 1981. Edited by P. David, T. Mayer-Kuckuk, and A. van der Woude. X, 462 pages. 1982.

Vol. 159: E. Seiler, Gauge Theories as a Problem of Constructive Quantum Field Theory and Statistical Mechanics. V, 192 pages. 1982.

Vol. 160: Unified Theories of Elementary Particles. Critical Assessment and Prospects. Proceedings, 1981. Edited by P. Breitenlohner and H.P. Dürr. VI, 217 pages. 1982.

Vol. 161: Interacting Bosons in Nuclei. Proceedings, 1981. Edited by J.S. Dehesa, J.M.G. Gomez, and J. Ros. V, 209 pages. 1982.

Vol. 162: Relativistic Action at a Distance: Classical and Quantum Aspects. Proceedings, 1981. Edited by J. Llosa. X, 263 pages. 1982.

Vol. 163: J. S. Darrozes, C. Francois, Mécanique des Fluides Incompressibles. XIX, 459 pages. 1982.

Vol. 164: Stability of Thermodynamic Systems. Proceedings, 1981. Edited by J. Casas-Vázquez and G. Lebon. VII, 321 pages. 1982.

Vol. 165: N. Mukunda, H. van Dam, L.C. Biedenharn, Relativistic Models of Extended Hadrons Obeying a Mass-Spin Trajectory Constraint. Edited by A. Böhm and J.D. Dollard. VI, 163 pages. 1982.

Vol. 166: Computer Simulation of Solids. Edited by C.R.A. Catlow and W.C. Mackrodt. XII, 320 pages. 1982.

Vol. 167: G. Fieck, Symmetry of Polycentric Systems. VI, 137 pages, 1982.

Vol. 168: Heavy-Ion Collisions. Proceedings, 1982. Edited by G. Madurga and M. Lozano. VI, 429 pages. 1982.

Vol. 169: K. Sundermeyer, Constrained Dynamics. IV, 318 pages. 1982.

Vol. 170: Eighth International Conference on Numerical Methods in Fluid Dynamics. Proceedings, 1982. Edited by E. Krause. X, 569 pages. 1982.

Vol. 171: Time-Dependent Hartree-Fock and Beyond. Proceedings, 1982. Edited by K. Goeke and P.-G. Reinhard. VIII, 426 pages. 1982.

Vol. 172: Ionic Liquids, Molten Salts and Polyelectrolytes. Proceedings, 1982. Edited by K.-H. Bennemann, F. Brouers, and D. Quitmann. VII, 253 pages. 1982.

Vol. 173: Stochastic Processes in Quantum Theory and Statistical Physics. Proceedings, 1981. Edited by S. Albeverio, Ph. Combe, and M. Sirugue-Collin. VIII, 337 pages. 1982.

Vol. 174: A. Kadić, D.G.B. Edelen, A Gauge Theory of Dislocations and Disclinations. VII, 290 pages. 1983.

Vol. 175: Defect Complexes in Semiconductor Structures. Proceedings, 1982. Edited by J. Giber, F. Beleznay, J. C. Szép, and J. László. VI, 308 pages. 1983.

Vol. 176: Gauge Theory and Gravitation. Proceedings, 1982. Edited by K. Kikkawa, N. Nakanishi, and H. Nariai. X, 316 pages. 1983.

Vol. 177: Application of High Magnetic Fields in Semiconductor Physics. Proceedings, 1982. Edited by G. Landwehr. XII, 552 pages. 1983.

Vol. 178: Detectors in Heavy-Ion Reactions. Proceedings, 1982. Edited by W. von Oertzen. VIII, 258 pages. 1983.

Vol. 179: Dynamical Systems and Chaos. Proceedings, 1982. Edited by L. Garrido. XIV, 298 pages. 1983.

Vol. 180: Group Theoretical Methods in Physics. Proceedings, 1982. Edited by M. Serdaroğlu and E. İnönü. XI, 569 pages. 1983.

Vol. 181: Gauge Theories of the Eighties. Proceedings, 1982. Edited by R. Raitio and J. Lindfors. V, 644 pages. 1983.

Vol. 182: Laser Physics. Proceedings, 1983. Edited by J. D. Harvey and D. F. Walls. V, 263 pages. 1983.

Vol. 183: J. D. Gunton, M. Droz, Introduction to the Theory of Metastable and Unstable States. VI, 140 pages. 1983.

Vol. 184: Stochastic Processes – Formalism and Applications. Proceedings, 1982. Edited by G. S. Agarwal and S. Dattagupta. VI, 324 pages. 1983.

Vol. 185: H. N. Shirer, R. Wells, Mathematical Structure of the Singularities at the Transitions between Steady States in Hydrodynamic Systems. XI, 276 pages. 1983.

Vol. 186: Critical Phenomena. Proceedings, 1982. Edited by F.J.W. Hahne. VII, 353 pages. 1983.

Vol. 187: Density Functional Theory. Edited by J. Keller and J.L. Gázquez. V, 301 pages. 1983.

Vol. 188: A. P. Balachandran, G. Marmo, B.-S. Skagerstam, A. Stern, Gauge Symmetries and Fibre Bundles. IV, 140 pages. 1983.

Vol. 189: Nonlinear Phenomena. Proceedings, 1982. Edited by K. B. Wolf. XII, 453 pages. 1983.

Vol. 190: K. Kraus, States, Effects, and Operations. Edited by A. Böhm, J. W. Dollard and W. H. Wootters. IX, 151 pages. 1983.

Vol. 191: Photon Photon Collisions. Proceedings, 1983. Edited by Ch. Berger. V, 417 pages. 1983.

Vol. 192: Heidelberg Colloquium on Spin Glasses. Proceedings, 1983. Edited by J. L. van Hemmen and I. Morgenstern. VII, 356 pages. 1983.

Vol. 193: Cool Stars, Stellar Systems, and the Sun. Proceedings, 1983. Edited by S. L. Balliunas and L. Hartmann. VII, 364 pages. 1984.

Vol. 194: P. Pascual, R. Tarrach, QCD: Renormalization for the Practitioner. V, 277 pages. 1984.

Lecture Notes in Physics

Edited by H. Araki, Kyoto, J. Ehlers, München, K. Hepp, Zürich
R. Kippenhahn, München, H. A. Weidenmüller, Heidelberg
and J. Zittartz, Köln
Managing Editor: W. Beiglböck, Heidelberg

229

Fundamentals of Laser Interactions

Proceedings of a Seminar
Held at Obergurgl, Austria
February 24 – March 2, 1985

Edited by F. Ehlotzky

Springer-Verlag
Berlin Heidelberg GmbH

Editor

Fritz Ehlotzky
Institute for Theoretical Physics, University of Innsbruck
A-6020 Innsbruck, Austria

ISBN 978-3-540-15640-6

CIP-Kurztitelaufnahme der Deutschen Bibliothek. Fundamentals of laser interactions : proceedings of a seminar held at the Bundessportheim in Obergurgl, Austria, February 24 – March 2, 1985 / ed. by F. Ehlotzky.
Springer, 1985.
(Lecture notes in physics: Vol. 229)
ISBN 978-3-540-15640-6 ISBN 978-3-540-39503-4 (eBook)
DOI 10.1007/978-3-540-39503-4
NE: Ehlotzky, Fritz [Hrsg.]: Bundessportheim <Obergurgl>; GT

Originally published by Springer-Verlag Berlin Heidelberg New York Tokyo in 1985

2153/3140-543210

FOREWORD

The Seminar on Fundamentals of Laser Interactions was the second Meeting on Laser Phenomena held at the Bundessportheim in Obergurgl. It was attended by 46 Physicists from Austria, Belgium, The Federal Republic of Germany, Finland, France, Hungary, Italy, Japan, The Netherlands, and the United States, who work actively in the rapidly developing field of laser interactions.

The Seminar presented an opportunity to discuss at leisure problems of mutual interest to theoreticians and experimentalists who are working on various aspects of the field of laser interactions. There was an attempt to bring together people who are doing research on multiphoton physics, on scattering phenomena, on many body problems, and on new methods of experimentation. In particular the following topics were chosen for discussion:

1) Multiphoton Spectroscopy
2) Electron Correlations in Multiphoton Transitions
3) Multiphoton Continuum Effects
4) Rydberg States in Strong Laser Fields
5) Laser Induced and Laser Assisted Scattering and Reactions
6) High Frequency Lasers
7) Laser Cooling and Trapping of Particles
8) Other Fundamental Interaction Processes

At the Seminar 18 Invited Lectures were given by:

G. Alber (JILA)
N. Andersen (Copenhagen)
W.E. Cooke (USC)
M. Crance (Orsay)
F.H.M. Faisal (Bielefeld)
G. Ferrante (Palermo)
M. Gavrila (Amsterdam)
J. Javanainen (Helsinki)
C.J. Joachain (Bruxelles)
H. Helm (Menlo Park)
H. Klar (Freiburg)
L.A. Lompré (Saclay)
C.K. Rhodes (Chicago)
F. Roussel (Saclay)
P.E. Toschek (Hamburg)
C.R. Vidal (MPI Garching)
H. Walther (MPI Garching)
K.H. Welge (Bielefeld)

In addition, there were 13 contributed papers presented at the meeting.

The following pages present the full text of the invited lectures and the abstracts of the contributed papers. The invited lecture of V.G. Minogin (Moscow) was not presented at the Seminar but has been accepted for publication in the Proceedings. The editor is grateful to the contributors for their collaboration in preparing their typescripts for rapid publication.

The active yet relaxed atmosphere of the Bundessportheim at Obergurgl, surrounded by the snow-capped peaks of the Ötztal Alps, supplied a congenial setting for a very stimulating and rewarding meeting. It is a pleasure to thank all participants for their interest and enthusiasm. The most valuable secretarial assistance of Miss G. Eder is gratefully acknowledged.

Innsbruck, April 1985

F. Ehlotzky

ACKNOWLEDGEMENTS

The Seminar on Fundamentals of Laser Interactions
has been supported by:

Bundesministerium für Wissenschaft und Forschung
Bundesministerium für Unterricht und Kunst
Amt der Tiroler Landesregierung
Magistrat der Stadt Innsbruck
Österreichische Forschungsgemeinschaft
Raiffeisen-Zentralkasse Tirol

CONTENTS

INVITED LECTURES

C O N T R I B U T E D P A P E R S
(ABSTRACTS)

I N V I T E D   L E C T U R E S

PART I: Collisions in Laser Fields

ELECTRON-ATOM INTERACTIONS IN INTENSE, HIGH FREQUENCY LASER FIELDS

M. Gavrila

FOM-Institute for Atomic and Molecular Physics,
Kruislaan 407, 1098 SJ Amsterdam, The Netherlands

I. INTRODUCTION

Substantial effort has been invested in the development of very intense lasers, yielding about 10^{16} W/cm^2, and operated in a range of frequencies extending from the IR to the VUV. At these high intensities atomic transitions abundantly involve multiphoton absorption and emission (for a review of these processes see ref. 1). The description by perturbation theory is no longer valid, and new methods of solution of the Schrödinger equation are needed. A nonperturbative theory was developed earlier by Kroll and Watson for the *low-frequency regime*[2,1], well suited for the range of the intense IR lasers. We have recently developed a nonperturbative approach to deal with the opposite case, of the *high-frequency regime*[3,4,1]. It applies to the intense excimer lasers already in operation in the VUV (e.g. see refs. 5,6), but extends beyond, into the XUV range.

In the following we shall present our theory for the high-frequency regime. We shall mainly deal with the case of electron-atom (ion) collisions in the radiation field, also termed *free-free transitions*. We will first describe the formalism (Sec.II), and then apply it to the case of a purely Coulomb potential (Sec.III). Further, in Sec.IV we will outline the extension of the method to encompass *atomic structure and multiphoton ionization*. Finally, in Sec.V we draw some conclusions.

II. FREE-FREE TRANSITIONS FORMALISM

A fully realistic description of the target atom is quite difficult. We shall represent it here by a potential model. (Very recently, however, we have extended the theory to take into account also the internal degrees of freedom). The potential will be taken to be of the central self-consistent type: Coulomb-like at the origin ($V(r) \simeq -Z/r$), short range or ionic ($V(r) \simeq -Z'/r$) at large distances, but unspecified otherwise.

The laser field will be represented by a monochromatic infinite plane wave, linearly polarized, in the dipole approximation. The plane-wave assumption is not critical, as the extension to a single-mode laser pulse of adiabatically varying intensity can subsequently be made[7]. Linear polarization is assumed in view of simplifying the algebra, and the dipole approximation is justified in the frequency range we are interested in (from the visible to the extreme ultraviolet). Consequently, we take the electrodynamic potentials of the wave in the form $\vec{A} = \vec{a}\cos\omega t$ (with $\vec{a}$ real) and $\phi = 0$. (Note that our premises are the same as those of Kroll and Watson[2].)

Application of the space translation transformation (Kramers[8], Henneberger[9]) to the Schrödinger equation gives

$$[\tfrac{1}{2}\vec{P}^2 + V(\vec{r} + \vec{\alpha}(t))]\psi = i(\partial\psi/\partial t) , \tag{1}$$

where

$$\vec{\alpha}(t) = -c^{-1} \int_0^t \vec{A}(t')dt' = \vec{\alpha}_0 \sin\omega t ,$$

$$\alpha_0 = -\alpha_0\vec{e}, \qquad \alpha_0 = a/\omega c = I^{\frac{1}{2}}\,\omega^{-2} , \tag{2}$$

and $\vec{e}$ and I are the real polarization vector and (time averaged) intensity of the plane wave. All our formulas are written in atomic units; the a.u. of (time averaged) intensity is $I_0 = 3{,}51 \times 10^{16}$ W/cm^2.

Eq.(1) should be solved by imposing the boundary conditions of our problem: an incoming current of particles of energy $E = p^2/2$, and radially outgoing currents of scattered particles of energies and momenta

$$E_n = E + n\omega, \quad E_n = p_n^2/2, \quad n = 0,\pm1,\pm2,\dots. \tag{3}$$

Equation (1) has periodic time-dependent coefficients. As usual, we seek a quasiperiodic solution of the form

$$\psi(\vec{r},t) = e^{-iEt}\,\Sigma_{n=-\infty}^{+\infty}\,\psi_n(\vec{r})e^{-in\omega t} . \tag{4}$$

Then, we Fourier analyze the potential:

$$V(\vec{r} + \vec{\alpha}(t)) = \Sigma_{n=-\infty}^{+\infty}\,V_n(\vec{\alpha}_0;\vec{r})e^{-in\omega t} . \tag{5}$$

By some algebraic manipulations the coefficients can be written as

$$V_n(\vec{\alpha}_0;\vec{r}) = (i^n/\pi) \int_{-1}^{+1} V(\vec{r} + \vec{\alpha}_0 u)T_n(u)(1 - u^2)^{-1/2}\,du , \tag{6}$$

where $T_n(u)$ are Chebyshev polynomials.

Insertion of Eqs.(4) and (5) into Eq.(1) leads to a system of coupled differential equations for the components $\psi_n(\vec{r})$, which we write

$$[\tfrac{1}{2}\vec{P}^2 + V_0 - (E+n\omega)]\psi_n = -\sum_{\substack{m=-\infty \\ (m\neq n)}}^{+\infty} V_{n-m}\psi_m . \tag{7}$$

The boundary conditions require that our solutions $\psi_n(\alpha_0,\omega;\vec{r})$ behave asymptotically as follows:

$$\psi_0(\vec{\alpha}_0,\omega;\vec{r}) \to \exp\left\{i[\vec{p}\vec{r} + \gamma_0 \ln(pr-\vec{p}\vec{r})]\right\} + \frac{1}{r} f_0(\vec{\alpha}_0,\omega;\hat{r})\exp[i(pr - \gamma_0 \ln 2pr)] , \quad (8)$$

$$\psi_n(\vec{\alpha}_0,\omega;\vec{r}) \to \frac{1}{r} f_n(\vec{\alpha}_0,\omega;\hat{r})\exp[i(p_n r - \gamma_n \ln 2p_n r)] \quad (n \neq 0), \quad (9)$$

with $\gamma_n = -Z'/p_n$ (for a short-range potential $Z' = 0$). Equation (8) contains the elastic scattering amplitude $f_0(\vec{\alpha}_0,\omega;\hat{r})$, and Eq.(9) that for absorption/emission $f_n(\vec{\alpha}_0,\omega;\hat{r})$. The associated scattering cross sections are

$$d\sigma_n/d\Omega = (p_n/p)|f_n(\vec{\alpha}_0,\omega;\hat{r})|^2 \quad (n = 0,\pm1,\pm2,\dots) . \quad (10)$$

For a single-mode laser pulse of adiabatically varying intensity, Eq.(10) should be time-averaged appropriately[7].

We shall now describe a method for handling the system Eq.(7). The left-hand side contains the Hamiltonian

$$H = \tfrac{1}{2}P^2 + V_0(\vec{\alpha}_0;\vec{r}) . \quad (11)$$

By use of the Green's operator $G(\Omega)$ associated to it, where Ω is the energy parameter, Eq.(7) may be formally solved as

$$\psi_n = \psi^{(+)}_{\vec{p}}\delta_{no} - G^{(+)}(E_n) \sum_{\substack{m \\ (m\neq n)}} V_{n-m}\psi_m . \quad (12)$$

Here $\psi^{(+)}_{\vec{p}}$ is the ($\vec{\alpha}_0$-dependent) solution of the equation

$$H\psi_{\vec{p}} = E\psi_{\vec{p}} , \quad (13)$$

satisfying the boundary condition Eq.(8) with an amplitude $f^{(0)}_0(\vec{\alpha}_0;\vec{r})$. It then follows from Eq.(12) that the ψ_n satisfy the boundary condition required by Eqs.(8) and (9) with the following expression for the scattering amplitudes:

$$f_n(\vec{\alpha}_0,\omega;\hat{r}) = f^{(0)}_0(\vec{\alpha}_0;\hat{r})\delta_{no} - (1-\delta_{no})\frac{1}{2\pi}\langle\psi^{(-)}_{\vec{p}_n}|V_n|\psi^{(+)}_{\vec{p}}\rangle +$$

$$+ \frac{1}{2\pi}\sum_{\substack{m \\ (m\neq n)}}\sum_{\substack{m' \\ (m'\neq m)}} \langle\psi^{(-)}_{\vec{p}_n}|V_{n-m}G^{+}(E_m)V_{m-m'}|\psi_{m'}\rangle . \quad (14)$$

Besides $\psi^{(+)}_{\vec{p}}$, Eq.(14) also contains $\psi^{(-)}_{\vec{p}_n}$, which is an incoming-wave solution of Eq.(13), as well as the unknown set of components $\psi_{m'}(r)$ satisfying Eq.(12).

By repeated insertion of Eq.(12) into Eq.(14) an expansion can be derived for f_n. Obviously, the iteration will have practical significance only if the successive terms decrease sufficiently rapidly. Since this will not be true in general, it is important to establish the conditions under which the first nonvanishing term of Eq.(14) represents a good approximation. For (a) $\omega \gg |E_0(\alpha_0)|$, where $E_0(\alpha_0)$ is the ground-state energy of the modified Hamiltonian Eq.(11) (Note that from Eq.(6) it follows that by *increasing* α_0, the potential V_0 becomes shallower, and therefore $|E_0(\alpha_0)|$ *decreases* from its unperturbed value at $\alpha_0 = 0$: $|E(\alpha_0)| < |E(o)|$.); (b) $\alpha_0^2\omega \gg 1$; (c) $\omega \gg E$, it was possible to extract the *exact* form of the dominant contribution to the last term of Eq.(14) (denoted below by $f^{(1)}$) for an *arbitrary* potential of the type discussed before. In the case of elastic scattering we find

$$\begin{pmatrix} \mathrm{Re}\, f_0^{(1)}(\hat{r}) \\ \mathrm{Im}\, f_0^{(1)}(\hat{r}) \end{pmatrix} =$$

$$= \frac{Z^2}{6\alpha_0\omega^2}\left[\psi_{\vec{p}_0}^{(-)*}(\vec{\alpha}_0)\psi_{\vec{p}}^{(+)}(\vec{\alpha}_0) + \psi_{\vec{p}_0}^{(-)*}(-\vec{\alpha}_0)\psi_{\vec{p}}^{(+)}(-\vec{\alpha}_0)\right]\begin{pmatrix} -(\ln\alpha_0^2\omega)^2 + O(\ln\alpha_0^2\omega) \\ \pi\ln\alpha_0^2\omega + O((\alpha_0^2\omega)^0) \end{pmatrix} \tag{15}$$

where $\vec{p}_0 = p\hat{r}$ is the *final* momentum (see Eq.(3) for $n = 0$), and the corrective terms O also depend on α_0, E, θ. Thus at fixed α_0 (this constraint appears also in the derivation of the Kroll and Watson result[2]), and sufficiently high ω (obeying conditions (a), (b), and (c) and the dipole-approximation assumption) it is possible to satisfy the inequality (d): $|f_0^{(1)}(\alpha_0,\omega;E,\theta)| \ll |f_0^{(o)}(\alpha_0;E,\theta)|$. Whereas this holds in general over wide ranges of parameter values, it should nevertheless be checked for each case separately because $f_0^{(o)}$ may become exceptionally small for certain angles. Conditions (a)-(d) together ensure the dominance of the first term in Eq.(14). However, this may hold under wider conditions than we were able to prove.

Thus, the *elastic amplitude* f_0 reduces, to lowest order (in the sense discussed above), to $f_0^{(o)}$, which is that calculated from the *time-independent* Schrödinger equation Eq.(13). This shows that in the high-frequency, high-intensity regime the incoming electron feels only the static distorted potential $V_0(\vec{\alpha}_0;\vec{r})$, the "dressed" potential associated to $V(r)$:

$$V_0(\vec{\alpha}_0,\vec{r}) = \frac{1}{\pi}\int_{-1}^{+1} V(\vec{r}+\vec{\alpha}_0 u)\,\frac{du}{\sqrt{1-u^2}}\ . \tag{16}$$

The dressed potential, Eq.(16), can be looked upon as created by a linear distribution of "charges" extending from $-\alpha_0$ to $+\alpha_0$ along $\vec{e}$, with density

$$\sigma(\xi) = \frac{1}{\pi\alpha_0}\left[1 - \left(\frac{\xi}{\alpha_0}\right)^2\right]^{-\frac{1}{2}}\ , \tag{17}$$

the unit of "charge" generating the potential $V(r)$. This behavior appears natural due to the rapid oscillations of the center of force in Eq.(1). In the regime we are considering, ω and I enter the scattering problem only through α_0.

For the *absorption amplitude* ($n > o$) we get to lowest order from Eq.(14)

$$f_n(\vec{\alpha}_0,\omega;\hat{r}) = -\frac{1}{2\pi}\, \langle\psi^{(-)}_{\vec{p}_n}|V_n|\psi^{(+)}_{\vec{p}}\rangle \,. \tag{18}$$

Thus, the Fourier component V_n acts in our regime as a transition operator between scattering states $\psi^{(\pm)}_{\vec{q}}$ of the dressed potential V_0. The amplitude f_n depends on ω via the final momentum $\vec{p}_n$ (see Eq.(3)). Because ω was assumed to be large at given α_0, all f_n will be small with respect to f_0. (This contrasts with the low-frequency case where many f_n may be larger than f_0.) Note that the condition (c) above precludes free-free emission ($n<0$).

III. ELASTIC SCATTERING FROM THE DRESSED COULOMB POTENTIAL

Since the original potential $V(r)$ is spherically symmetric, $V_0(\vec{\alpha}_0;\vec{r})$ has axial symmetry around $\vec{\alpha}_0$ (we have assumed linear polarization). In the case of the pure Coulomb potential $V(r) = -Z/r$, Eq.(16) yields

$$V_0(\vec{\alpha}_0;\vec{r}) = -(2Z/\pi)(r_+r_-)^{-1/2}\; K(2^{-1/2}(1-\hat{r}_+r_-)^{1/2}) \,, \tag{19}$$

where $\vec{r}_\pm = \vec{r} \pm \vec{\alpha}_0$, and K is a complete elliptic integral of the first kind. V_0 has a logarithmic singularity along the distribution of charges, and $r_\pm^{-1/2}$ singularities at its end points (all weaker than the original Coulomb singularity). The dressed Coulomb potential Eq.(19) is represented in Fig. 1.

The axial symmetry of V_0 complicates the elastic scattering problem, as the azimuthal quantum number ℓ is no longer conserved. Computations taking this into account have been performed by Van de Ree, Kaminski and Gavrila (to be published). In this section we shall describe a simplified approach[4] in which $V_0(\vec{\alpha}_0,\vec{r})$ is modelled by its spherical average $\bar{V}_0(\alpha_0,r)$. The problem is thus reduced to a phase shift calculation. This should give the right order of magnitude for the elastic scattering, particularly at low energies, when the electron wavelength is larger than the extension of the line of charges, i.e. $p\alpha_0 \lesssim 1$ (p is the electron momentum). On the other hand, interesting features related to the dependence of the cross section on the polarization vector $\vec{e}$ (or $\vec{\alpha}_0$) will thus be lost.

By averaging Eq.(19) over all directions of $\vec{\alpha}_0$, one finds:

$$\begin{aligned}\bar{V}_0(\alpha_0,r) &= \frac{-Z}{\pi\alpha_0\rho}\left[2\arcsin\rho - \rho\ln\frac{1-(1-\rho^2)^{\frac{1}{2}}}{1+(1-\rho^2)^{\frac{1}{2}}}\right] , \quad \rho \leq 1 \\ &= -\frac{Z}{\alpha_0\rho} , \quad \rho \geq 1 ,\end{aligned} \tag{20}$$

where $\rho = r/\alpha_0$. For $r \geq \alpha_0$, $\bar{V}_0$ coincides with the original Coulomb potential. For $r < \alpha_0$, $\bar{V}_0$ can be expanded as

$$\bar{V}_0(\alpha_0,r) = -\frac{2Z}{\pi\alpha_0}\left(\ln\rho + \sum_{n=0}^{\infty} A_n\rho^{2n}\right) , \tag{21}$$

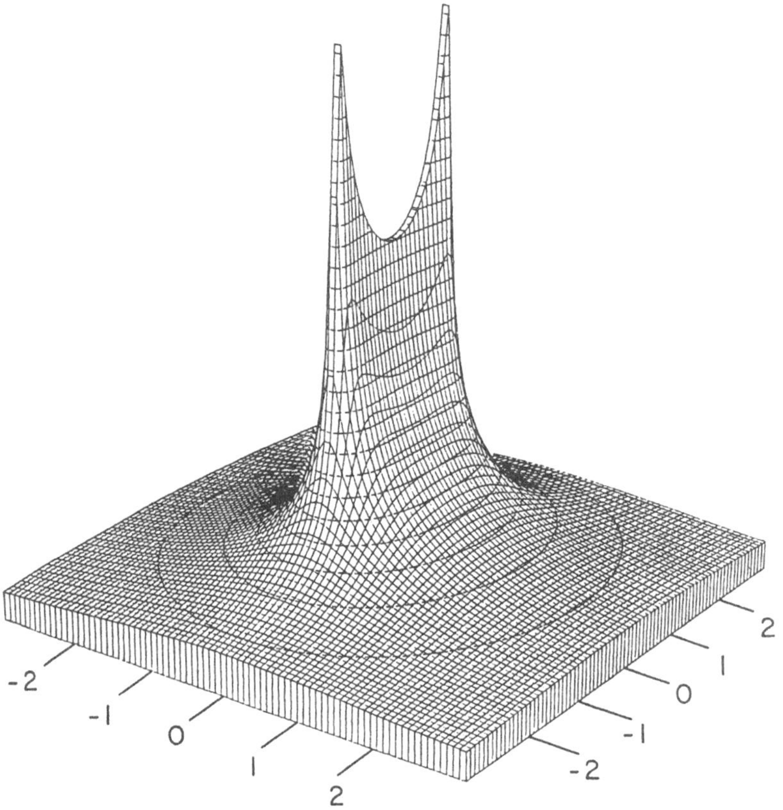

Fig. 1. Values of the dressed Coulomb potential V_o in a plane passing through the symmetry axis defined by $\vec{\alpha}_o$. In this plane, taken as the horizontal base of the figure, distances are measured in units of α_o. Along the vertical we represent $(-\alpha_o V_o/Z)$ in atomic units, up to the value 10. The saddle of the figure reflects the rise in V_o near the line of singularities (extending from $-\vec{\alpha}_o$ to $\vec{\alpha}_o$) and their increasing strength towards the end points. At radial distances from the origin larger than α_o, the distortion of the Coulomb potential fades away (the level lines become circular).

which displays a *logarithmic singularity* at the origin. This is much weaker than that of the Coulomb potential.

As is easily seen, $\bar{V}_0(\alpha_0,r)$ is the potential energy due to a spherically symmetric distribution of electric charges extending up to $r=\alpha_0$, of radial density:

$$\tau(r) = 2Z\,\sigma(r)\ , \tag{22}$$

with the function σ defined by Eq.(17). Our scattering problem resembles thus that of an electron probing an extended charge nucleus[10] of density $\tau(r)$ and radius α_0; in contrast to this, however, we are dealing with a nonrelativistic case.

The determination of the continuum solutions of the potential $\bar{V}_0$ of Eqs.(20), (21) raises some interesting mathematical points on account of its logarithmic singularity at the origin. A number of results have been obtained lately for the bound state problem of the purely logarithmic case (all $A_n = 0$ in Eq.(21)), e.g. see Gesztesy and Pittner[11], and Müller-Kirsten and Bose[12]. By expanding the ℓ-th reduced radial partial wave as follows

$$P_\ell(\rho) = \rho^{\ell+1} \sum_{m=0}^{\infty} (\rho^2 \ln \rho)^m \sum_{n=0}^{\infty} a_{mn}^{(\ell)} \rho^{2n} , \tag{23}$$

we have uniquely determined the coefficients $a_{mn}^{(\ell)}$ by recurrence relations. We have checked numerically the convergence of the series for small ρ. On account of its behaviour for $\rho \to 0$, it represents the physically acceptable solution. It can be used to start the numerical solution in the vicinity of the origin. Beyond $\rho = 1$ the solution goes over into a linear combination of the regular and irregular Coulomb functions, and the extra phase shifts δ_ℓ, due to the deviation of $\bar{V}_0$ from a Coulomb potential, can be determined.

For an attractive potential ($Z > 0$) δ_ℓ is negative, and for given ℓ and E at small values α_0, its dominant behaviour reads:

$$\delta_\ell = \gamma \frac{2^{2\ell} e^{-\pi\gamma} |\Gamma(\ell+1+i\gamma)|^2}{[(2\ell+1)!]^2} \cdot \frac{\Gamma(\ell+\frac{3}{2})}{(\ell+1)(2\ell+3)\sqrt{\pi}} (p\alpha_0)^{2\ell+2} \cdot (1+\mathcal{O}(\alpha_0)) , \tag{24}$$

in which $\gamma = -Z/p$. For values $\alpha_0 \leq 10^{-2}$, Eq.(24) agrees quite well with the numerical computation.

For large values of α_0, on the other hand, we find analytically (at given ℓ and E) the behaviour:

$$\delta_\ell = \gamma \ln \alpha_0 + \chi_\ell(p) + \mathcal{O}(1/\alpha_0) , \tag{25}$$

where $\chi_\ell(p)$ is independent of α_0. Eq.(25) can be derived from an extension of the usual JWKB phase shift expression to the case of a modified Coulomb potential like that of Eq.(20). It is limited by the condition: $p\alpha_0 >> \ell + \frac{1}{2}$. Note that, although p is fixed and can be small in our case, the JWKB approximation still applies, because at large α_0 the magnitude of the potential $\bar{V}_0$ (and its derivatives) becomes small, while its range increases as α_0. The logarithmic build-up of δ_ℓ with α_0 displayed by Eq.(25), is peculiar to the Coulomb long-range behaviour of $\bar{V}_0$. For $\alpha \geq 10^2$, this was quite well checked numerically.

Because of the long range behaviour of the potential, Eq.(20), the scattering amplitude is given by the theory for *modified Coulomb scattering*:

$$f(\theta) = f_c(\theta) + f'(\theta) , \tag{26}$$

$$f_c(\theta) = \frac{(-\gamma)}{2p \sin^2 \frac{\theta}{2}} \exp(-i\gamma \ln \sin^2 \frac{\theta}{2} + 2i\sigma_0) , \tag{27}$$

$$f'(\theta) = \frac{1}{2ip} \sum_{\ell} (2\ell+1) e^{2i\sigma_\ell} (e^{2i\delta_\ell} - 1) P_\ell(\cos\theta) . \tag{28}$$

Here θ is the scattering angle, f_c is the Coulomb amplitude, f' is the extra contribution due to $\bar{V}_0 - V$, and σ_ℓ are the Coulomb phases. The modified cross section is

$$\frac{d\sigma}{d\Omega} = \frac{d\sigma_c}{d\Omega} + 2\,\mathrm{Re}\, f_c^* f' + |f'(\theta)|^2 , \tag{29}$$

where the first term at the right is the Rutherford cross section and the second represents the interference of the Coulomb and short-range amplitudes. Eqs.(26)-(29) are similar to those used in the analysis of the classical proton-proton collision experiments at low (nuclear) energies with the short range nuclear force taken into account; see Ref. 13, chapter 10, §§ 5 and 9 (exchange effects are absent here).

For small α_0, Eq.(24) shows that only the $\ell = 0$ phase should be considered; to lowest order, $\delta_0 = O(\alpha_0^2)$. The cross section, Eq.(29), can then be cast into the form used for proton-proton scattering (with exchange neglected), see Ref. 13, chapter 10, equation (5.1).

In Fig. 2 we present the angular dependence of the ratio $R = (d\sigma/d\Omega)/(d\sigma_c/d\Omega)$ of the modified cross section, Eq.(29), to the Rutherford cross section at $Z = 1$ for a number of electron energies E and values of α_0. The energies E chosen satisfy the condition of validity (b) of our theory given in Sec.II, for ω attainable with some existing high-frequency lasers (see Luk *et al.*[5], Rhodes[6]). The values considered for α_0 have also been achieved experimentally. (When the laser of Luk *et al.*[5] is operated at $I = 10^{15}$ W/cm^2, we have $\alpha_0 = 3.1$ and at $I = 10^{16}$ W/cm^2, $\alpha_0 = 9.8$).

In all cases $R \to 1$ as $\theta \to 0$. This is due to the fact that $f_c(\theta)$ and $d\sigma_c/d\Omega$ become infinite as $\theta \to 0$, whereas $f'(\theta)$ stays finite. Moreover, for $\theta \to 0$, R has infinitely many oscillations of increasing frequency and decreasing amplitude. These are due to the presence of the Coulomb phase factor, see Eq.(27), in the interference term of Eq.(29) (but not in $d\sigma_c/d\Omega$). Indeed, by setting $\gamma = 0$ in the phase factor, the oscillations disappear and R either becomes a monotonically decreasing function of θ (for small α_0), or has a broad maximum and then decreases (for larger α_0).

The amplitude of these *Coulomb-interference oscillations* decreases for $\theta \to 0$ because the relative importance of the interference term in Eq.(29) diminishes compared to $d\sigma_c/d\Omega$. Note that, as seen in Fig. 2, the Coulomb-interference oscillations occur in an accessible range of experimental parameters. (They were not detected in the proton-proton collision experiments, since there the energy had to be high enough, $\gamma \simeq 0$, to overcome the Coulomb repulsion so that the particles could approach within the range of nuclear forces).

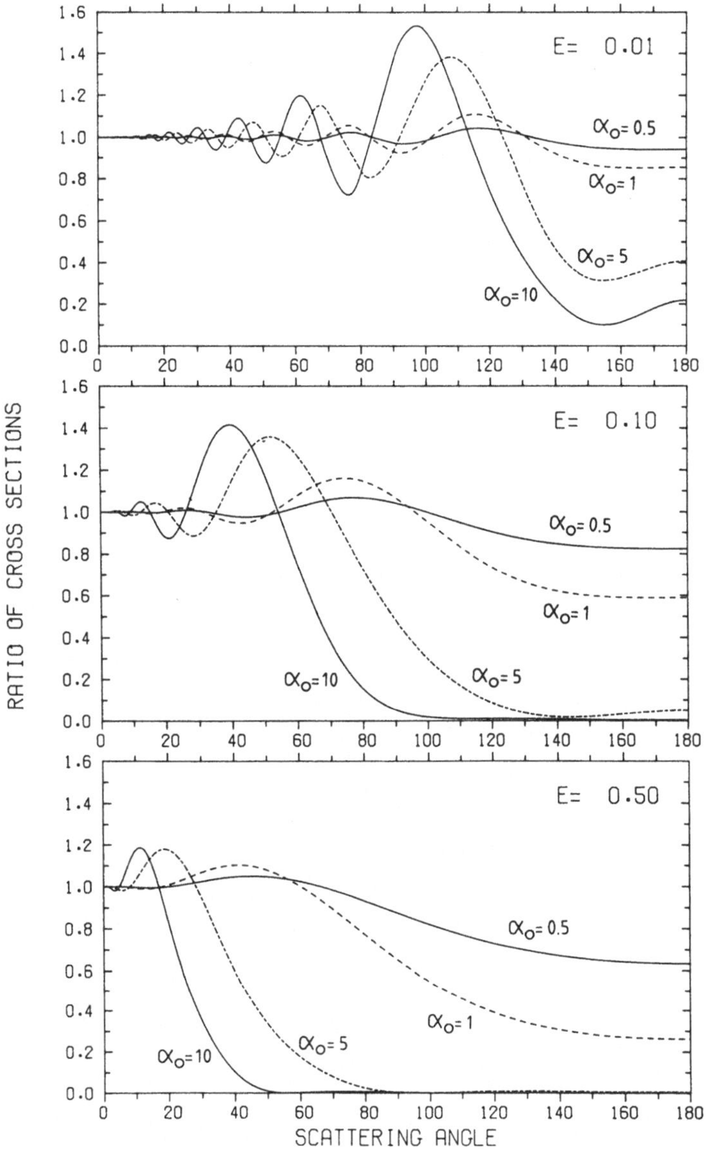

Fig. 2. Ratio R of the elastic scattering cross section for the averaged dressed Coulomb potential, Eq.(20), to the Rutherford cross section. Nuclear charge $Z = 1$. Values of the electron energy E (in Ry) and of α_o, as indicated (for the definition of α_o see Eq.(2)).

For large scattering angles Fig. 2 shows that the modified cross section is considerably reduced with respect to the original value, because of destructive interference in Eq.(29).

In Fig. 3 we display the effect on R of the increase of the nuclear charge Z, at given E and α_0. It is apparent that the Coulomb oscillations will extend to higher angles, and become quite regular. For $Z = -1$ (positron-proton scattering) and the parameters of Fig. 3, $R = 1$ for all angles (not drawn). This happens because in this case the positron wave function does not penetrate into the region ($r \leq \alpha_0$) where the potential is modified by the laser field, and the scattering remains purely Coulombic.

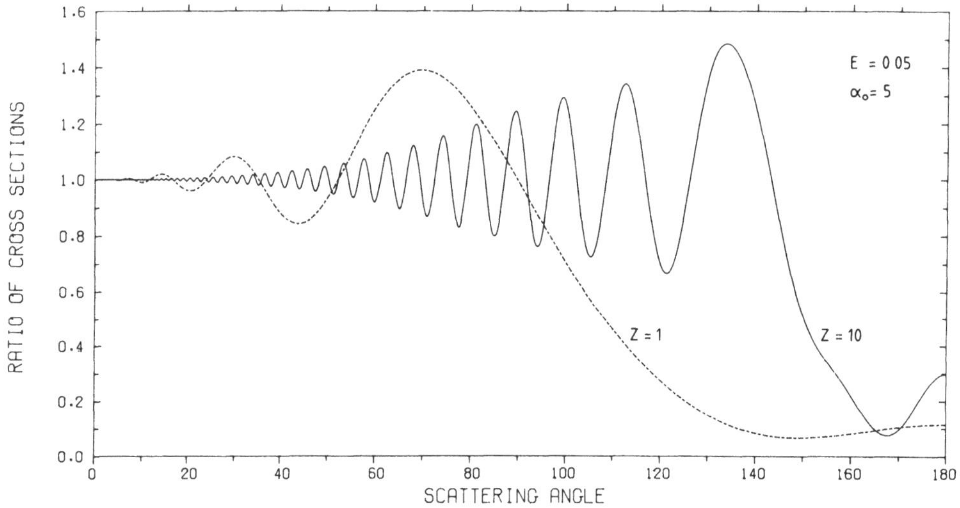

Fig. 3. *Ratio of cross sections R for Z = 1 and Z = 10, at E = 0.05 Ry and α_0 = 5.*

IV. ATOMIC STRUCTURE AND MULTIPHOTON IONIZATION

An atom driven by an intense laser field undergoes distortion, while decaying by multiphoton ionization. We are interested in analyzing these phenomena in the high-frequency regime.

We shall consider here the simplest case of a one-electron atom, with a binding potential of the general form described in Sec.II. The assumptions concerning the radiation field will be the same as there.

We start again from the space-translated Schrödinger equation Eq.(1), and seek a *quasienergy solution*, Eq.(4), where the quasienergy E remains to be determined (for the background of this type of approach see Refs. 14,15). As before one obtains the system of coupled equations Eq.(7). However, in the present case we want to solve it with decaying bound state, asymptotic conditions:

$$\begin{aligned} &\psi_n(\vec{\alpha}_0,\omega;\vec{r}) \to \frac{1}{r} f_n(\vec{\alpha}_0,\omega;\hat{r}) \exp[i(p_n r - \gamma_n \ln 2p_n r)] , \quad n > 0 , \\ &\psi_n(\vec{\alpha}_0,\omega;\vec{r}) \to 0 \text{ (exponentially)}, \quad n \leq 0 , \end{aligned} \tag{30}$$

where $f_n(\vec{\alpha}_0,\omega;\hat{r})$ is the n-photon ionization amplitude. For $n > 0$ we have the open channels, and for $n \leq 0$ the closed ones. The energies E_n and the corresponding momenta p_n are given by Eq.(3) in terms of the real part of E (denoted in Eq.(31) as E_r).

The existence of the open channels forces the quasienergy E to be complex. As usual, we write it in the form

$$E = E_r - i \frac{\Gamma}{2} , \tag{31}$$

where Γ is the total width of the state considered. The theory of decaying states shows that the quasienergies are poles of the S matrix (resonances)[14].

Using Eq.(30) one finds for the angular decay rate by n-photon ionization the expression:

$$\frac{d\Gamma^{(n)}}{d\Omega} = p_n |f_n(\vec{\alpha}_0,\omega;\hat{r})|^2 , \tag{32}$$

if the solution Eq.(4) is properly normalized. The sum of all (angle integrated) multiphoton decay rates is equal to the total Γ:

$$\sum_{n>0} \Gamma^{(n)} = \Gamma . \tag{33}$$

The system of coupled equations Eq.(7) can be solved similarly to Eq.(12):

$$\psi_n = \psi_0 \delta_{no} - (1-\delta_{no}) G^{(+)}(E_n) \sum_{\substack{m \\ (m \neq n)}} V_{n-m} \psi_m , \tag{34}$$

where, for $n=0$, the equation reduces to an identity. ψ_n satisfies the boundary conditions Eq.(30). By iteration of Eq.(34) ψ_n can be expressed as a series in terms of ψ_o and E. Upon inserting it into the system of coupled equations one finds:

$$[\tfrac{1}{2}\vec{P}^2 + V_o + \sum_{\substack{m \\ (m\neq o)}} V_{-m}\, G^{(+)}(E+m\omega)V_m + \cdots]\,\psi_o = E\psi_o\,. \tag{35}$$

This determines the complex quasienergy E together with ψ_o.

Assuming ω to be sufficiently large, Eq.(35) yields *to lowest order* in $1/\omega$:

$$H\psi_o = E\psi_o \tag{36}$$

where H is the (Hermitian) Hamiltonian defined by Eq.(11). Taking into account the closed channel boundary condition for $n=0$, Eq.(30), one obtains $\psi_o = u_\lambda$ and $E = W$, where u_λ is an eigenfunction of H, and W_λ the corresponding real eigenvalue.

This shows that in the high-frequency limit ($\omega\to\infty$) *the atom is stable* no matter the intensity of the radiation (the associated total decay width $\Gamma=0$), but may be strongly distorted as a consequence of the "dressing" of its potential by the field.

To *next order* in $1/\omega$, Eq.(35) can be solved by perturbation theory. By denoting

$$E = W_\lambda + E_\lambda^{(1)} + \cdots\,, \tag{37}$$

one finds for the complex $E_\lambda^{(1)}$:

$$E_\lambda^{(1)} = \Delta_\lambda - i\,\frac{\Gamma_\lambda}{2}\,; \tag{38}$$

$$\Delta_\lambda = \sum_{\substack{m \\ (m\neq o)}} \langle u_\lambda|V_{-m}\,\mathcal{P}(H - E_\lambda^{(o)} - m\omega)V_m|u_\lambda\rangle\,, \tag{39}$$

$$\Gamma_\lambda = -2\pi \sum_{\substack{m \\ (m\neq o)}} \langle u_\lambda|V_{-m}\,\delta(H - E_\lambda^{(o)} - m\omega)V_m|u_\lambda\rangle\,. \tag{40}$$

By comparing with Eq.(31) it follows that:

$$E_r = W_\lambda + \Delta_\lambda + \cdots\,,\qquad \Gamma = \Gamma_\lambda + \cdots\,. \tag{41}$$

The eigenvalue equation Eq.(36) for the high frequency limit, together with Eqs.(38)-(40), was obtained earlier by Gersten and Mittleman[16] by a different method (see also Ref. 17, § 3.6).

From the asymptotic behavior of the *correction* term to the $\psi_o = u_\lambda$ approximation, we obtain for the n-photon ionization amplitude

$$f_{\vec{p}_n\lambda} = -\frac{1}{2\pi}\langle\psi^{(-)}_{\vec{p}_n}\,|\,V_n|u_\lambda\rangle\,. \tag{42}$$

This is similar in form to the free-free transition amplitude Eq.(18). Both decrease rapidly with ω.

We are now in the process of solving Eq.(36) numerically for a number of model potentials (Coulomb, Yukawa, etc.), to study the α_0 dependence of the levels.

V. CONCLUSIONS

The theory presented here for the high-intensity, high-frequency regime reveals new features, different from those known at lower frequencies. The atomic structure is strongly distorted (because of the high intensity), while the decay by multiphoton ionization is strongly quenched (because of the high frequency). The atom is thus quasi "frozen" in a distorted state. In the case of free-free transitions the elastic scattering dominates, while again multiphoton absorption is strongly suppressed. These features appear in ranges of frequency and intensity now opening up to experiment.

REFERENCES

1) Gavrila M, in *Atomic Physics* 9 (Proceedings of the 9th ICAP, Seattle, USA, 1984), Ed. N Fortson (Plenum, 1985).
2) Kroll NM and Watson KM, Phys.Rev. A 8, 804 (1973).
3) Gavrila M and Kaminski JZ, Phys.Rev.Lett. 52, 613 (1984), and to be published.
4) Offerhaus MJ, Kaminski JZ and Gavrila M, submitted for publication.
5) Luk TS, Pummer H, Boyer K, Shahidi M, Egger H and Rhodes CK, Phys.Rev.Lett. 51, 110 (1983).
6) Rhodes CK, Second Topical Meeting on Laser Techniques in the XUV, Boulder, Colo, USA (Optical Society of America vol. 84.2, 1984, p. MB1-1).
7) Krüger H and Jung Ch, Phys.Rev. A 17, 1706 (1978).
8) Kramers HA, *Collected Scientific Papers* (North-Holland, Amsterdam, 1956), p.262.
9) Henneberger WC, Phys.Rev.Lett. 21, 838 (1968).
10) Elton LR, *Nuclear Sizes* (Oxford University Press, 1961).
11) Gesztesy F and Pittner L, J.Phys. A 11, 679 (1978).
12) Müller-Kirsten HJ and Bose SK, J.Math.Phys. 20, 2471 (1979).
13) Evans R, *The Atomic Nucleus* (McGraw-Hill, 1955).
14) Baz AI, Zeldovich YaB and Perelomov AM, *Scattering, Reactions and Decay in Nonrelativistic Quantum Mechanics* (translated from the Russian by the Israel Program for Scientific Translations, Jerusalem 1969), Chap. 5.
15) Zeldovich YaB, Sov.Phys.Usp. 16, 427 (1974).
16) Gersten JI and Mittleman MH, J.Phys. B 9, 2561 (1976).
17) Mittleman MH, *Introduction to the theory of Laser-Atom interactions* (Plenum, 1982).

STRONG-COUPLING THEORY OF ELECTRON SCATTERING FROM ATOMS IN A RADIATION FIELD

F.H.M Faisal
Fakultät für Physik
Universität Bielefeld
Federal Republic of Germany

1. Introduction

Recent developments in the electron-atom collision in an external laser field shows the need[1] for systematic analysis of the effect of simultaneous coupling of the laser field with the target atom as well as with the scattering electron. This requirement may be fulfilled by a strong-coupling theory in which the field-interactions and the collision-interactions enter on the same footing. Such a theory permits one for example to study systematically all new channels associated with the emission and absorption of photons, both by the electron as well as by the atom, during the course of the collision.

Here we first give the derivation of a generalized strong-coupling theory of radiative scattering obtained in close analogy with the powerfull close-coupling theory of electron scattering in the absence of the field. This permits one to use the solution techniques and if desired even known data from the field-free case, toward the solution of the more difficult problem of radiative e-atom scattering. Finally, results of specific calculations for low-energy electron collisions are used to predict a number of new scattering phenomena which occur only in the presence of the field.

2. The Hamiltonian Of The System

The Hamiltonian of the entire system of "electron+atom+field", in which each sub-system is allowed to interact with the other two, is

$$H = H_a + \omega a^+ a + \frac{i}{c} \vec{\nabla}_{\vec{x}} \vec{\varepsilon}(a^+ + a)$$

$$- \frac{1}{2} \nabla^2_{\vec{r}} + \frac{i}{c} \vec{\nabla}_{\vec{r}} \vec{\varepsilon} (a^+ + a) + V(\vec{r},\vec{x}) \tag{1}$$

where the first line of (1) describes the Hamiltonian of the isolated atom, H_a, that of the mono-mode field, $\omega a^+ a$, and the atom-field interaction. The second line of (1) gives the kinetic energy operator of the electron and its interaction with the field and with the atom.

In analogy with the field-free close coupling theory in which the eigenstates of the bare atom plays the central role in the development of the total wave-function we shall expand the wave-function of the present system in terms of the dynamical polarised states (also to be called the Floquet-states or the dressed-states) of the target atom in the field. But before this is done, the problem of definition of the "asymptotic states" which is central to a consistent treatment of the radiative electron-atom scattering problems, will be resolved.

3. Notion Of The Asymptotic States In Radiative Electron-Atom Scattering

It is first necessary to recall that the collisional interaction is generally of such a range (e.g. several Bohr radii) that the asymptotic region for the electron motion is reached as, $r \to \infty$, in the atomic scale (e.g. several hundred Bohr radii). On the other hand, the electron-field interaction in (1) is of macroscopic extend (e.g. laser beam width of about 0.1 mm $\approx 10^6$ a_0) and as such do not vanish "asymptotically" in the atomic scale. Hence the following two condi-

tions prevail:
(i) the scattering electron keeps interacting with the field long after (before) it ceases (begins) to interact with the target atom,
(ii) the scattered electron in all probability must leave the field region when it interacts with the detector.

These two conditions pose a new problem of definition of the asymptotic scattering states, and hence of the reference states of the transition associated with the radiative scattering, which does not arise in the field-free case. This circumstance is partly reminiscent of the problem of the asymptotic states in the long-range (in atomic scale) coulomb scattering and partly of the quantum field-theoretical problems in which the interaction "electron+(vacuum) field" does not "switch-off" at any distance whatever.

For the Hamiltonian (1) the reference states for the scattering are naturally defined by the eigenstates of the Hamiltonian in the absence of the scattering potentials i.e.

$$H_R \equiv H - V(\vec{c},\vec{x}) \tag{2}$$

Furthermore, the condition (ii) must be met by requiring that the reference eigen-states also be such that as the electron emerges from the field region, they must go over to the usual (plane-wave) scattering states. Because of the generally slower motion of the target atom with respect to the electron motion, the state of the target-atom will remain, in general, dressed by the field as the electrons leave the field region and attain the plane wave scattering states.

4. The Reference Eigen-Functions

The eigen-functions of the reference Hamiltonian (2) are not, as may be thought of as first, simply the product of a dressed-state of the atom and a dressed-state (e.g. Volkov-state) of the electron.

This is because the electron and the atom interact with each other via the field (see eq.(1)) even in the absence of the collisional potential. We may, however, derive the eigen-functions of (2) as appropriate linear combinations of such product states. It may be verified by substitution of (3) given below that the eigen-states of (2) are

$$|\vec{k},\lambda_p\rangle = \sum_{mn} \sum_{j} a_{jm}(\lambda_p) \; |j(\vec{x})\rangle \; J_{n-m}(\vec{k}\cdot\vec{\alpha}_0) \; e^{i\vec{k}\cdot\vec{r}} \; |n\rangle \tag{3}$$

which belong to the eigenvalues

$$E = \frac{k^2}{2} + \lambda_p \tag{4}$$

where $a_{jm}(\lambda_p)\rangle$ are the Floquet-coefficients satisfying the periodicity relation

$$a_{jm}(\lambda_p) = a_{j,m+n}(\lambda_p^n) \tag{5}$$

with

$$\lambda_p^n = \lambda_p + n\omega \tag{6}$$

$|j(x)\rangle$ are the bare states of the target atom, $j=1,2,3,\ldots J$, and $|n\rangle$ is a number state; $(n,m)=0,\pm1,\ldots\pm\infty$. The Floquet-coefficients and the eigenvalues λ_p^n are obtained from the solution of the algebraic equations

$$E - \varepsilon_j - m\omega) \; a_{jm} = i\omega \; \frac{\vec{\alpha}_0}{2} \; \vec{\nabla}_{\vec{x}} \; (a_{j,m-1} + a_{j,m+1}) \tag{7}$$

where

$$\vec{\alpha}_0 = \vec{\varepsilon} \; \frac{F_0}{\omega^2} \; ,$$

with unit polarization vector $\vec{\varepsilon}$, field amplitude F_0 and frequency ω.

For the development of the strong-coupling theory we shall rewrite (3) as

$$|\vec{k},\lambda_p\rangle = \sum_{n=-\infty}^{\infty} J_n(\vec{k}\cdot\vec{a}_0)\, e^{i\vec{k}\cdot\vec{r}}\, |\lambda_p^n;\vec{x}\rangle \tag{8}$$

where

$$|\lambda_p^n;\vec{x}\rangle \equiv \sum_{jm} a_{jm}(\lambda_p^n)\,|j(\vec{x})\rangle\,|m\rangle \tag{9}$$

are the "dressed-states" which incorporate fully the field induced dynamical polarization of the target-atom. Expression (8) is derived from (3) by making use of the periodicity relation (5) and subsequently shifting the infinite summation indices. The states (3) or (8) properly satisfy the asymptotic requirement (ii). This is because as α_0, the electron-field interaction parameter, in (8) goes to zero adiabatically across the field boundary,

$$|\vec{k},\lambda_p\rangle \rightarrow e^{i\vec{k}\cdot\vec{r}}\,|\lambda_p^n;\vec{x}\rangle \tag{10}$$

which is the desired product state of the scattering electron in the detector region and the target atom in the field region. It should be noticed that in case the target atom should leave the field-region then again (8) will take the approppriate asymptotic product state of the free-atom ⊗ free-electron, automatically. The latter result follows from the limiting properties of the Floquet-coefficients as the "atom-field" coupling goes to zero adiabatically.

$$a_{jn}(\lambda_p) \longrightarrow \delta_{j,p}\,\delta_{n,0}$$

and (11)

$$\lambda_p \longrightarrow \varepsilon_p$$

5. The Strong-Coupling Equations

In analogy with the field-free close coupling theory in which the eigenstates of the bare atom provides the basis of expansion of the total wave-function, we now expand the total wavefunction of the total Hamiltonian (1) as

$$\psi(\vec{r},\vec{x}) = \sum_{p=1}^{J} \sum_{n=-\infty}^{\infty} \{ F_{pn}(\vec{r}) \; |\lambda_p^n;\vec{x}> \quad F_{pn}(\vec{x}) \; |\lambda_p^n;\vec{r}> \} \qquad (12)$$

where for the sake of notational economy and of clarity we explicitly consider the two-electron system. The scattering functions $F_{pn}(r)$ give rise to the radiative scattering amplitude for the singlet (+) and the triplet (-) scattering; the exchange symmetry being incorporated explicitly in this expansion. To find the equations for the unknown scattering wavefunctions F_{pn} we substitute (12) into the Schrödinger equation of the total system:

$$[E - H] \; \psi(\vec{r},\vec{x}) = 0 \qquad (13)$$

We now make use of the following two relations which can be derived without difficulty by appropriately shifting the summation indexes and using the periodicity property (5) and the Bessel-identity,

$$\sum_{n=-\infty}^{\infty} J_{n-m}(\vec{x}) \; J_{n-m'}(\vec{x}) = \delta_{m,m'} \qquad (14)$$

$$H(\vec{r},\vec{x}) \sum_{pn} F_{pn}(\vec{r}) \; |\lambda_p^n;\vec{x}>$$

$$= \sum \; |\lambda_p^n;\vec{x}> [- \tfrac{1}{2} \nabla^2 + \lambda_p^n + \beta_{\vec{r}} \; (s_n^- + s_n^+) + V(\vec{r},\vec{x})] \; F_{pn}(\vec{r}) \qquad (15)$$

where

$$s_n^{\pm} \; F_{pn}(\vec{r}) \equiv F_{p,n\pm1}(\vec{r}) \qquad (16)$$

and

$$\beta_{\vec{r}} \equiv \frac{i}{c} \vec{A}_0 \cdot \vec{\nabla}_{\vec{r}} \qquad (17)$$

Using the symmetry of the Hamiltonian (1), under interchange of coordinates, $H(\vec{r},\vec{x}) = H(\vec{x},\vec{r})$, and interchanging $\vec{r} \longleftrightarrow \vec{x}$ in (15) one also finds

$$H(\vec{r},\vec{x}) \quad \sum_{pn} \quad F_{pn}(\vec{x}) \; |\lambda_p^n;\vec{r}>$$

$$= \sum_{pn} \; |\lambda_p^n;\vec{r}> \; [-\tfrac{1}{2}\,\nabla^2 + \lambda_p^n \; + \beta_{\vec{x}}(s_n^- + s_n^+) + V(\vec{x},\vec{r})] \; F_{pn}(\vec{x}) \qquad (18)$$

with

$$\beta_{\vec{x}} \equiv \frac{i}{c}\,\vec{A}_0 \cdot \vec{\nabla}_{\vec{x}}$$

We then use (15) and (18) in (13) and project onto the orthonormal dressed-states $|\lambda_p{}^n,x>$. This yields, after slight rearrangements of terms the dressed strong coupling equations for the $F_{pn}(\vec{r})$:

$$[\nabla^2 + k_{pn}^2 - 2\beta_{\vec{r}}\,(s_n^- + s_n^+)] \; F_{pn}^{(\pm)}(\vec{r})$$

$$= \sum_{p'n'} \; U_{pn}^{p'n'}(\vec{r}) \; F_{p'n'}^{(\pm)}(\vec{r}) \qquad \sum_{p'n'} \; \int \; d\vec{r}' \; W_{pn}^{p'n'}(\vec{r},\vec{x}) \; F_{p'n'}^{(\pm)}(\vec{x})$$

$$(19)$$

where

$$k_{pn} = (2(E - \lambda_p^n)^{1/2}$$

are the channel-momenta in the scattering-channel { pn} and

$$U_{pn}^{p'n'}(\vec{r}) \quad \text{and} \quad W_{pn}^{p'n'}(\vec{r},\vec{x})$$

are the generalized direct potentials and the exchange-potentials respectively. More explicitly the direct potentials are found to be

$$U_{pn}^{p'n'}(\vec{r}) = 2 \sum_{jj'm} \; a_{j'm}^{*}(\lambda_{p'}^{n'}) \; a_{jn}(\lambda_p^n) \; <j'(\vec{x})|V(\vec{r},\vec{x})|j(\vec{x})> \qquad (20)$$

and the exchange-potential operator is

$$W_{pn}^{p'n'}(\vec{r},\vec{x}) = -\,2 \sum_{jj'm} \; |j(\vec{r})> \; a_{j'm}^{*}(\lambda_{p'}^{n'}) \; a_{jm}(\lambda_p^n)$$

$$<j'(\vec{x})| \; [(E - \lambda_{p'}^{n'} - \varepsilon_{j'}) - \beta_{\vec{x}}(s_n^+ + s_n^-) + V(\vec{x},\vec{r}) - V_c(x)] \qquad (21)$$

where $V_c(\vec{r})$ is the potential interaction of the electron with the "core".

The direct and the exchange potentials differ from the corresponding potentials in the field free case due to the appearence of the Floquet-coefficients which incorporates the persistent influence of the dynamical polarization of the target by the field. It shows that the modification of the scattering potentials due to the field is not entirely a transient phenomena but is rather reminescent of the persistent effect of the perturbation in the field theory.

6. The Scattering Amplitudes

To obtain the radiative scattering amplitudes the strong coupling equations (19) are to be solved under the outgoing wave boundary condition as $r\to\infty$. In this limit the exchange and the direct potentials will tend to zero (the latter vanish mostly exponentially and the former vanish as inverse powers of the distance). The equations (19) then reduce to

$$[\nabla^2 + k_{pn}^2 - 2\beta_{\vec{r}}(s_n^- + s_n^+)]\ F_{pn}(\vec{r}) \to 0 \tag{22}$$

which have required solutions of the form

$$J_{n-N}(\vec{k}_{pN}\cdot\vec{\alpha}_0)\ e^{i\vec{k}_{pN}\cdot\vec{r}} \tag{23}$$

for any n,N and p. Consistent with the asymptotic conditions for the radiative scattering, therefore, the boundary conditions on the channel wave function $F_{pn}(\vec{r})$ take the form

$$F_{pn}(\vec{r}) \underset{\lim r\to\infty}{=} J_{n-N}(\vec{k}_{pN}\cdot\vec{\alpha}_0)\ e^{i\vec{k}_{pN}\cdot\vec{r}}\ \delta_{p,i}\ \delta_{N,0} + \sum_{N=-\infty}^{\infty} J_{n-N}(\vec{k}_{pN}\cdot\vec{\alpha}_0)\ \frac{e^{ik_{pN}r}}{r}\ f^{(\pm)}_{i,0\to p,N}(\Omega_0,\Omega) \tag{24}$$

where

$$f^{(\pm)}_{i,0\to p,N}(\Omega_0,\Omega)$$

finally yield the singlet and the triplet radiative electron-atom scattering amplitudes for the scattering of the electron from the initial momentum state $\vec{k}_{i,0}$, in direction Ω_0, to the final momentum $\vec{k}_{pN}$, in direction Ω accompanied by the emission (N>0) or absorption of (N<0) of N-photons.

In the rest of this lecture I shall discuss some problems of much current interest and give numerical evidence of a number of new scattering phenomena which we predict to occur for radiative electron-atom collisions in a strong field.

7. New Phenomena In Low-Energy Radiative Scattering

Perhaps the most interesting aspects of radiative electron-atom scattering processes can occur at electron energies E_i in the region $E_i < \hbar\omega$ and in high field strength. In this situation neither the Born-like approximation[2] nor the low-frequency approximations[3] may be applicable.

Calculations are performed here using a multichannel pseudo-potential for scattering in the absence of the field, whose parameters are defined by fitting to the known electron-H-atom scattering information in the field-free case[4].

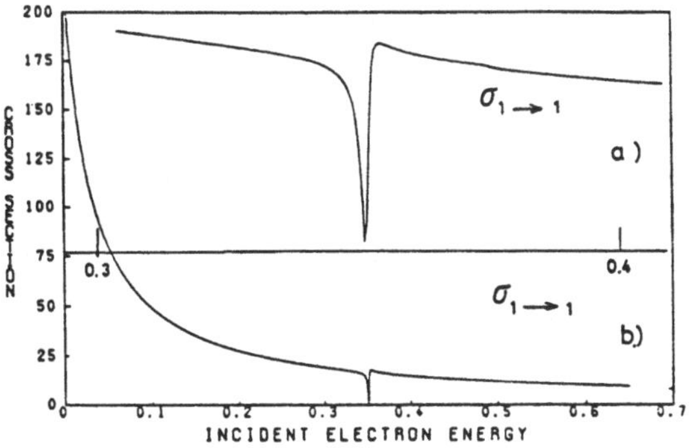

Fig.1. - Low energy e-H elastic scattering cross section showing the ^1S resonance at ≈ 9.558 eV (width ≈ 0.04 eV), obtained from the pseudo-potential model, Scales are in a.u..

Fig.1 shows the kind of result one obtains in this way when the field is off. The elastic 1s-1s cross section at low scattering energies is shown. It also reproduces the famous e-H scattering resonance at 9.6 eV (width 0.04 eV see magnification in the inset)[4] The excellent agreement in the position, width and the shape with the experimental resonance[5] and the present pseudopotential model is seen in Fig.2.

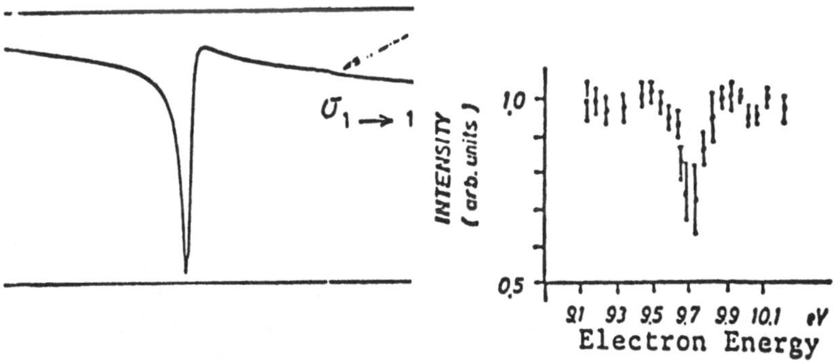

Fig. 2. - $\sigma_{1\to1}$ on the left hand side is the calculated 1s-1s resonance at ≈9.6 eV. This is compared on the right handside with the experimental result obtained by Kleinpoppen and Riable (ref. 5.) for the same resonance.

The same multichannel-potentials yield the H^- (negative-ion) bound state at -0.75 eV (not shown in the figure). The cross-section at zero-energy is seen to compare with the standard two-state close-coupling calculation which gives $\sigma(E=0) \approx 254.75\ a_0^2$ (although the actual value could be somewhat less[4]). It is gratifying that the 2-state pseudo-potential model reproduces the known low energy scattering data physically correctly. The model is therefore extended to incorporate and investigate the effect of the external field on the scattering process.

7.1 Field Induced Scattering Resonance

Consider the e-H scattering with the field on, at an electron energy far below the inelastic threshold (E_{th} = 10.2 eV) and at a photon frequency ω = .05 au. The result is shown in figure 3.

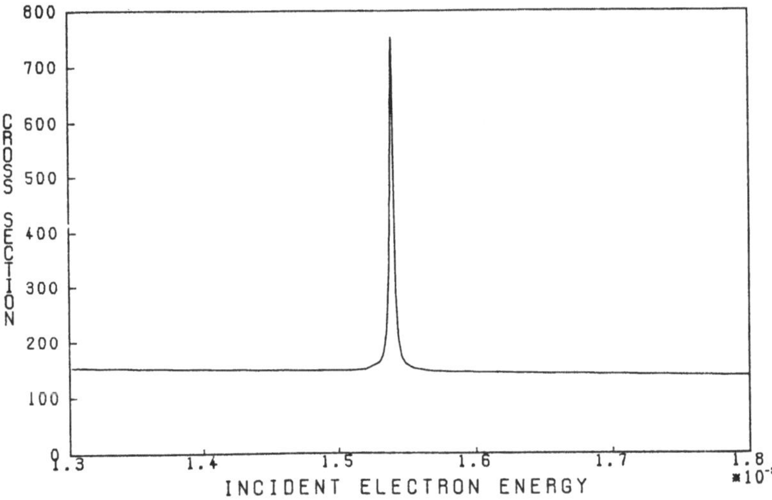

Fig.3 - The field modified e-H elastic cross section at F_0 = .0005 a.u. ω = 1.17 eV. Sharp structure is an induced one photon "capture-escape" resonance with respect to the H^- state at -0.75 eV. The scales are in a.u..

We see a dramatic change in the field modified elastic cross section $\sigma_{1,0\to1,0}$ (with no net emission or absorption of photons); it exhibits a sharp structure which "reflects" the bound H^- negative ion-state into the elastic channel. At $E_i \approx .0154$ (au) the photon energy $\hbar\omega = .043$ (au) matches approximately with the binding energy of H^-, $E_B = -.0276$ (au) and we have a resonant capture of the elastic electron in the H^- state (as it loses the energy of one-photon by the stimulated emission). But the captured electron can also reabsorb a laser photon and return to its positive energy elastic channel; the delay caused by the temporary "capture and escape" episodes shows up as a new resonance in the elastic channel. We note that a smaller two-photon "capture-escape" resonance is also discernable at an electron energy which is $\hbar\omega = .043$ (au) above the primary one-photon resonance. Clearly the main resonance may be used to determine the electron-affinity by elastic scattering in the field.

7.2 Inelastic Excitation Of The Target By Sub-Threshold Electrons

Another interesting phenomenon, at electron energies below the first inelastic excitation threshold of the H-atom, is the resonant excitation of the n=2 level, by the sub-threshold electrons. This process may permit one to probe experimentally a portion of the off-shell electron scattering amplitude. "Off-shell" calculations[7-9] up till now, have been confined to the above threshold electron energies where the effect occurs along with the direct collisional excitation but can be separated. With the sub-threshold electrons, on the other hand, the dominant process itself is the off-shell one.

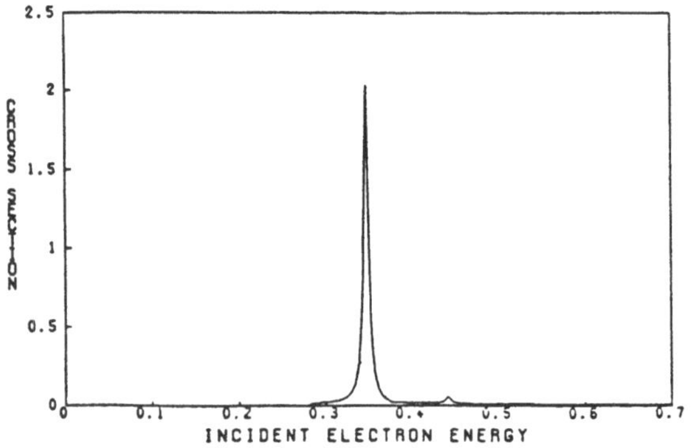

Fig.4 - Resonant sub-threshold excitation of H(n=2) by low energy electrons. F_0 = 0.01 a.u., ω = 0.094 a.u.. The scales are in a.u..

Figure 4 exhibits a typical result of such excitation cross-section, $\sigma_{1,0\to 2,-1}$, in which the electron borrows a photon from the field and excites the upper 2s state of the atom by the collisional transfer of its enhanced kinetic energy to the atom. What is more, the process becomes resonant in nature due to the presence of the elastic scattering resonance (at ≈ 9.6 eV). We also note the existence of a threshold cusp[10-12] (at ≈ 10.2 eV) and a secondary resonance at ≈ 0.35 (au). We should emphasise that the resonant sub-threshold excitations should occur in other systems as well. Excitation of vibrational states of diatomic molecules, for example H_2 and N_2, by very low energy electrons in the presence of a strong infrared laser (e.g. a CO_2-laser at $\hbar\omega$ = .117 eV) should also exhibit the resonant sub-threshold excitation phenomena.

7.3 Breaking The Spherical Symmetry Of The Low Energy Angular Distribution

The field can affect the differential scattering cross sections due to both symmetry and dynamical reasons. In the absence of the field, the scattering process maintains its cylindrical symmetry about the incident beam direction, which leads to the usual azimuthal invariance of the angular distributions of the scattered electrons. The very presence of the field destroyes this invariance, since in general it provides an extra quantization direction in the laboratory (except perhaps if the polarization direction lies along the incident electron beam direction). Besides, the distribution with respect to the polar angle, in a given azimuthal plane also can be overwhelmingly modified. Below we show the result of calculation indicating the modification of the low-energy angular distribution of the scattered electrons, which are spherically symmetric in the field-free case.

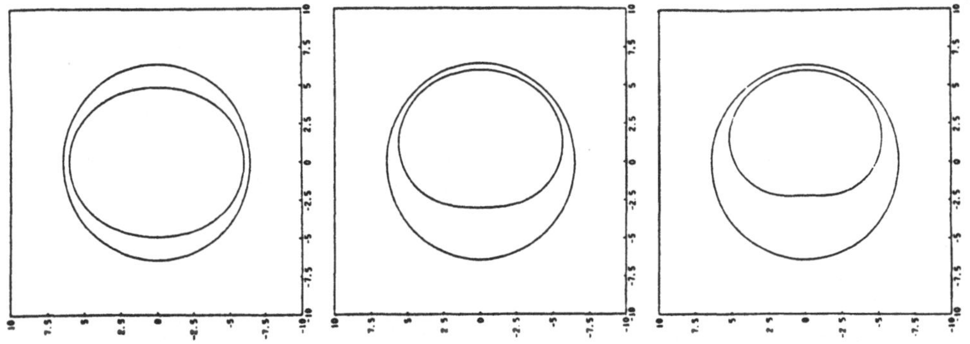

Fig.5 - Low energy e-H_0 differential scattering cross section in the azimuth plane $\varphi = 45^o$ modified by a circularly polarized field. Electrons energy E_i = 1.36 eV; F_0 = .005 a.u., ω = 0.05 a.u..

Figure 5 shows an example of the change in the low-energy (E_i = .05 a.u.) angular distribution for the e-H scattering, due to the presence of the field ($\hbar\omega$ = .05 a.u., peak field strength F_0 = .005a.u.) when a circularly polarized photon-beam is directized a) along the incident electron-beam direction (left-pannel) b) along 45° (middle-pannel), and c) perpendicular (right-pannel), to the electron-beam direction. The outer circles in these pannels correspond to the usual spherical distribution of the scattered low-energy electrons (when the field is off).

7.4 Photon Amplification By Scattering Of Electrons.

Just as the electron can borrow energy from the field, the field may also gain energy from the electron during the collision. Is it possible that at specific electron energies and or photon frequencies the field can gain more energy from the electron than the other way around, which could actually lead to an amplification of the field intensity? The answer may be given in positive in view of the result obtained below.

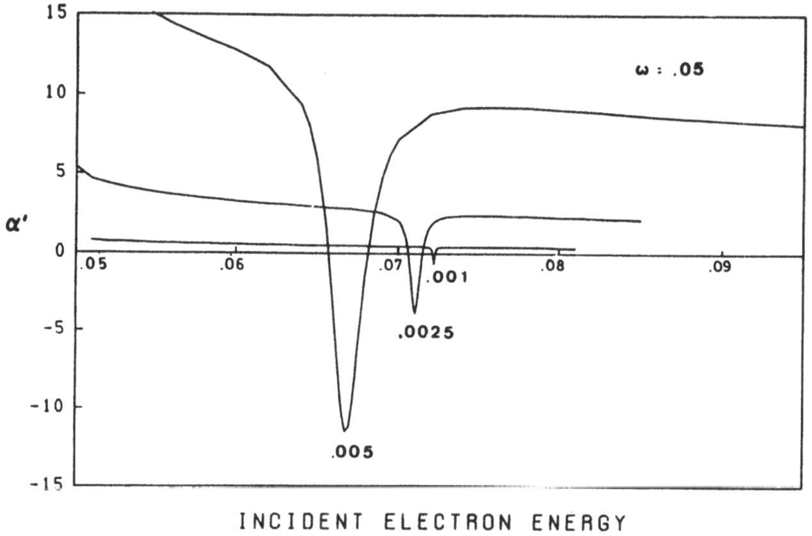

Fig.6: The photon-absorption coefficient $\alpha(E_i)$ vs. the incident electron energy E_i, for the radiative scattering of electrons with H atoms at $\omega = 0.05$ a.u. and three different field strengths $F_0 = 0.001$, 0.0025 and 0.005 a.u. Note that the negative absorption or the gain occurs resonantly (see text for the discussion). The scales are in a.u.

In figure 6 we show the coefficient, $\alpha(E)$, of the net absorption of photons by the scattering system (as a function of the incident electron energy E_i, for a fixed field frequency, ω = .05 au) where both the elastic scattering and the atomic excitation are allowed for. It clearly exhibits how at preferred electron energies the absorption coefficient becomes negative and hence changes into a "gain-coefficient". Observe also that the "gain-resonance" deepens, broadens, as well as shifts towards the lower energies, as the field strength increases, from F_0 = .001 (a.u.) = 5.1 10^6 V/cm, through .0025 (a.u.) to .005 (a.u.). Here we have a case of photon-amplification by the radiative scattering of electrons, in the absence of any undulators.

8. Coherent Scattering By Dynamic Polarization

The scattering signal from a dynamically polarized target created by a near (or on)-resonant laser pulse depends on the duration (and, as we shall see, also on the condition of the application) of the field. Thus if the pulse is strong it must be short enough, t(pulse) < t(decay), for the dynamically polarized state is in essence a coherent superposition-state (during its natural life-time). If, further, the pulse rise time t(rise) < 1/Ω(Rabi) (where $\Omega(\mathrm{Rabi}) = (\Delta^2 + \beta^2)^{1/2}$ is the two-level generalized Rabi-frequency with the detuning Δ, and the coupling strength β), then the pulse will effectively switch on "suddenly" (at a time say, t=0).

The dynamically polarized state will evolve (in the resonant two-level approximation) as

$$\psi_{sud}(t) = e^{-i\lambda_1 t} |\lambda_1\rangle\langle\lambda_1|\psi(0)\rangle + e^{-i\lambda_2 t} |\lambda_2\rangle\langle\lambda_2|\psi(0)\rangle \quad (25)$$

where

$$|\lambda_1\rangle = \cos\theta\ |1,n_0\ \rangle - \sin\theta\ |2,n_0-1\rangle; \quad (26)$$

and

$$|\lambda_2\rangle = \sin\theta\ |1,n_0\ \rangle + \cos\theta\ |2,n_0-1\rangle; \quad (27)$$

are the two dominant dressed states; $-\pi/4 \leq \theta \leq \pi/4$, $\tan 2\theta = \beta/\Delta$ and $|1,n_0\rangle$, $|2,n_0\rangle$ are the product states, $|\mathrm{atom}\rangle \otimes |\mathrm{field}\rangle$. The perturbed eigen-frequencies are

$$\lambda_1 = \omega_1 + \frac{\Delta}{2} - \frac{\Omega}{2} \qquad \to \quad \omega_1 \text{ (as } \beta \to 0,\ \Delta > 0\text{)}$$

and (28)

$$\lambda_2 = \omega_1 + \frac{\Delta}{2} + \frac{\Omega}{2} \qquad \to \quad \omega_2 - \omega \text{ (as } \beta \to 0,\ \Delta > 0\text{)}$$

(and they are reversed for $\Delta<0$).
where ω_1 and ω_2 are the Bohr-frequencies of the levels $|1\rangle$ and $|2\rangle$; $\Delta \equiv \omega_2 - \omega_1 - \omega$; $\lambda_2 - \lambda_1 = \Omega$.

Note that if initially $|\psi(0) = |1, n_0\rangle$ then

$$\langle\lambda_1|\psi(0)\rangle = \cos\theta \quad \text{and} \quad \langle\lambda_2\ |\psi(0)\rangle = \sin\theta \tag{29}$$

8.1 Reduction Of The Pure State Into A Mixture (In λ-Representation)

Consider now that the on-set of the collision, between a pair of target and projectile atoms, occurs at $t=t_0$. The density matrix of the target at this time is (from eq. (25))

$$\varrho(t_0) = \begin{pmatrix} \cos^2\theta & e^{i\Omega t_0}\cos\theta\sin\theta \\ e^{-i\Omega t_0}\cos\theta\sin\theta & \sin^2\theta \end{pmatrix} \tag{30}$$

Clearly, the target is in a pure-state at this time ($[\varrho(t_0)]^2 = \varrho(t_0)$). But, the instant t_0, for all pairs of colliding atoms, is essentially a random quantity; so the observable signal depends on the average of (30) over the distribution of t_0. Assuming an uniform distribution we get

$$\bar{\varrho} \equiv \lim_{T\to\infty} \frac{1}{T} \int_0^T \varrho(t_0)\, dt_0$$

$$= \lim_{T\to\infty} \begin{pmatrix} \cos^2\theta & , & -i\frac{1}{\Omega T}(e^{i\Omega T}-1)\cos\theta\sin\theta \\ & & \\ i\frac{1}{\Omega T}(e^{-i\Omega T}-1)\cos\theta\sin\theta & , & \sin^2\theta \end{pmatrix} \quad (31)$$

One sees from (31) that the off-diagonal correlation between the dressed states $|\lambda_1>$ and $|\lambda_2$ is important but for a few Rabi-periods, $T = 0\,(1/\Omega)$.

In the limit the correlation is negligible and we get,

$$\bar{\varrho} = \begin{pmatrix} \cos^2\theta & 0 \\ & \\ 0 & \sin^2\theta \end{pmatrix} \quad (32)$$

8.2 The Scattering Signal In The Vicinity Of The Incident Energy.

Thus the scattering signal becomes effectively that from a mixture of states $([\varrho]^2 \neq [\varrho])$ in the $|\lambda>$-representation:

$$\frac{ds}{d\Omega} = \mathrm{Tr}\,[f\,\bar{\varrho}\,f^+] \quad (33)$$

where the amplitude matrix

$$f = \begin{pmatrix} f_{\lambda_1\to\lambda_1}(k_0,k) & \left(\frac{k_{21}}{k_0}\right)^{1/2} f_{\lambda_2\to\lambda_1}(k_0,k_{21}) \\ & \\ \left(\frac{k_{12}}{k_0}\right)^{1/2} f_{\lambda_1\to\lambda_2}(k_0,k_{12}) & f_{\lambda_2\to\lambda_2}(k_0,k) \end{pmatrix} \quad (34)$$

the scattered wave vectors are $\vec{k}$, $\vec{k}_{12}$ and $\vec{k}_{21}$ with

$$k_{21} = \left(k_0^2 + \frac{2\mu}{\hbar}\,\Omega\right)^{1/2}$$

$$k_{12} = \left(k_0^2 - \frac{2\mu}{\hbar}\,\Omega\right)^{1/2}$$

$$k = k_0$$

where k_0 is the incident (relative) wave-vector and μ is the reduced mass. Hence from (33) and (34) one finds that the "elastic signal" splits up into three components, separated by the Rabi-energy $h\Omega$, from each other:

$$\frac{ds}{d\Omega} = \frac{ds_{el}}{d\Omega} + \frac{ds_{\lambda_1\to\lambda_2}}{d\Omega} + \frac{ds_{\lambda_2\to\lambda_1}}{d\Omega} \tag{35}$$

where the central peak has the intensity

$$\frac{ds_{el}}{d\Omega} = \cos^2\theta\ \frac{d\sigma_{\lambda_1\to\lambda_1}(\vec{k}_0,\vec{k})}{d\Omega} + \sin^2\theta\ \frac{d\sigma_{\lambda_2\to\lambda_2}(\vec{k}_0,\vec{k})}{d\Omega}$$

and the intensities of the side bands are

$$\frac{ds_{\lambda_1\to\lambda_2}}{d\Omega} = \cos^2\theta\ \frac{d\sigma_{\lambda_1\to\lambda_2}(\vec{k}_0,\vec{k}_{12})}{d\Omega}$$

and

$$\frac{ds_{\lambda_2\to\lambda_1}}{d\Omega} = \sin^2\theta\ \frac{d\sigma_{\lambda_2\to\lambda_1}(\vec{k}_0,\vec{k}_{21})}{d\Omega}$$

with

$$\frac{d\sigma_{\lambda_i\to\lambda_j}(\vec{k}_0,\vec{k}_{ij})}{d\Omega} = \frac{k_{ij}}{k_0}\,\left|f_{\lambda_i\to\lambda_j}(\vec{k}_0,\vec{k}_{ij})\right|^2 \quad ; \qquad i,j = 1,2 \tag{37}$$

To isolate them clearly the energy resolution would require to be better than $\hbar\Omega$.

8.3 INTERFERENCE-SCATTERING

It would appear that the signal (35) is merely an incoherent sum of cross-sections. This is indeed so but in the $|\lambda\rangle$-representation. The signal (35), when analysed in the representation of the asymptotically free-states, reveals the hidden interference scattering between the scattering amplitudes directly arising from the atomic states $|1\rangle$ and $|2\rangle$.

Consider a typical amplitude in the $|\lambda\rangle$-representation,

$$f_{\lambda_i\to\lambda_j}(\vec{k}_0,\vec{k}_{ij}) = -\frac{\mu}{2\pi\hbar^2}\langle\vec{k}_{ij}|\langle\lambda_j|\ T\ |\lambda_i\rangle\ |\vec{k}_0\rangle\ , \qquad (i,j=1,2) \qquad (38)$$

where T is the transition operator and $|k_0\rangle$, $|k_{ij}\rangle$ are plane waves.

It is sufficient to show the existence of the interference scattering in the simplest (FBA) approximation of $T \simeq V$. (For the heavy-particle beam collisions of interest, with k_0 large, this is often a sufficient approximation.) In this case, clearly,

$$\begin{aligned}
&\langle 1,n_0|\ T\ |n_0,1\rangle = \langle 1|\ V\ |1\rangle = V_{11}(\vec{R})\\
&\langle 2,n_0-1|\ T\ |n_0-1,2\rangle = \langle 2|\ V\ |2\rangle = V_{22}(\vec{R}) \qquad (39)\\
&\langle 1,n_0|\ T\ |n_0-1,2\rangle = \langle 2,n_0-1|\ V\ |n_0,1\rangle = 0
\end{aligned}$$

where the interaction potential $V=V(\vec{R},\vec{x})$; $\vec{R}$ is the relative coordinate and $\vec{x}$ are the target coordinates.
Using eqs. (37)-(39) in (35) one easily finds the following results,

$$\frac{ds_{el}}{d\Omega} = (\cos^6\theta + \sin^6\theta)\frac{d\sigma_{1\to1}}{d\Omega} + \cos^2\theta\sin^2\theta\frac{d\sigma_{2\to2}}{d\Omega} + 2\cos^2\theta\sin^2\theta\ \mathrm{Re}(f^*_{1\to1}\ f_{2\to2}) \qquad (40)$$

$$\frac{ds_{\lambda_1\to\lambda_2}}{d\Omega} = \cos^4\theta\sin^2\theta\left[\frac{d\sigma_{1\to1}}{d\Omega} + \frac{d\sigma_{2\to2}}{d\Omega} - 2\ \mathrm{Re}\ (f^*_{1\to1}\ f_{2\to2})\right] \qquad (41)$$

$$\frac{ds_{\lambda_2\to\lambda_1}}{d\Omega} = \sin^4\theta\cos^2\theta\left[\frac{d\sigma_{1\to1}}{d\Omega} + \frac{d\sigma_{2\to2}}{d\Omega} - 2\ \mathrm{Re}\ (f^*_{1\to1}\ f_{2\to2})\right] \qquad (42)$$

Thus, not only the central peak (40), but also the splitted components (41) and (42), depend on the direct interference scattering between the two elastic amplitudes $f_{1\to 1}(\vec{k}_0\to\vec{k})$ and $f_{2\to 2}(\vec{k}_0\to\vec{k})$.

ACNOWLEDGEMENTS

I would like to thank Mr.L.Dimou for his collaboration in this work and for preparing the graphical results. This research has been partially supported by Deutsche Forschungsgemeinschaft Sonderforschungs Bereich 216 Teilprojekt M2.

REFERENCES

1. F.H.M. Faisal, Comments In At. Mol. Phys. 15 (1984) 119

2. F.V. Bunkin and M.V. Fedorov, JETP 22, (1966) 844,
N.K. Rahman, Phys. Rev. A8, (1974) 804,
I.V. Hertel and L. Hahn, J.phys. B5, (1972) 1995.

3. N.M. Kroll and K.M. Watson, Phys. Rev. A8, (1973) 804.
H.Krüger and C. Jung, Phys. Rev. A17, (1978) 1706.

4. N.F. Mott and H.S.W. Massey, The Theory of Atomic Collisions, 3rd. Ed. Oxford (1965) p.530.

5. H.Kleinpoppen and V. Raible, Phys. Lett. 18 24 (1965)

6. E.P. Wigner, Phys. Rev. 98, (1965) 145.

7. N.K. Rahman and F.H.M. Faisal, J. Phys. B9, (1976) L275 ;
Phys. Lett. 57A, 426 (1976); J. Phys. B 11, (1978) 2003.

8. F.H.M. Faisal, in Coherence and Correlation in Atomic Collisions, ed. H. Kleinpoppen and J.F. Williams, Plenum Press, n.Y. (1980), p.479.

9. S. Jetzke, F.H.M. Faisal, R. Hippler and H.O. Lutz, Z. Physik A 315, (1984) 271.

10. L.D. Landau and E.M. Lifshitz, Quantum Mechanics 2nd Ed. Pergamon Press, Oxford (1965) pp. 518, 565.

11. C. Jung and H.S. Taylor, Phys. Rev. A23, (1981) 1115.

12. F.H.M. Faisal, in Laser Assisted Collision and Related Topics, ed. N.K. Rahman and C. Guidotti, Harwood Acad. Publishers, New York (1982), p. 287;

ELECTRON-ATOM COLLISIONS IN A STRONG LASER FIELD

C.J. Joachain
Physique Théorique, Faculté des Sciences,
Université Libre de Bruxelles, Belgium,
and
Institut de Physique Corpusculaire
Université de Louvain, Louvain-la-Neuve, Belgium

1. INTRODUCTION

Electron-atom collisions in the presence of a strong laser field have attracted considerable theoretical attention in recent years, not only because of the importance of these processes in applied areas (such as plasma heating or laser-driven fusion), but also in view of their interest in fundamental atomic collision theory. Reviews of various aspects of the subject may be found in the references [1-4]. The problem is in general very complex, since in addition to the difficulites associated with the treatment of electron-atom collisions, the presence of the laser introduces new parameters (for example the laser photon energy $\hbar\omega$ and intensity I) which may influence the collision. Moreover, the laser photons can play the role of a "third body" during the collision, and "dress" the atomic states.

It is therefore of interest to begin the theoretical analysis by considering the simpler problem of the scattering of an electron by a potential, in the presence of a strong laser field. Section 2 is devoted to a brief survey of this topic. In Section 3, we shall consider the more realistic case of a genuine atomic target. In particular, we shall study the "elastic" scattering of fast electrons by atoms in the presence of a laser field which, although strong by laboratory standards, is weak when measured in atomic units. We recall in this connection that the atomic unit of electric field strength is $e/a_0^2 \simeq 5 \times 10^9$ V cm^{-1}, the corresponding intensity being

$I_o \simeq 3.5 \times 10^{16}$ W cm^{-2}. The "dressing" of the target due to the laser field will be shown to produce important effects on the scattering at small momentum transfers, especially when photons are exchanged with the field.

2. POTENTIAL SCATTERING

Let us start our discussion by considering the non-relativistic scattering of an electron, of mass m and charge (-e) by a potential $V(\underset{\sim}{r})$ in the presence of a laser field. We assume that this laser field is treated classically as a spatially homogeneous, monochromatic, linearly polarized electric field $\underset{\sim}{\mathcal{E}}(t) = \underset{\sim}{\mathcal{E}}_o \sin(\omega t)$, the corresponding vector potential in the Coulomb gauge being $\underset{\sim}{A}(t) = \underset{\sim}{A}_o \cos(\omega t)$, with $\underset{\sim}{A}_o = c\, \underset{\sim}{\mathcal{E}}_o/\omega$. After removal of the A^2 term by means of a unitary transformation, the time-dependent Schrödinger equation for the state vector $\Psi(\underset{\sim}{r},t)$ is given by

$$i\hbar \frac{\partial}{\partial t} \Psi(\underset{\sim}{r},t) = [-\frac{\hbar^2}{2m} \nabla^2 - \frac{ie\hbar}{mc} \underset{\sim}{A}.\underset{\sim}{\nabla} + V(\underset{\sim}{r})]\Psi(\underset{\sim}{r},t) \qquad (1)$$

Let us first consider the motion of the electron in the presence of the laser field, but without scattering potential (V = 0). The corresponding Schrödinger equation for this "free" electron

$$i\hbar \frac{\partial}{\partial t} \chi(\underset{\sim}{r},t) = [-\frac{\hbar^2}{2m} \nabla^2 - \frac{ie\hbar}{mc} \underset{\sim}{A}.\underset{\sim}{\nabla}]\chi(\underset{\sim}{r},t) \qquad (2)$$

is readily solved to give the Volkov wave function

$$\chi_{\underset{\sim}{k}}(\underset{\sim}{r},t) = (2\pi)^{-3/2} \exp\{i[\underset{\sim}{k}.\underset{\sim}{r} - \underset{\sim}{k}.\underset{\sim}{\alpha}_o \sin(\omega t) - E_k\, t/\hbar]\} \qquad (3)$$

where $\underset{\sim}{k}$ is the electron wave vector, $E_k = \hbar^2 k^2/2m$ its kinetic energy, $\underset{\sim}{\alpha}_o = e\, \underset{\sim}{\mathcal{E}}_o/m\omega^2$ and we have normalized $\chi_{\underset{\sim}{k}}$ to a delta function. We remark that the effect of the field on the "free" electron states is characterized by the dimensionless parameter $\beta = k\alpha_o = ke\mathcal{E}_o/m\omega^2$ [5]. It is important to note that for strong, low-frequency fields and fast electrons (such that $ka_o > 1$, where a_o is the first Bohr radius of atomic hydrogen), the parameter β is always larger than unity. For example, in the case of a CO_2 laser with photon energy $\hbar\omega = 0.12$ eV and intensity $I = 10^8$ W cm^{-2} (so that $\mathcal{E}_o = 2.7 \times 10^5$ V cm^{-1}), corresponding to the one used in the experiments of Weingartshofer et al [6], one has $\beta \simeq 4ka_o$. For a laser with photon energy $\hbar\omega = 2$ eV and electric field strength $\mathcal{E}_0 = 10^8$ V cm^{-1} (so that $I = 1.3 \times 10^{13}$ W cm^{-2}) one has again $\beta \approx 4ka_0$.

In order to solve the full Schrödinger equation (1), we first recall that the causal propagator $G_0^{(+)}(\mathbf{r},t;\mathbf{r}',t')$ satisfying the equation

$$[i\hbar \frac{\partial}{\partial t} + \frac{\hbar^2}{2m} \nabla^2 + \frac{ie\hbar}{mc} \mathbf{A}.\nabla] G_0^{(+)}(\mathbf{r},t;\mathbf{r}',t') = \delta(\mathbf{r}-\mathbf{r}')\delta(t-t') \quad (4)$$

is given by

$$G_0^{(+)}(\mathbf{r},t;\mathbf{r}',t') = - \frac{i}{\hbar} \Theta(t-t') \int \chi_{\mathbf{k}}(\mathbf{r},t) \chi_{\mathbf{k}}^{*}(\mathbf{r}',t') d\mathbf{k} \quad (5)$$

where $\Theta(x) = 1$ for $x > 0$ and 0 for $x < 0$. The corresponding causal solution of the Schrödinger equation (1), which we denote by $\Psi_{\mathbf{k}}^{(+)}(\mathbf{r},t)$, satisfies the integral equation

$$\Psi_{\mathbf{k}}^{(+)}(\mathbf{r},t) = \chi_{\mathbf{k}}(\mathbf{r},t) + \int_{-\infty}^{t} dt' \int d\mathbf{r}' \, G_0^{(+)}(\mathbf{r},t;\mathbf{r}',t') V(\mathbf{r}') \Psi_{\mathbf{k}}^{(+)}(\mathbf{r}',t') \quad (6)$$

and the S-matrix element for a transition $\mathbf{k}_i \to \mathbf{k}_f$ in the presence of the laser field is given by

$$S_{\mathbf{k}_f,\mathbf{k}_i} = - \frac{i}{\hbar} (\chi_{\mathbf{k}_f}, V \Psi_{\mathbf{k}_i}^{(+)}) \quad (7)$$

where the parentheses indicate an integration over all space and time.

Let us solve equation (6) by iteration starting with the Volkov wave function (3). We then obtain for $S_{\mathbf{k}_f,\mathbf{k}_i}$ the Born series

$$S_{\mathbf{k}_f,\mathbf{k}_i} = \sum_{n=1}^{\infty} S^{Bn}_{\mathbf{k}_f,\mathbf{k}_i} = - \frac{i}{\hbar} \sum_{n=1}^{\infty} (\chi_{\mathbf{k}_f}, V[G_0^{(+)}V]^{n-1} \chi_{\mathbf{k}_i}) \quad (8)$$

The time integration in equation (8) can be performed by using the relation

$$\exp(i\, x \sin u) = \sum_{\ell=-\infty}^{+\infty} J_\ell(x) \exp(i\, \ell\, u) \quad (9)$$

where $J_\ell(x)$ is an ordinary Bessel function of order ℓ. Thus

$$S_{\mathbf{k}_f,\mathbf{k}_i} = - 2\pi i \sum_{\ell=-\infty}^{+\infty} \delta(E_{k_f} - E_{k_i} - \ell\hbar\omega) \, T^{\ell}_{\mathbf{k}_f,\mathbf{k}_i} \quad (10)$$

The delta function in the above equation ensures energy conservation,

$$\frac{\hbar^2 k_f^2(\ell)}{2m} = \frac{\hbar^2 k_i^2}{2m} + \ell\hbar\omega \; , \qquad \ell = 0, \pm 1, \pm 2, \ldots \quad (11)$$

where positive values of ℓ correspond to photon absorption (inverse bremsstrahlung), negative ones to photon emission (stimulated bremsstrahlung) and $\ell = 0$ corresponds to pure elastic scattering in the presence of the laser field.

The T-matrix element $T^{\ell}_{\mathbf{k}_f,\mathbf{k}_i}$ for scattering with the exchange of ℓ photons which appears in equation (10) has the Born series expansion

$$T^{\ell}_{\mathbf{k}_f,\mathbf{k}_i} = \sum_{n=1}^{\infty} T^{Bn,\ell}_{\mathbf{k}_f,\mathbf{k}_i} \tag{12}$$

whose terms $T^{Bn,\ell}_{\mathbf{k}_f,\mathbf{k}_i}$ are readily written down. In first Born approximation we have

$$T^{B1,\ell}_{\mathbf{k}_f,\mathbf{k}_i} = J_\ell(\mathbf{\Delta}.\mathbf{\alpha}_o)\ \tilde{V}(\mathbf{\Delta}) \tag{13}$$

where $\mathbf{\Delta} = \mathbf{k}_i - \mathbf{k}_f$ is the wave vector transfer and

$$\tilde{V}(\mathbf{\Delta}) = (2\pi)^{-3} \int \exp(i\mathbf{\Delta}.\mathbf{r})\ V(\mathbf{r})\ d\mathbf{r} \tag{14}$$

For $n \geqslant 2$,

$$T^{Bn,\ell}_{\mathbf{k}_f,\mathbf{k}_i} = \sum_{L_1,\ldots L_{n-1}} \int d\mathbf{k}_1 \ldots d\mathbf{k}_{n-1} \frac{\tilde{V}(\mathbf{K}_1)\ \ldots\ \tilde{V}(\mathbf{K}_n)}{(D_1+L_1\hbar\omega)\ \ldots\ (D_{n-1}+L_{n-1}\hbar\omega)}$$
$$J_{\ell-L_1}(\mathbf{K}_1.\mathbf{\alpha}_o)\ J_{L_1-L_2}(\mathbf{K}_2.\mathbf{\alpha}_o)\ \ldots\ J_{L_{n-1}}(\mathbf{K}_n.\mathbf{\alpha}_o) \tag{15}$$

where $\mathbf{K}_1 = \mathbf{k}_1 - \mathbf{k}_f$, $\mathbf{K}_2 = \mathbf{k}_2 - \mathbf{k}_1$, ... $\mathbf{K}_n = \mathbf{k}_i - \mathbf{k}_{n-1}$ and $D_s = E_{k_i} - E_{k_s} + i\epsilon$ with $E_{k_s} = \hbar^2 k_s^2/2m$ and $\epsilon \to 0^+$.

The differential cross section for the scattering process $\mathbf{k}_i \to \mathbf{k}_f$ with the exchange of ℓ photons is given by

$$\frac{d\sigma^{\ell}}{d\Omega} = (2\pi)^4 \frac{m^2}{\hbar^4} \frac{k_f(\ell)}{k_i} |T^{\ell}_{\mathbf{k}_f,\mathbf{k}_i}|^2 \tag{16}$$

In general, the evaluation of the T-matrix element $T^{\ell}_{\mathbf{k}_f,\mathbf{k}_i}$ is a very difficult task, but there are two limiting cases in which considerable simplifications occur. The first one arises when the first Born approximation provides an accurate description of the scattering by $V(\mathbf{r})$. Roughly speaking, this means (for most interaction potentials) that the potential is weak enough or that the electron kinetic energy is large compared to its potential energy. In this case the differential cross section (16) is given to good approximation by the first Born result

$$\frac{d\sigma^{B1,\ell}}{d\Omega} = (2\pi)^4 \frac{m^2}{\hbar^4} \frac{k_f(\ell)}{k_i} J_\ell^2(\mathbf{\Delta}.\mathbf{\alpha}_o)\ |\tilde{V}(\mathbf{\Delta})|^2 \tag{17}$$

obtained by Bunkin and Fedorov [7]. If $E_{k_i} = \hbar^2 k_i^2/2m \gg \ell\hbar\omega$, we see from equation (11) that $k_f(\ell)/k_i \simeq 1$, so that

$$\frac{d\sigma^{B1,\ell}}{d\Omega} \simeq J_\ell^2(\underset{\sim}{\Delta}\cdot\underset{\sim}{\alpha}_0)\,\frac{d\sigma^{B1}}{d\Omega} \tag{18}$$

where $d\sigma^{B1}/d\Omega$ is the field-free (no laser) first Born differential cross section. We remark that in the above equation the laser parameters enter only through the Bessel function. From equation (18) we deduce with the help of the sum rule

$$\sum_{\ell=-\infty}^{+\infty} J_\ell^2(x) = 1 \tag{19}$$

that

$$\sum_{\ell=-\infty}^{+\infty} \frac{d\sigma^{B1,\ell}}{d\Omega} \simeq \frac{d\sigma^{B1}}{d\Omega} \tag{20}$$

The second limiting case for which matters simplify is the low frequency (soft photon) limit, such that the laser photon energy $\hbar\omega$ is small compared to the projectile energy E_{k_i}. Kroll and Watson [8] have shown that in this case the T-matrix element $T^\ell_{\underset{\sim}{k}_f,\underset{\sim}{k}_i}$ is given by

$$T^\ell_{\underset{\sim}{k}_f,\underset{\sim}{k}_i} = J_\ell(\underset{\sim}{\Delta}\cdot\underset{\sim}{\alpha}_0)\langle \chi_{\underset{\sim}{k}_f^*} | \mathcal{T}(E_{k_i^*}) | \chi_{k_i^*}\rangle + O(\omega^2) \tag{21}$$

where $\underset{\sim}{k}_i^*$ and $\underset{\sim}{k}_f^*$ are the shifted wave vectors

$$\underset{\sim}{k}_i^* = \underset{\sim}{k}_i + \frac{\ell m\omega}{\hbar\underset{\sim}{\Delta}\cdot\underset{\sim}{\alpha}_0}\underset{\sim}{\alpha}_0 , \qquad \underset{\sim}{k}_f^* = \underset{\sim}{k}_f + \frac{\ell m\omega}{\hbar\underset{\sim}{\Delta}\cdot\underset{\sim}{\alpha}_0}\underset{\sim}{\alpha}_0 \tag{22}$$

and $\mathcal{T}(E_{k_i^*})$ is the T-operator in the absence of the laser, corresponding to the energy $E_{k_i^*} = \hbar^2 k_i^{*2}/2m$. Setting $E_{k_f^*} = \hbar^2 k_f^{*2}/2m$ and using equation (11) yields $E_{k_i^*} = E_{k_f^*}$. The T-matrix element on the right of equation (21) is therefore on shell, so that the scattering in the presence of the laser field can be completely described in terms of scattering in the absence of the laser field. This is a generalization of a result derived by Low [9,10] in the weak field limit for single photon processes. Using equation (21), the differential cross section for transfer of ℓ photons is

$$\frac{d\sigma^\ell}{d\Omega} = \frac{k_f(\ell)}{k_i} J_\ell^2(\underset{\sim}{\Delta}\cdot\underset{\sim}{\alpha}_0)\,\frac{d\sigma}{d\Omega}(\underset{\sim}{k}_f^*,\underset{\sim}{k}_i^*) + O(\omega^2) \tag{23}$$

where $d\sigma(\underset{\sim}{k}_f^*,\underset{\sim}{k}_i^*)/d\Omega$ refers to the transition $\underset{\sim}{k}_i^* \to \underset{\sim}{k}_f^*$ in the absence of the laser. If ω is small enough to neglect the ℓ-dependence in k_f,

k_i^* and k_f^*, then the sum rule (19) may again be used to give

$$\sum_{\ell=-\infty}^{+\infty} \frac{d\sigma^\ell}{d\Omega} \approx \frac{d\sigma}{d\Omega} \tag{24}$$

where $d\sigma/d\Omega$ is the field-free differential cross section.

The Kroll-Watson result may be understood by noting that in the low frequency regime the electron cannot completely exchange a photon during the collision since the collision time (about $\hbar/E_{k_i}$) is small with respect to the time (roughly $2\pi/\omega$) necessary for a photon to be exchanged. Thus, if $\hbar\omega \ll E_{k_i}$, the laser field has only a small effect on the scattering of the electron by the potential, and the cross section for scattering with the exchange of photons "decouples" into the product of a field-free scattering cross section and a factor depending only on the coupling of the laser field with a free electron. We also remark that when scattering resonances exist in the absence of the laser the Kroll-Watson results must be modified [11-13].

Although a number of formal derivations of the low-frequency results have been given - see for example the references [2,3,13-16] - there are very few explicit calculations in which the range of validity of the Kroll-Watson soft-photon approximation has been investigated. Shakeshaft [17] has developed a method for treating electron scattering from a potential in a radiation field, which he applied to the case of a separable potential, and more recently to a Gaussian potential of the form $V(r) = V_0 \exp(-r/a)^2$. The Kroll-Watson approximation was found to be surprisingly accurate, even for cases such that $\hbar\omega/E_{k_i} = 0.5$. A similar conclusion was reached by Joachain, Piraux and Nedeljkovic [18], who studied the scattering of an electron by a screened Coulomb (Yukawa) potential of the form $V(r) = V_0 r^{-1} \exp(-r/a)$, in the presence of a laser field, using second order perturbation theory in the electron-potential interaction V.

Finally, we mention that the scattering of an electron by a potential in the presence of an intense, high frequency laser field has been investigated by Gavrila and Kaminski [19].

3. ELECTRON-ATOM COLLISIONS

Let us now consider the scattering of electrons by "real" atomic targets - having an internal structure - in the presence of a laser field. As an example, we shall treat the case of the "elastic" scattering (accompanied by the transfer of ℓ photons) of fast electrons by atoms in the presence of a strong laser field, following the recent work of Byron and Joachain [20]. We shall again treat the laser field classically as a spatially homogeneous electric field, linearly polarized and single mode. We also assume that the laser photon energy $\hbar\omega$ is low with respect to typical atomic excitation energies.

Our first task is to obtain the "dressed" states of the target atom embedded in the laser field. We must therefore solve the Schrödinger equation

$$i\hbar \frac{\partial}{\partial t} \Phi(\underset{\sim}{r}_1, \underset{\sim}{r}_2, \ldots, \underset{\sim}{r}_N, t) = H_T \Phi(\underset{\sim}{r}_1, \underset{\sim}{r}_2, \ldots, \underset{\sim}{r}_N, t) \tag{25}$$

where $\underset{\sim}{r}_1, \underset{\sim}{r}_2, \ldots, \underset{\sim}{r}_N$ are the target electron coordinates and H_T is the Hamiltonian of the target atom in the presence of the field. As stated in the introduction, we shall consider laser fields which, although strong by laboratory standards, are nevertheless such that $\mathscr{E}_o \ll e/a_o^2 \simeq 5 \times 10^9$ V cm^{-1}. The Schrödinger equation (25) can then be solved by using first-order time-dependent perturbation theory. For example, in the case of an hydrogen atom target one finds in this way for the "dressed" ground state wave function, to first order in $\mathscr{E}_o$

$$\Phi_o(\underset{\sim}{r}_1, t) = \exp(-\frac{i}{\hbar} w_o t) \exp(-i\underset{\sim}{a} \cdot \underset{\sim}{r}_1)[\psi_o(r_1) - \sin(\omega t) \sum_n \frac{M_{np,o}}{\hbar\omega_{no}} \psi_{np}(\underset{\sim}{r}_1)$$

$$+ i \cos(\omega t) \sum_n (\frac{\omega}{\omega_{no}}) \frac{M_{np,o}}{\hbar\omega_{no}} \psi_{np}(\underset{\sim}{r}_1)] \tag{26}$$

where ψ_o is the "undressed" ground state wave function, ψ_{np} is the nth "undressed" p-state, w_o and w_n are the corresponding eigenenergies, $\hbar\omega_{no} = w_n - w_o$ and the summation includes an integration over the continuum states. Moreover, we have written $\underset{\sim}{a} = e\underset{\sim}{A}/\hbar c$ and $M_{np,o} = \mathscr{E}_o \cdot \langle \psi_{np}|e\underset{\sim}{r}_1|\psi_o\rangle$. We remark that in equation (26) the effect of the laser field is characterized by the two dimensionless parameters $\underset{\sim}{a} \cdot \underset{\sim}{r}_1 \simeq aa_o = e \mathscr{E}_o a_o/\hbar\omega$ and $|M_{np,o}|/\hbar\omega_{no} \simeq e \mathscr{E}_o a_o/\hbar\omega_{no}$. The former can be of order unity even for field strengths $\mathscr{E}_o \ll e/a_o^2$; the latter is always small when $\mathscr{E}_o \ll e/a_o^2$.

We now turn to the electron-atom collision problem in the presence of the laser field, starting with direct (no exchange) collisions. The Hamiltonian of the electron-atom system in the presence of the laser field may be written in the direct (initial) arrangement channel as

$$H = H_F + H_T + V_d \tag{27}$$

where H_F is the Hamiltonian of the "free" electron in the laser field, H_T is the "dressed" target Hamiltonian appearing in equation (25) and V_d is the electron-atom interaction in the direct (initial) arrangement channel. The Schrödinger equation describing the non-relativistic motion of the "free" electron in the laser field, governed by H_F, is of course equation (2) and its solutions are the Volkov wave functions (3). Thus, in the absence of electron-atom interaction ($V_d = 0$) the wave functions of the system are products of Volkov wave functions $\chi_{\underset{\sim}{k}}(\underset{\sim}{r},t)$ for the "free" electron times "dressed" target wave functions $\Phi(\underset{\sim}{r}_1,\underset{\sim}{r}_2,...,\underset{\sim}{r}_N,t)$.

To first order in V_d, i.e. in first Born approximation, the S-matrix element for direct "elastic" scattering ($\underset{\sim}{k}_i \rightarrow \underset{\sim}{k}_f$) from a "dressed" target state Φ_o, in the presence of the laser field, is given by

$$S_{e\ell}^{B1} = -\frac{i}{\hbar}\int_{-\infty}^{+\infty} dt \langle \chi_{\underset{\sim}{k}_f}(\underset{\sim}{r},t)\Phi_o(\underset{\sim}{r}_1,\underset{\sim}{r}_2,..,\underset{\sim}{r}_N,t)\,|V_d|\,\chi_{\underset{\sim}{k}_i}(\underset{\sim}{r},t)\Phi_o(\underset{\sim}{r}_1,\underset{\sim}{r}_2,..,\underset{\sim}{r}_N,t)\rangle \tag{28}$$

The t integration is readily performed by using equation (9). Working from now on in atomic units, we have

$$S_{e\ell}^{B1} = (2\pi)^{-1}\, i \sum_{\ell=-\infty}^{+\infty} \delta(E_{k_f}-E_{k_i}-\ell\omega)\, f_{e\ell}^{B1,\ell} \tag{29}$$

where $f_{e\ell}^{B1,\ell}$ is the first Born approximation to the "elastic" scattering amplitude with the transfer of ℓ photons. For the case of an atomic hydrogen target, we have $V_d = |\underset{\sim}{r}-\underset{\sim}{r}_1|^{-1} - r^{-1}$ and we find with the help of equations (3) and (26) that

$$\begin{aligned} f_{e\ell}^{B1,\ell}(\underset{\sim}{\Delta}) &= J_\ell(\underset{\sim}{\Delta}\cdot\underset{\sim}{\alpha}_o)\, f_{e\ell}^{B1}(\Delta) \\ &\quad - i\, J_\ell'(\underset{\sim}{\Delta}\cdot\underset{\sim}{\alpha}_o) \sum_n \frac{f_{o,np}^{B1}(\underset{\sim}{\Delta})\, M_{np,o} + M_{o,np}\, f_{np,o}^{B1}(\underset{\sim}{\Delta})}{w_n - w_o} \end{aligned} \tag{30}$$

where we recall that $\underset{\sim}{\Delta} = \underset{\sim}{k}_i - \underset{\sim}{k}_f$ and the first Born amplitudes $f_{e\ell}^{B1}$, $f_{np,o}^{B1}$ and $f_{o,np}^{B1}$ refer, respectively, to the scattering processes o → o (elastic scattering), o → np and np → o in the absence of the laser. We note that because $J_{-\ell}(x) = (-1)^\ell J_\ell(x)$, it follows that $f_{e\ell}^{B1,-\ell} = (-1)^\ell f_{e\ell}^{B1,\ell}$.

The first term on the right of equation (30) is of the "potential scattering" type. Taken alone, it would lead to the Bunkin and Fedorov result (17). The second term represents the effect of the dipole distortion of the atomic target by the laser field, which "mixes in" the p states. We emphasize that although $|M_{np,o}|/(w_n-w_o)$ is small, the amplitudes $f^{B1}_{o,np}(\underset{\sim}{\Delta})$ and $f^{B1}_{np,o}(\underset{\sim}{\Delta})$ are very large for small values of Δ. Hence at small momentum transfers this second term is very important. In fact, except for the case $\ell = 0$ it dominates over the first term at small angles.

In the case of scattering by atomic hydrogen, the summation in equation (30) can be performed exactly. However, for more complex target atoms, this procedure would be very difficult, so that the closure approximation (in which the target energy differences $w_n - w_o$ are replaced by an average excitation energy $\bar{w}$) suggests itself. Using the closure approximation we see that equation (30) reduces to

$$f^{B1,\ell}_{e\ell}(\underset{\sim}{\Delta}) = J_\ell(\underset{\sim}{\Delta}\cdot\underset{\sim}{\alpha}_0) f^{B1}_{e\ell}(\Delta) + \frac{4}{\Delta^2 \bar{w}} J'_\ell(\underset{\sim}{\Delta}\cdot\underset{\sim}{\alpha}_0)\, \mathcal{E}_0 \cdot \nabla_{\underset{\sim}{\Delta}} \langle o|\exp(i\underset{\sim}{\Delta}\cdot\underset{\sim}{r}_1)|o\rangle \qquad (31)$$

and the generalization to complex target atoms is obvious.

As an illustration of the above results, we show in Fig. 1 the first Born differential cross section

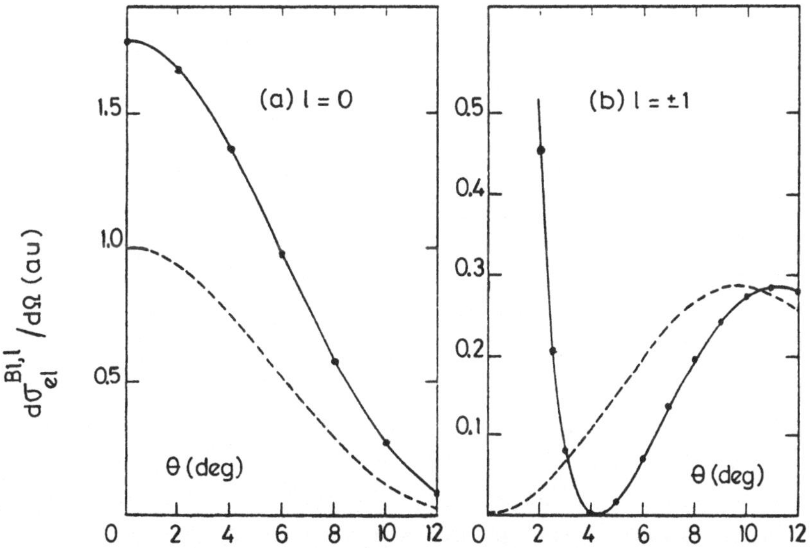

Fig. 1. The first Born differential cross section (32), in atomic units, for "elastic" electron-atomic hydrogen scattering with the transfer of (a) no photons and (b) one photon, at an incident electron energy of 100 eV. The laser photon energy is $\hbar\omega = 2$eV, the electric field strength $\mathcal{E}_0 = 10^8$ V cm^{-1} and $\hat{\underset{\sim}{\mathcal{E}}}_0$ is parallel to $\underset{\sim}{\Delta}$. —— : calculation including the "dressing" of the target, i.e. using equation (30) for $f^{B1,\ell}_{e\ell}$; • same calculation, using the closure approximation of equation (31), with $\bar{w} = 4/9$; --- : results obtained by neglecting the "dressing" of the target, i.e. by omitting the second term on the right of equation (30). Taken from [20].

$$\frac{d\sigma_{e\ell}^{B1,\ell}}{d\Omega} = \frac{k_f(\ell)}{k_i} |f_{e\ell}^{B1,\ell}|^2 \tag{32}$$

for "elastic" electron-atomic hydrogen scattering at an incident electron energy of 100 eV, accompagnied by no transfer of photons ($\ell = 0$, Fig. 1(a)) and by the transfer of one photon ($\ell = \pm 1$, Fig. 1(b)). We have chosen an electric field strength $\mathcal{E}_o = 10^8$ V cm^{-1} and a laser photon energy $\hbar\omega = 2$eV. The direction $\hat{\mathcal{E}}_o$ of the electric field was taken to be parallel to the momentum transfer. In Fig. 1(b) the enhancement of the differential cross section due to the "dressing" of the target is seen to be very important at small angles. In fact, it is a simple matter to show that for any atom one has in this geometry ($\hat{\mathcal{E}}_o$ parallel to Δ), for $\ell \neq 0$

$$f_{e\ell}^{B1,\ell}(\theta=o) = J_\ell(\Delta\,\alpha_o)[f_{e\ell}^{B1}(\theta=o) - \frac{4\alpha\,E_{k_i}}{\ell}] \tag{33}$$

where α is the dipole polarizability of the target atom. Also displayed in Fig. 1 are the results obtained by using the closure approximation of equation (31) with the value $\bar{w} = 4/9$ giving the correct dipole polarizability $\alpha = 4.5$ of atomic hydrogen. The agreement of these closure results with the values calculated from equation (30) is excellent.

We now consider higher order contributions (in the electron-atom interaction V_d) to $f_{e\ell}^{\ell}$, the full direct amplitude for "elastic" scattering with the transfer of ℓ photons. A complete analysis of this problem is very difficult, but several important conclusions can be reached without performing detailed calculations. Firstly, the term $f_{e\ell}^{B1}$ in equation (30) will be replaced by $f_{e\ell}$ (the full field-free direct amplitude) in which, according to the low frequency method of Kroll and Watson [8], the electron wave vectors (i.e. momenta in atomic units) k_i and k_f are replaced respectively by the shifted wave vectors k_i^* and k_f^* given by equation (22). Similarly, the amplitudes $f_{np,o}^{B1}$ and $f_{o,np}^{B1}$ will also be modified by higher order contributions. However, because these terms arising from target "dressing" effects are important only at small angles, and since s-p amplitudes are known to be dominated in the small angle region by the first Born term, we shall neglect these higher-order modifications to $f_{np,o}^{B1}$ and $f_{o,np}^{B1}$. Moreover, we shall also neglect the shift of momenta $k_i \to k_i^*$ and $k_f \to k_f^*$ in calculating $f_{e\ell}$ since k_i and k_f are large for fast electrons and therefore the shift is significant only at small momentum transfers, where the second term of equation (30) dominates for $\ell \neq 0$ [21]. For fast incident electrons a good approximation to the field-free direct amplitude

$f_{e\ell}$ is provided by the eikonal-Born series (EBS) direct amplitude $f_{e\ell}^{EBS}$ [22-24], so that we shall write the full direct amplitude $f_{e\ell}^{\ell}$ for "elastic" scattering accompanied by the transfer of ℓ photons as

$$f_{e\ell}^{\ell}(k_i,\underset{\sim}{\Delta}) = J_\ell(\underset{\sim}{\Delta}\cdot\underset{\sim}{\alpha}_o)\ f_{e\ell}^{EBS}(k_i,\Delta) - i\ J_\ell'(\underset{\sim}{\Delta}\cdot\underset{\sim}{\alpha}_o)\ \sum_n \frac{f_{o,np}^{B1}(\underset{\sim}{\Delta})\ M_{np,o} + M_{o,np}\ f_{np,o}^{B1}(\underset{\sim}{\Delta})}{w_n - w_o} \tag{34}$$

Let us now examine exchange effects in the presence of a laser field, a problem which has been studied by several authors [25-28]. Since we are dealing here with fast electrons we shall, in the spirit of the EBS approach, consider only the leading term of $g_{e\ell}^{\ell}$, the exchamge amplitude for "elastic" scattering with the transfer of ℓ photons. In addition, because field-free s-p exchange amplitudes are small in comparison with field-free elastic exchange amplitudes at small angles, "dressing" effects are small in this angular region. Moreover, the momentum shift $\underset{\sim}{k}_i \to \underset{\sim}{k}_i^*$, $\underset{\sim}{k}_f \to \underset{\sim}{k}_f^*$ may again be neglected since at small angles the second term on the right of equation (34) is dominant. Thus we may write to good approximation, for fast electrons and small angles,

$$g_{e\ell}^{\ell}(k_i,\underset{\sim}{\Delta}) = J_\ell(\underset{\sim}{\Delta}\cdot\underset{\sim}{\alpha}_o)\ g_{Och}(k_i,\Delta) \tag{35}$$

where g_{Och} is the Ochkur elastic exchange amplitude [29].

The full differential cross section for the "elastic" scattering of unpolarized electrons by unpolarized hydrogen atoms, accompanied by the transfer of ℓ photons, is given by

$$\frac{d\sigma_{e\ell}^{\ell}}{d\Omega} = \frac{k_f(\ell)}{k_i}\left(\frac{1}{4}\ |f_{e\ell}^{\ell}+g_{e\ell}^{\ell}|^2 + \frac{3}{4}\ |f_{e\ell}^{\ell}-g_{e\ell}^{\ell}|^2\right) \tag{36}$$

In Fig. 2 we show the values we have obtained for $d\sigma_{e\ell}^{\ell}/d\Omega$ by using respectively equation (34) for $f_{e\ell}^{\ell}$ and (35) for $g_{e\ell}^{\ell}$ (full curves), at an incident electron energy of 100 eV, and for the cases $\ell = 0$ (Fig. 2(a)) and $\ell = \pm 1$ (Fig. 2(b)). The laser field parameters are the same as in Fig. 1. We also display in Fig. 2 the results obtained by using the closure approximation with $\bar{w} = 4/9$ (dots) to calculate the second term on the right of equation (34); these closure results are seen to be in excellent agreement with the full results obtained from equation (34). Also shown in Fig. 2 are the results obtained by neglecting the "dressing" of the target, i.e. by omitting the second term on the right of equation (34). Comparing the results of Fig. 2(a) with the corresponding first Born ones shown in Fig. 1(a), we remark that the small angle differential cross section is considerably modified. This

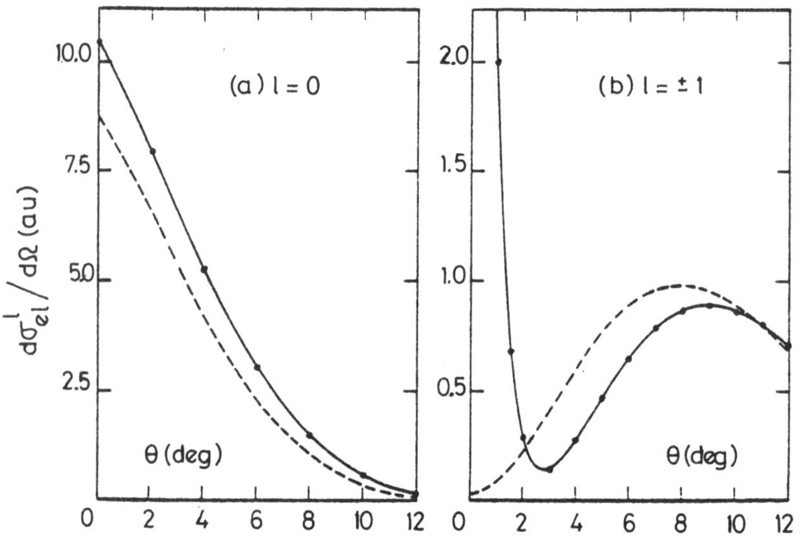

Fig. 2. The full differential cross section (36), in atomic units, for "elastic" electron-atomic hydrogen scattering with the transfer of (a) no photons and (b) one photon, at an incident electron energy of 100 eV. The laser field parameters are the same as in Fig. 1. —— : calculation using equation (34) for $f_{e\ell}^{\ell}$ and equation (35) for $g_{e\ell}^{\ell}$, which includes the "dressing" of the target states via the second term on the right of equation (34); • same calculation, using the closure approximation with $w = 4/9$ to evaluate this second term; --- results obtained by neglecting the "dressing" of the target, i.e. by omitting this second term. Taken from [20].

is mainly due to polarization and absorption effects coming from the second order contribution to the field-free amplitude $f_{e\ell}$. It is also apparent from Fig. 2(a) that the relative importance of the "dressing" of the target is rather small in the $\ell = 0$ case. By contrast, the results displayed in Fig. 2(b) for the $\ell = \pm 1$ case are very similar to those shown in Fig. 1(b); in both cases we see that there is a spectacular enhancement of the differential cross section at small momentum transfers, due to the target "dressing" effects, which are identical in equations (30) and (34). Important "dressing" effects are also present at small Δ for other values of $\ell(\neq 0)$.

We have therefore shown that in "elastic" electron-atom collisions accompanied by the transfer of $\ell(\neq 0)$ photons, dramatic modifications of the angular distribution are predicted at small momentum transfers, due to the "dressing" of atomic states by a strong laser field. Since experiments of this kind involving atomic hydrogen targets would be difficult, it is sensible to look for these "dressing" effects in electron collisions with noble gas atoms. We have studied electron-helium "elastic" collisions, and found that in this case the "dressing" effects are reduced because of the smaller helium dipole polarizability, as can be understood from equation (33). Thus a more favorable target for obtaining the experimental confirmation of the "dressing" effects discussed above would be a noble gas atom with a large dipole polarizability, such as argon.

The "elastic" scattering of fast electrons by atoms in the presence of a non-resonant strong laser field is only one interesting aspect of laser-assisted electron-atom collisions. Other problems which have attracted a great deal of theoretical attention include low-energy electron-atom radiative scattering [4], electron-impact excitation [4, 30, 31] and ionization [32, 33] of atoms in the presence of a laser field, electron-atom scattering in a resonant laser [2] and electron-atom collisions in intense, high frequency laser fields [34]. Hopefully, this wealth of theoretical calculations should stimulate new experiments, which are very much needed in the field of laser-assisted electron-atom collisions.

REFERENCES

1. M. Gavrila and M. Van der Wiel, Comments At. Mol. Phys. 8, 1 (1978); M. Gavrila, in Electronic and Atomic Collisions, ed. by G. Watel (North-Holland, Amsterdam, 1978), p. 165.
2. M.H. Mittleman, Introduction to the Theory of Laser-Atom Interactions (Plenum Press, New York, 1982).
3. L. Rosenberg, Adv. At. Mol. Phys. 18, 1 (1982).
4. F.H.M. Faisal, Comments. At. Mol. Phys. 15, 119 (1984).
5. According to L. Rosenberg [Phys. Rev. A 23, 2283 (1981)] a measure of the effective coupling between the electron and the field is given by the ratio δ_1/δ_2, where $\delta_1 = eA_0/cp = eA_0/c\hbar k$ is a dimensionless parameter providing a measure of the field strength and $\delta_2 = \hbar\omega/E_k$. This is consistent with the parameter $\beta = k\alpha_0$ which we have introduced, since $\beta = 2\delta_1/\delta_2$.
6. A. Weingartshofer, E.M. Clarke, J.K. Holmes and C. Jung, Phys. Rev. A 19, 2371 (1979); A. Weingartshofer, J. Holmes, J. Sabbagh and S. Chin, J. Phys. B 16, 1805 (1983).
7. F.V. Bunkin and M.V. Fedorov, Sov. Phys. JETP 22, 844 (1966).
8. N.M. Kroll and K.M. Watson, Phys. Rev. A 8, 804 (1973).
9. F.E. Low, Phys. Rev. 110, 974 (1958).
10. L.S. Brown and R.L. Goble, Phys. Rev. 173, 1505 (1968).
11. H. Kruger and C. Jung, Phys. Rev. A 17, 1706 (1978).

12. M.H. Mittleman, Phys. Rev. A 20, 1965 (1979).
13. L. Rosenberg, Phys. Rev. A 23, 2283 (1981).
14. G. Ferrante and C. Leone, Nuovo Cimento B 48, 35 (1978); G. Ferrante, Nuovo Cimento B 52, 229 (1979).
15. M.H. Mittleman, Phys. Rev. A 19, 134 (1979).
16. L. Rosenberg, Phys. Rev. A 20, 275 (1979); Phys. Rev. A 20, 457 (1979); Phys. Rev. A 20, 1352 (1979); Phys. Rev. A 21, 157 (1980); Phys. Rev. A 21, 1939 (1980).
17. R. Shakeshaft, Phys. Rev. A 28, 667 (1983); R. Shakeshaft, Phys. Rev. A 29, 383 (1984).
18. C.J. Joachain, B. Piraux and N.N. Nedeljkovic, to be published.
19. M. Gavrila and J.Z. Kaminski, Phys. Rev. Letters 52, 613 (1984).
20. F.W. Byron, Jr. and C.J. Joachain, J. Phys. B 17, L 295 (1984).
21. When $\ell = 0$ there is of course no approximation in neglecting the shift $\mathbf{k}_i \rightarrow \mathbf{k}_i^*$, $\mathbf{k}_f \rightarrow \mathbf{k}_f^*$, as seen from equation (22).
22. F.W. Byron, Jr. and C.J. Joachain, Phys. Rev. A 8, 1267 (1973).
23. C.J. Joachain, Quantum Collision Theory (3rd ed., North-Holland, Amsterdam, 1983), Chapter 19.
24. F.W. Byron, Jr. and C.J. Joachain, Phys. Reports 34, 233 (1977).
25. M.H. Mittleman, Phys. Rev. A 21, 79 (1980).
26. G. Ferrante, C. Leone and F. Trombetta, J. Phys. B 15, L 475 (1982).
27. M.A. Prasad and K. Unnikrishnan, J. Phys. B 16, 3443 (1983).
28. F. Trombetta, C.J. Joachain and G. Ferrante, Proc. 13th ICPEAC (Berlin, 1983) p. 678.
29. V.I. Ochkur, Zh. Eksp. Teor. Fiz. 45, 734 (1963) [Sov. Phys. JETP 18, 503 (1964)].
30. N.K. Rahman and F.H.M. Faisal, J. Phys. B 9, L 275 (1976); J. Phys. B 11, 2003 (1978).
31. H.S. Brandi, B. Koiller and H.G.P. Lins de Barros, Phys. Rev. A 19, 1058 (1979).
32. P. Cavaliere, G. Ferrante and C. Leone, J. Phys. B 13, 4495 (1980).
33. M. Zarcone, D.L. Moores and M.R.C. McDowell, J. Phys. B 14, L 11 (1983).
34. M. Gavrila, in this volume.

ELECTRON SCATTERING IN STOCHASTIC RADIATION FIELDS

R. Daniele[+], G. Ferrante[+], F. Morales[o] and F. Trombetta[+]

+ Istituto di Fisica, via Archirafi 36, 90123 Palermo, Italy.

o Istituto di Fisica, Facoltà di Ingegneria, Parco d'Orleans, 90128 Palermo, Italy.

1. Introduction

This Lecture is concerned with the problem of particle-atom collisions in the presence of strong radiation fields with randomly distributed parameters. The rationale of this problem may be formulated as follows.

The last years have witnessed an increasing interest in the field-assisted particle-atom collisions, though restricted mainly to theoretical investigations (For references and reviews, see, for instance, ref.s 1-4 and contributions to this volume). A distinguishing feature of the theoretical efforts has been to consider fields of arbitrary strength, in order to be able to treat strong field situations. In fact, only in the latter case, a nonresonant radiation field is likely to yield significant modifications of the collision parameters.

In most of the investigations, in extending the conventional scattering theory methods to include a strong radiation field, the latter has been generally considered within the simplest model of a pure, spatially homogeneous field, in dipole approximation. However, the intense real lasers required to appreciably modify the collision event, are likely to be poorly described by the single mode field model. So, to make more close contact with laboratory conditions and more reliable predictions, there is the need to improve the models to describe more realistically the strong assisting radiation field. This is the first, basic scope in considering this problem here.

In general, in constructing the theory of collision events in the presence of strong radiation fields, it is of interest in its own right to consider differents laser models, corresponding to different possible real situations, and to establish how they modify the scattering parameters. In the related area of multiphoton processes in atoms and molecules the analogous problem has been the subject of several investigations in recent years[5]; and to some extent our task below will be to extend to field-assisted collisions the analysis now familiar in the realm of multiphoton processes, borrowing in particular models originally worked out for the latter. Of course, it is needless to say that the physical situations encountered in collisions may be quite different compared with those considered in processes like multiphoton excitation, ionization, and similar. Or the information obtained within the field-assisted collisions may will be different from what is currently known from the area of multiphoton processes. As shown below, it is just what happens. As a matter of fact, in some cases of field-assisted collisions (potential scattering, relatively energetic particle-atom scattering and so on) the field is included in the theoretical treatment exactly, to all orders, through exact solutions for the asymptotic scattering states; or accurately enough through solutions obtained within the formalism of the quasienergies. Thus field-assisted collisions offer the possibility to investigate rather accurately the properties of strong radiation fields up to consider very high intensity domains. It is expected to produce new information on the properties of strong radiation fields, which may significantly complement what is presently known from research areas, more directly concerned with these aspects of radiation-matter interaction. This possibility forms the second scope of the present paper. Investigations on the field coherence role in field-assisted collisions are at their beginning, and the information available in the literature is still quite incomplete[6-12].

Below we will be specifically concerned with the theory of charged particle scattered by a static potential V(r) in the presence of a strong radiation field, considered within three models, familiar in the literature.

The first is the phase diffusion model (PDM), in which the field is assumed to undergo only phase fluctuations, and corresponds to an intensity stabilized single mode laser field. The second is that of a chaotic field (CH), in which the field is assumed to undergo both amplitude and phase fluctuations, and corresponds to a pulsed multimode laser field with a large number of uncorrelated modes. The third model is that of a gaussian-amplitude field (G), in which the field undergoes only amplitude fluctuations. This last model is an example of a field which undergoes stronger amplitude fluctuations than a chaotic field, and its consideration is useful to isolate those effects due solely to amplitude fluctuations.

The limited space of this Lecture does not allow to give details of the statistical descriptions of the above radiation field models, and it is considered a pitiful flaw. However, they may be found in several places (see, for instance, ref.s 13 and 14).

In Section 2 we review briefly the theory of nonrelativistic potential scattering in the presence of an arbitrary strong radiation field, treated classically and in dipole approximation. In Section 3 we restrict the theory to First Born Approximation (FBA), take the averages of the squared modulus of the S-matrix appropriate to the field models mentioned above; and obtain the corresponding transition probabilities and differential cross sections. Section 4 considers some limiting cases. In Section 5 and 6 we present a number of results obtained either in analytical form or by numerical calculations. They are mainly meant to compare the scattering parameters, obtained within different field models, and to point out similarities and differences with pertinent results from other research areas. Section 7 concludes the paper with some short remarks.

2. Potential Scattering in the Presence of a Strong Radiation Field

We start by considering a particle of mass m and charge e being scattered by a static, local potential V(r) in the presence of a radiation field, represented by its vector potential $\underset{\sim}{A}(t)$. The corresponding Schrodinger equation is

$$\frac{1}{2m}\left[\frac{\hbar}{i}\nabla - \frac{e}{c}A(t)\right]^2\psi(\underline{r},t) + V(r)\psi(\underline{r},t) = i\hbar\dot{\psi}(\underline{r},t) \tag{2.1}$$

Assuming V(r) as the perturbation responsible for the transition, the unperturbed initial and final states are the solution of the equation

$$\frac{1}{2m}\left[\frac{\hbar}{i}\nabla - \frac{e}{c}A(t)\right]^2\chi(\underline{r},t) = i\hbar\dot{\chi}(\underline{r},t) \tag{2.2}$$

i.e., the Volkov waves

$$\chi_{\underline{k}}(\underline{r},t) = \exp\left\{i\,\underline{k}\cdot\underline{r} - \frac{i\hbar}{2m}\int^t\left[k^2 - \frac{2e}{\hbar c}\,\underline{k}\cdot\underline{A}(\tau)\right]d\tau\right\}\times$$
$$\times\exp\left\{-\frac{i}{\hbar}\int^t\frac{e^2}{2mc^2}A^2(\tau)\,d\tau\right\} \tag{2.3}$$

where $\hbar\underline{k}$ is the particle momentum averaged over the field period or statistics, as the case requires. The exact S-matrix for the transition $i \to f$ is given by

$$S = (-i/\hbar)(\chi_f, V\psi_i^+) \tag{2.4}$$

where χ_f is given by (2.3) for the final state and ψ_i^+ is the exact formal solution of (2.1), with standard causal boundary conditions, for the incident channel. The round brackets (...,...) in (2.4) denotes both space and time integrations. Formally, the solution of (2.1) is written as

$$\psi_i^+ = \chi_i + G^+V\chi_i \tag{2.5}$$

where G^+ is the retarded full Green function (i.e., in the presence of both the static potential and the radiation field), which, expanded in powers of the scattering potential, is

$$G^+ = G_0^+ + G_0^+VG_0^+ + G_0^+VG_0^+VG_0^+ + \cdots \tag{2.6}$$

G_0^+ being the retarded Green function in the absence of V(r). Using (2.5)-(2.6) in (2.4), the S-matrix is obtained as

$$S = (-i/\hbar)\left[(\chi_f, V\chi_i) + (\chi_f, VG_0^+V\chi_i) + \cdots\right] = \sum_{\nu=1}^{\infty} S^{(\nu)} \tag{2.7}$$

where the ν-th order contains ν times V(r) and (ν -1) times G_o^+. Let us explicitly remark that in (2.7) at each order in V(r), the field enters exactly, through the unperturbed states $\chi_{\underset{\sim}{k}}$. The first two orders are explicitly given by

$$S^{(1)} \underset{T\to\infty}{=} (-i/\hbar) \int_{-T}^{T} dt \int d^3r \; \chi_f^*(\underline{r},t)\, V(r)\, \chi_i(\underline{r},t) , \tag{2.8}$$

$$S^{(2)} \underset{T\to\infty}{=} (-i/\hbar)^2 \int_{-T}^{T} dt_1 \int_{-T}^{T} dt_2 \sum_m \int d^3r_1 \; \chi_f^*(\underline{r}_1,t_1)\, V(r_1)\, \chi_m(\underline{r}_1,t_1) \times$$

$$\times \int d^3r_2 \; \chi_m^*(\underline{r}_2,t_2)\, V(r_2)\, \chi_i(\underline{r}_2,t_2) \tag{2.9}$$

Finally, in order to account for the field fluctuations, we have to make the average

$$\langle |S|^2 \rangle = \langle |\sum_\nu S^{(\nu)}|^2 \rangle \tag{2.10}$$

properly accounting for the specific statistics of the assisting field. The average of the entire S-matrix (2.10) appears to be a formidable task for each case under consideration here, and for the time being we content ourselves with the averages of the first term of (2.10) only.

3. First-Order Theory of the Potential Scattering in the Presence of a Strong Stochastic Field

Using (2.3), (2.9) and (2.10), the averaged first-order squared S-matrix is given by

$$\langle |S^{(1)}|^2 \rangle = \lim_{T\to\infty} \int_{-T}^{T} dt \int_{-T}^{T} dt' \; \exp\{i\omega_{fi}(t-t')\} \times$$

$$\times F(t,t')\, C(\underline{Q}) \tag{3.1}$$

where

$$C(\underset{\sim}{Q}) = \hbar^{-2} \left| \int d^3r \; V(r) \exp(i\, \underset{\sim}{Q}\cdot \underset{\sim}{r}) \right|^2 \tag{3.2}$$

$$F(t,t') = \langle \exp\{i\alpha_{fi}\int_{t'}^{t} d\tau\, A(\tau)\}\rangle \tag{3.3}$$

$$\omega_{fi} = (\varepsilon_f - \varepsilon_i)/\hbar \;, \qquad \varepsilon_\alpha = \hbar^2 K_\alpha^2/2m \;, \qquad (\alpha = i,f)\;, \tag{3.4}$$

$$\alpha_{fi} = (e/mc)\,\hat{e}_L\cdot\underset{\sim}{Q} \;, \qquad \hat{e}_L = \underset{\sim}{A}(t)/A(t) \;, \qquad \underset{\sim}{Q} = \underset{\sim}{K}_i - \underset{\sim}{K}_f \tag{3.5}$$

In the present problem, with V(r) treated in the first-order, the radiation field enters only the exponential (3.3), whatever be its statistical properties. If $A(\tau)$ is a gaussian process (as is the case of the CH and the G fields), F(t,t') can be calculated using the generating functional property

$$\langle \exp\{i\alpha_{fi}\int_{t'}^{t} A(\tau)\, d\tau\}\rangle = \exp\{-\frac{\alpha_{fi}^2}{2}\int_{t'}^{t} d\tau\int_{t'}^{t} d\tau' \langle A(\tau)A(\tau')\rangle\} \tag{3.6}$$

For these field models, F(t,t') and, consequently, $\langle |S^{(1)}|^2\rangle$, are completely characterized by the first-order correlation function of $A(\tau)$. If, additionally, the gaussian process $A(\tau)$ is stationary (depends only on time difference), as it happens for the CH field, the function F(t,t') further simplifyies, becoming

$$F(|t-t'|) = \exp\{-\frac{1}{2}\alpha_{fi}^2 \int\!\!\int_0^{|t-t'|} d\tau\, d\tau' \langle A(\tau)A(\tau')\rangle\} \tag{3.7}$$

(3.7) allows the calculation of the time integrals of (3.1) without much difficulty but some lengthy algebra[9,15]. The correlation function $\langle A(\tau)A(\tau')\rangle$ for the vector potential

$$A(t) = \bar{A}(t)\, e^{-i\omega t} + \bar{A}^*(t)\, e^{i\omega t}$$

within a given model may be obtained either by assuming directly a first-order correlation function for the fluctuating complex amplitude $\bar{A}(t)$, or by deducing it from that of the electric field

$$E(t) = \bar{\mathcal{E}}(t)\, e^{-i\omega t} + \bar{\mathcal{E}}^*(t)\, e^{i\omega t}$$

with the same statistics, through the usual relation $A(t) = -c\int^{t} E(\tau)\,d\tau$. We wish only to point out that the results will be the same only for vanishing bandwidth. Below, for all the models, the first-order correlation function of the complex amplitudes will be assumed to be an exponential

$$\langle \bar{f}(t_1)\bar{f}^*(t_2)\rangle = f_0^2 \exp\left\{-\frac{1}{2}\Delta\omega|t_1-t_2|\right\}, \tag{3.8}$$

and accordingly, the resulting field energy spectrum will be necessarily a Lorentzian. Thus no consideration will be given here to the important problem of nonLorentzian spectra. In spite of this, the reported results are expected to be generally significant, except for those concerning the tails of the scattering lineshapes. In (3.8) f(t) may stand, as the case requires, for $\bar{A}(t)$ or $\bar{\mathcal{E}}(t)$; $\Delta\omega$ is the full width at half maximum (FWHM) of the Lorentzian spectrum; and f_0^2 is the variance of f(t). Inserting now an explicit expression for $\langle A(\tau)A(\tau')\rangle$ and following usual procedures, the transition probability per unit time and the cross section are eventually arrived at[9,10,15]. For the PDM field, the property (3.6) does not hold, and an alternative treatment is required. With this aim, we write the vector potential as

$$\underset{\sim}{A}(t) = \hat{e}_L A_0 \cos[\omega t + \varphi(t)], \qquad (A_0 = cE_0/\omega), \tag{3.9}$$

$\varphi(t)$ being the stochastic phase of the field, and define the function

$$X(t) = \exp\left\{i\,\alpha_{fi}\int_0^t A(\tau)\,d\tau\right\} \tag{3.10}$$

which obeys the differential equation

$$\dot{X}(t) = i\,\alpha_{fi}\,A_0\cos[\omega t + \varphi(t)]\,X(t) \tag{3.11}$$

with X(0)=1. Then, it may be shown[16] that F is given by

$$F(0,t) = \langle X(t)\rangle = \int_0^{2\pi} d\varphi\, X(\varphi,t), \tag{3.12}$$

where the "marginal average" $X(\varphi, t)$, in which now φ and t are uncoupled variables, obeys the equation

$$\left[\frac{\partial}{\partial t} - \Delta\omega\frac{\partial^2}{\partial\varphi^2}\right]X(\varphi,t) = i\,\alpha_{fi}\,A_0\cos(\omega t+\varphi)\,X(\varphi,t) \tag{3.13}$$

with $X(\varphi, t=0)=1/2\pi$ and periodic boundary condition on φ.

A method of solving eq.(3.13) is reported in ref. 11, and for the averaged transition probability per unit time it gives

$$W = 2\,\mathrm{Re}\left\{i/[\omega_{fi} + \xi(Q_0^+ + Q_0^-)]\right\}C(\underset{\sim}{Q}) \tag{3.14}$$

with $\xi = \alpha_{fi} A_0/2$ and Re $\{.....\}$ has the usual meaning of real part of the bracketed fraction. Q_0^+ and Q_0^- are the continued fractions with

$$Q_0^+ = \cfrac{1}{\beta_1 \xi^{-1} + \cfrac{1}{\beta_2 \xi^{-1} - \cdots}} \quad ; \quad Q_0^- = \cfrac{1}{\beta_{-1} \xi^{-1} - \cfrac{1}{\beta_{-2} \xi^{-1} - \cdots}} \tag{3.15}$$

with

$$\beta_m = -\omega_{fi} + m\omega - i\, m^2 \Delta\omega \tag{3.15a}$$

The "double" differential cross section (DDCS) is given by

$$\left(\frac{d^2\sigma}{d\Omega\, d\varepsilon_f}\right)^{PDM} = \left(\frac{K_f}{K_i}\right) \frac{1}{\hbar\pi} \mathrm{Re}\left\{ i/[\omega_{fi} + \xi(Q_0^+ + Q_0^-)]\right\} \left(\frac{d\sigma}{d\Omega}\right)^{FBA} \tag{3.16}$$

where

$$(d\sigma/d\Omega)^{FBA} = (m/2\pi\hbar)^2\, C(\underline{Q}) \tag{3.17}$$

is the First Born Approximation (FBA) to the potential scattering. The "single" differential cross section (DCS) is given by

$$(d\sigma/d\Omega) = \int_0^\infty d\varepsilon_f\, (d^2\sigma/d\Omega\, d\varepsilon_f) \tag{3.18}$$

The total cross section (TCS) follows in the usual way.

The general expression for the cross sections in the other two radiation field models (CH and G) are too involved to be reported here[9,15], and below only particular cases will be considered. We now proceed to briefly present a number of results.

4. Limiting Cases

We consider first two limiting cases.

4.1 Zero Bandwidth and Arbitrary Field Intensity

In this case, the PDM field reduces to a homogeneous, single mode field (SM) without fluctuating parameters, i.e. becomes a pure, coherent field. From eq.s (3.10)-(3.13) we recover a now familiar result

$$\left(\frac{d^2\sigma}{d\Omega\, d\varepsilon_f}\right)^{PDM} = \left(\frac{d^2\sigma}{d\Omega\, d\varepsilon_f}\right)^{SM} = \sum_{n=-\infty}^{\infty} \left(k_f/k_i\right) J_n^2(\lambda)\, \delta(\varepsilon_f - \varepsilon_i - n\hbar\omega) \left(\frac{d\sigma}{d\Omega}\right)^{FBA} \qquad (4.1)$$

with $J_n(\lambda)$ the Bessel function of integer index and real argument, and

$$\lambda = 2\xi/\omega = (e/mc\omega)\, A_0\, \hat{e}_L \cdot \underline{Q} \qquad (4.2)$$

the basic field coupling parameter for this kind of processes.

Comparison of (4.1) and (3.16) yields the interesting equality[11]

$$\lim_{\Delta\omega \to 0} \mathrm{Re}\left\{ i / \left[\omega_{fi} + \xi (Q_0^+ + Q_0^-) \right] \right\} = \hbar\pi \sum_{m=-\infty}^{\infty} J_m^2(\lambda)\, \delta(\varepsilon_f - \varepsilon_i - m\hbar\omega), \qquad (4.3)$$

which gives a representation of the r.h.s. in terms of continued fractions. Thanks to its properties of fast convergency even for very high values of λ , the representation (4.3) is expected to prove very useful in many multiphoton calculations within unperturbative schemes, either for calculations of processes with a given n, or when all the n contributions need to be summed, as in plasma heating problems.

In the same zero bandwidth limit, the DCS for a CH field is given by[6,7]

$$\left(\frac{d\sigma}{d\Omega}\right)^{CH} = \sum_n \left(k_f(n)/k_i\right) \exp\left(-\frac{\bar{\lambda}^2}{2}\right) I_n\left(\frac{\bar{\lambda}^2}{2}\right) \left(\frac{d\sigma(n)}{d\Omega}\right)^{FBA} \qquad (4.4)$$

with $k_f(n)$ fixed by the energy conservation condition for the n-th channel

$$\frac{\hbar^2 k_f^2}{2m} = \frac{\hbar^2 k_i^2}{2m} + n\hbar\omega \qquad (4.5)$$

and

$$\bar{\lambda} = (e/m\omega^2)\, \mathcal{E}_0\, \hat{e}_L \cdot \underline{Q} = \left(\frac{\mathcal{E}_0}{E_0}\right) \lambda, \qquad (4.6)$$

$\mathcal{E}_0$ being the variance of the chaotic field. I_n is the Bessel function of imaginary argument. We observe that (4.4) may be also obtained as and average of (4.1) over the distribution function of the CH field amplitudes, or within a model of a field with N equal-frequency randomly distributed modes, when N becomes very large[7].

For the G field one has[15]

$$\left(\frac{d\sigma}{d\Omega}\right)^{G} = \sum_{m} \left(k_f(m)/k_i\right) G_n(\bar{\lambda}) \left(\frac{d\sigma(m)}{d\Omega}\right)^{FBA} \qquad (4.7)$$

with

$$G_n = \frac{1}{\pi} \frac{\Gamma(m+\frac{1}{2})}{(\nu!)^2} \left(\frac{\bar{\lambda}^2}{2}\right)^{\nu} {}_2F_2\left(m+\frac{1}{2};\, m+\frac{1}{2};\, 2m+1,\, m+1;\, -\frac{\bar{\lambda}^2}{2}\right) \qquad (4.8)$$

$(\nu = |m|)$

In (4.8) Γ is the gamma function, and ${}_2F_2$ the generalized hypergeometric series. (4.8) may be cast in a more instructive form, namely as[15]

$$G_m(\bar{\lambda}) = \frac{1}{\pi}\int_0^{\pi} d\vartheta \, \exp\{-\bar{\lambda}^2 \cos^2\vartheta\} \, I_n(\bar{\lambda}^2 \cos^2\vartheta), \qquad (4.9)$$

clearly showing the connections between a model in which only the amplitude fluctuates and a model in which fluctuate both the amplitude and the phase. In fact the CH field result is contained in (4.9) as a limiting case, when instead of averaging over ϑ all the expression under the integral sign we substitute in it the average value of $\cos^2\vartheta$:

$$\frac{1}{\pi}\int_0^{\pi} d\vartheta \, \cos^2\vartheta = \frac{1}{2} \qquad (4.10)$$

$$G_n(\bar{\lambda}) \longrightarrow \exp\left(-\frac{\bar{\lambda}^2}{2}\right) I_n\left(\frac{\bar{\lambda}^2}{2}\right) \qquad (4.10a)$$

and

$$\left(\frac{d\sigma}{d\Omega}\right)^{G} \longrightarrow \left(\frac{d\sigma}{d\Omega}\right)^{CH} \qquad (4.10b)$$

(4.9)-(4.10b) show that the phase fluctuations act to reduce the fluctuations of the amplitude. Mathematically, this is expressed in the fact that the quantity $\bar{\lambda}^2\cos^2\vartheta$ instead of varying from 0 to $\bar{\lambda}^2$ as in (4.9), in (4.10a) is held fixed at the value $\bar{\lambda}^2/2$ and this is solely due to the phase fluctuations.

In order to get some preliminary information on the role of fluctuating field parameters in determining the shape of the scattering lines, we now turn our attention to another limiting case.

4.2 Weak Particle-Field Coupling ($\lambda \ll 1$) and Nonzero Bandwidth

In this limit, for the PDM field, at lowest order, with the first correction included, one finds[11]

$$\left(\frac{d^2\sigma}{d\Omega\, d\varepsilon_f}\right)^{PDM} = \sum_m \left(K_f(m)/K_i\right)\left(\frac{d\sigma(m)}{d\Omega}\right)^{FBA} \frac{1}{(\nu!)^2}\left(\frac{\lambda}{2}\right)^{2\nu} \mathcal{L}_n^{PDM}(\omega), \tag{4.11}$$

$$\mathcal{L}_m^{PDM}(\omega) = \frac{1}{\pi}\frac{\hbar(\nu^2+\lambda^2/2)\Delta\omega}{[\hbar(\nu^2+\lambda^2/2)\Delta\omega]^2 + [\varepsilon_f - \varepsilon_i - n\hbar\omega]^2} \tag{4.11a}$$

With $\lambda=0$, eq. (4.11a) reproduces the lineshape predicted by perturbation theory, when the broadening is provided only by a PDM field spectrum[17,18] through its bandwidth. The term containing λ^2 gives the first correction to it, is proportional, as it must be, to the field intensity, and act to further broaden the scattering lineshape.

For a CH field, within the same degree of approximation as for the PDM field, one has[9]

$$\left(\frac{d^2\sigma}{d\Omega\, d\varepsilon_f}\right)^{CH} = \sum_m \left(K_f(m)/K_i\right)\left(\frac{d\sigma(m)}{d\Omega}\right)^{FBA} \frac{1}{\nu!}\left(\frac{\bar{\lambda}^2}{2}\right)^{2\nu} \mathcal{L}_m^{CH}(\omega), \tag{4.12}$$

$$\mathcal{L}_m^{CH}(\omega) = \frac{1}{\pi}\frac{\hbar(\nu+3\bar{\lambda}^2/2)\Delta\omega}{[\hbar(\nu+3\bar{\lambda}^2/2)\Delta\omega]^2 + [\varepsilon_f - \varepsilon_i - m\hbar\omega]^2} \tag{4.12a}$$

Again, eq. (4.12a) with $\lambda=0$ reproduces the result predicted by perturbation theory[17,18].

We have considered in this subsection the weak coupling limit $\lambda \ll 1$ just to show that the present treatment, which is exact in the radiation field, correctly reproduces the weak field results, known in the literature. At the same time, it must be emphasized that (4.11) and (4.12) are only extreme limiting cases, and that the situation changes drastically when we consider cases for which λ is no more restricted to be small. As an example, Fig. 1 gives an idea how the overall scattering linewidth is broadened by the field fluctuations, when the particle field coupling is not small[11]. Fig. 1 shows the ratio of the total scattering linewidth Γ to the field energy width $\hbar\Delta\omega$ for various multiphoton exchanges (n=1,3,5) and two field models (CH and PDM) as a function of the amplitude of the electric field (with $E_o^{PDM} = \mathcal{E}_o^{CH}$). The curves refer to differential cross sections of scattering by a screened coulomb potential, calculated with the following parameters: $\mathcal{E}_i$ =100 eV; scattering angle ϑ =45°, $\hbar\omega$ =1.17 eV, $\hbar\Delta\omega = 10^{-4}$ eV. From Fig. 1 it is easily seen that only for the lowest values of electric field, the linewidth is controlled by the photon multiplicity, according to the predictions of perturbation theory. Increasing the field, the broadening becomes independent of n, depending only on E_o. The CH field is found to broaden the line much more effectively than the PDM field, and Γ becomes orders of magnitude broader than $\hbar\Delta\omega$ ($\Delta\omega$ being the only width considered in the present theoretical treatment). The results for the G field in the limit considered in this subsection are omitted for the sake of brevity. They will be presented elsewhere[15]

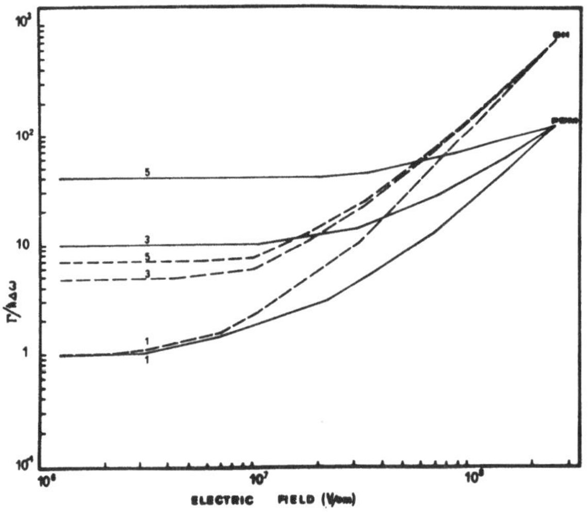

Fig. 1 Ratio $\Gamma/\hbar\Delta\omega$ of the overall width of the scattering lines to the field bandwidth vs the electric field amplitude for various numbers of exchanged photons (numbers over the curves). Full curves: PDM field; dashed curves: Chaotic field. See main text for the values of the other pertinent parameters (after ref.11).

5. The Coherence Factors

We now proceed to consider for a couple of cases the ratios of the DCS calculated within the various field models. For simplicity, we restrict ourselves to the case of vanishing field bandwidths $\Delta\omega/\omega \approx 0$, and take, of course, $\mathcal{E}_0 = E_0$. When these ratios are considered within perturbation theory and the comparison is with the single mode (SM) field results, they are commonly named "Enhancement Factors". In fact, the perturbation theory (PT) and the factorization properties of the n-th order correlations of the various field models generally predict enhancement[17,18]

$$(R_n^{CH})_{PT} = D_m^{CH} / D_m^{SM} = m! \tag{5.1}$$

$$(R_m^{G})_{PT} = D_m^{G} / D_m^{SM} = m!! = 1\cdot 3 \cdots (2m-1) \tag{5.2}$$

with D_n^X the DCS of absorption (or emission) of n photons in the presence of a field treated within the model X=CH, G, PDM, SM. (For particular case of $\Delta\omega = 0$, an exception is the PDM field for which $(R_n^{PDM})_{PT} = 1$). As it will become soon apparent, enhancement occurs not always, so below the ratios R_n^X are more neutrally termed "Coherence Factors". The analysis of the coherence factors given below is based on simple analytical estimates in which two quantities play a major role: 1) λ, the particle-field coupling parameter, and 2) $|n|$, the number of exchanged photons (the photon multiplicity). Loosely speaking, below we will consider that the inequality $|m| \ll \lambda$ refers to processes in which (relatively) few photons are exchanged (low multiplicity domain), while $|m| \gg \lambda$ identify the opposite situation (high multiplicity domain). As a result of the exact inclusion of the field in our treatment, (5.1) and (5.2) are replaced, for $\Delta\omega = 0$, and $E_0 = \mathcal{E}_0$, respectively by

$$R_m^{CH} = \exp\left\{-\frac{\lambda^2}{2}\right\} I_n\left(\frac{\lambda^2}{2}\right) / J_n^2(\lambda) \tag{5.3}$$

and

$$R_m^{G} = \frac{1}{\pi}\int_0^{\pi} d\vartheta \exp\left\{-\lambda^2 \cos^2\vartheta\right\} I_n(\lambda^2 \cos^2\vartheta) / J_n^2(\lambda) \tag{5.4}$$

Using known asymptotic expressions of the Bessel functions appearing in (5.3) and (5.4) and specializing to $n > 0$, a number of limiting forms of R_n^X may be given. For instance, for the chaotic coherence factor:

i) $\lambda \ll 1$ and $n \gg \lambda$

$$R_n^{CH} \approx n!$$

The n! behaviour is recovered in another limiting case. Namely,

ii) $\lambda \gg 1$ and $n \gg \lambda$

$$R_n^{CH} \approx n!$$

A completely different behaviour is instead found in the limit of strong coupling ($\lambda \gg 1$) and photon multiplicity n smaller or comparable to λ :

iii) $\lambda \gg 1$, $n \lesssim \lambda$

$$R_n^{CH} \approx \sqrt{\pi}\, \exp\{-n^2/\lambda^2\}\,(1-n^2/\lambda^2)^{1/2} \tag{5.5}$$

It is easy to see that: $R_n^{CH} \rightarrow \sqrt{\pi}$ for $n \ll \lambda$; is much smaller than unity for $n \approx \lambda$; and approaches zero as $n \rightarrow \lambda$. As a whole, we see that the n! behaviour is restricted to only given ranges of the pertinent parameters, and that for the other ranges either D_n^{CH} is only slightly larger than D_n^{SM}, or even $D_n^{CH} \ll D_n^{SM}$. Exact numerical evaluation of (5.3) fully confirms the predictions based on the analytical estimates[10,11]. Similar conclusions are arrived at for the gaussian coherence factor. Namely, we have

iv) $\lambda \ll 1$ and $n \gg \lambda$; or $\lambda \gg 1$ and $n \gg \lambda$

$$R_n^G \approx n!!$$

v) $\lambda \gg 1$, $n \lesssim \lambda$

$$R_n^G \approx \frac{K_0(n^2/4\lambda^2)}{\sqrt{2\pi}}\left(1-n^2/\lambda^2\right)^{1/2} \exp\left\{-\frac{n^2}{4\lambda^2}\right\} \tag{5.6}$$

In (5.6) $K_0(z)$ is the cylindrical function of imaginary argument. Again it is easy to see that R_n^G becomes much smaller than unity for $n \approx \lambda$; and tends to zero as $n \rightarrow \lambda$. It becomes, instead increasingly large for $n \ll \lambda$ (where the chaotic coherence factor was only about $\sqrt{\pi}$). This last result is an indication that the larger amplitude fluctuations of the G field are more effective, as compared to those of a CH field, in raising the cross sections of processes in the ranges $\lambda \gg 1$ and $n \ll \lambda$.

Finally, is of some interest to compare the chaotic and gaussian coherence factors in the critical range ($\lambda \gg 1$, $n \to \lambda$), where individually they tend to become very small. In this range, we have

$$R_m^G / R_m^{CH} \approx K_0\left(\tfrac{1}{4}\right) \exp\left(\tfrac{3}{4}\right)\left(\pi\sqrt{2}\right)^{-1}$$

implying that for $\lambda \gg 1$ and $n \to \lambda$, the gaussian coherence factor decreases faster than R_n^{CH}.

Numerical evaluation of the coherence factors for $\Delta\omega \neq 0$ in the case of PDM and CH fields may be found in ref. 11.

6. Differential and Total Cross Sections

Here we report very briefly on selected calculations of DCS and TCS vs the scattering angle and the field intensity for the PDM, SM and CH fields. Unless otherwise stated, the physical parameters are chosen as follows. Initial particle energy $\mathcal{E}_i$=100 eV; photon energy $\hbar\omega$ =1.17 eV; field energy width $\hbar\Delta\omega=10^{-3}$ eV; mean field intensity $I=10^{13}$ W/cm^2; scattering angle ϑ=45°; screened coulomb potential with Z=1 and screening parameter $r_o=0.02\ a_o^{-1}$. n=1 for DCS and n=0 - 5 for TCS.

The reported results are meant to illustrate, in some representative way, i) the nonlinear behaviour of the DCS and TCS vs the field intensity; and ii) the dependence of the DCS and TCS on the field statistics. A more complete analysis will be given elsewhere. Concerning the specific numerical values used in the following calculations, it is perhaps worth to remark that they are chosen merely for purposes of an illustration, and not as values maximizing particular field effects we wish to stress. We point out only that decreasing the photon energy (to $\hbar\omega$=0.117 eV, for instance), or changing the scattering angle, or some other parameter, it is possible to have the same effects reported below or even stronger with field intensities of orders of magnitude smaller. Fig. 2a shows the DCS for n=1 vs the scattering angle ϑ for the PDM and CH fields; Fig. 2b shows the same for the CH and SM fields. Fig. 3a shows the DCS for n=1 vs the field intensity for the CH and SM fields, while

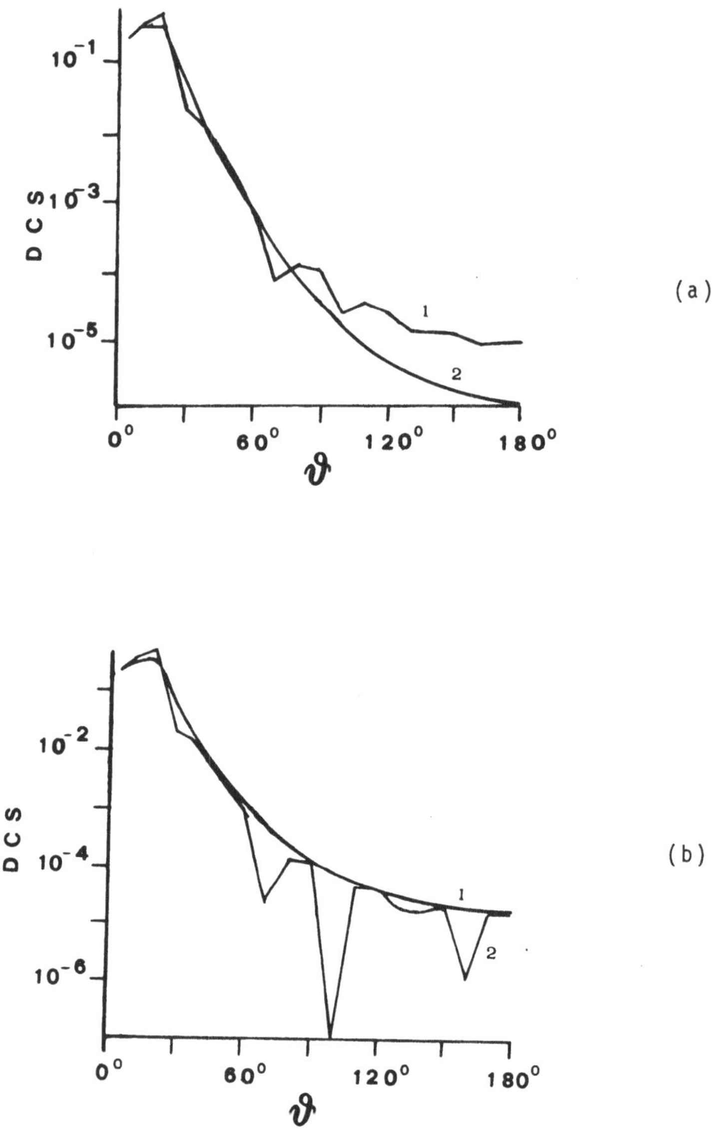

Fig. 2 Differential Cross Sections (in units of πa_0^2 ster^{-1}) vs the scattering angle for different field models. (a) curve 1-PDM, curve 2-CH. (b) curve 1-CH, curve 2-SM. $\mathcal{E}_i$=100 eV; $\hbar\omega$=1.17 eV; $\hbar\Delta\omega=10^{-3}$ eV; $I=10^{13}$ W/cm^2; V(r) screened coulomb potential with Z=1 and screening parameter r_0=0.02 a_0^{-1}; n=1 (one photon absorption).

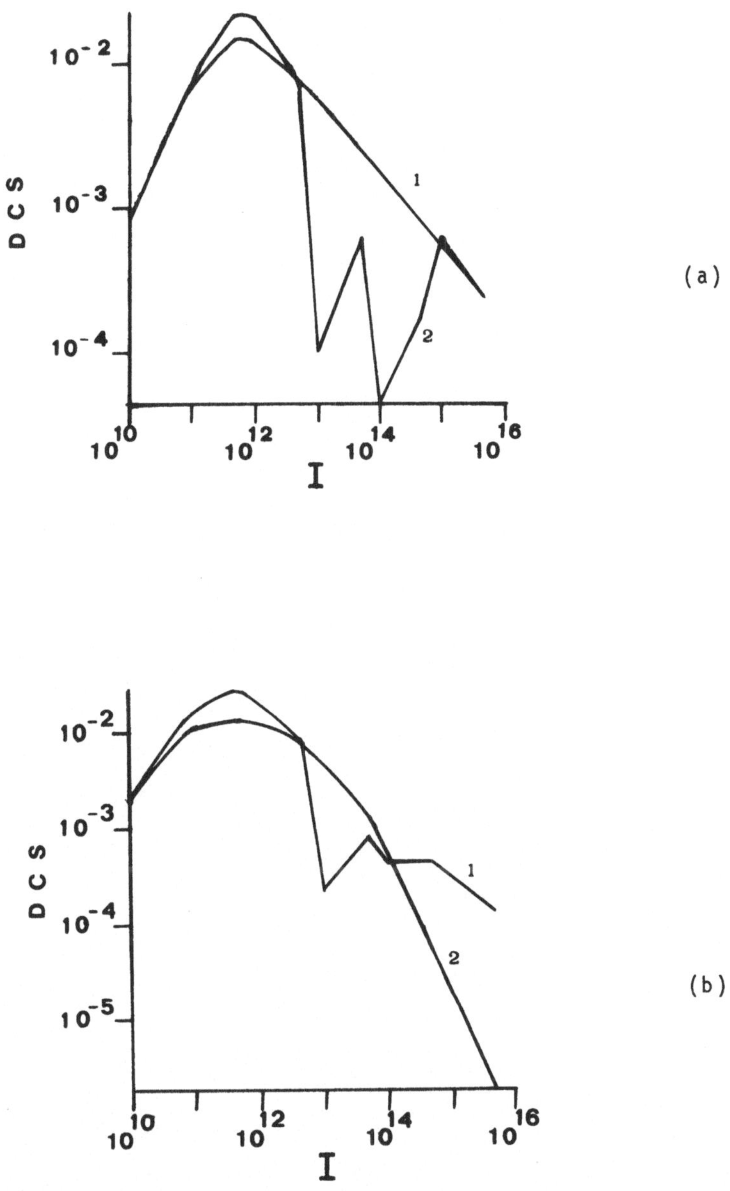

Fig. 3 DCS (in units of πa_o^2 ster^{-1}) vs the field intensity (in W/cm^2). (a) curve 1-CH; curve 2-SM. (b) curve 1-PDM; curve 2-CH. Scattering angle ϑ=45°. The others parameters as in Fig.2.

Fig. 3b shows the same for the PDM and CH fields. Fig. 4 and 5 reports on TCS *vs* the intensity for n=0 -5 , $\Delta\omega=0$ and the SM and CH fields.

7. Concluding Remarks

As a conclusion, we remark that both the DCS and TCS are significantly affected by the field statistics, each model modifying them in a peculiar way, which however can not be simply predicted for all the values of the field and collision parameters. This because of the highly nonlinear dependence of the DCS on the field parameters and on particle-field coupling. It is likely that considerably new information may be obtained on both the cross sections of the processes in the presence of

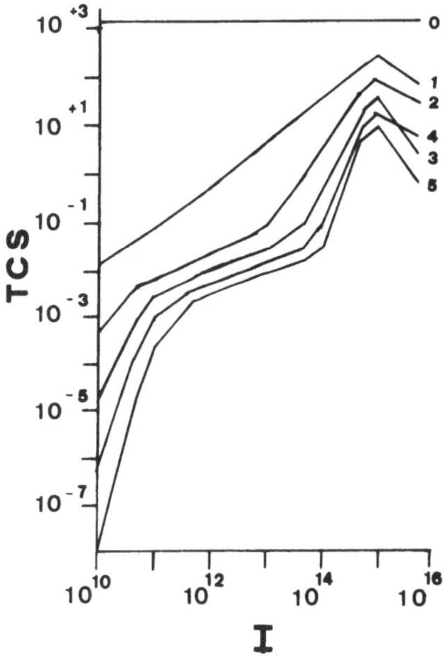

Fig. 4 Total Cross Sections (in units of πa_o^2) *vs* the field intensity (in W/cm^2) with absorption of n photons (n=0-5) for a single mode field (SM). The parameters as in Fig. 2.

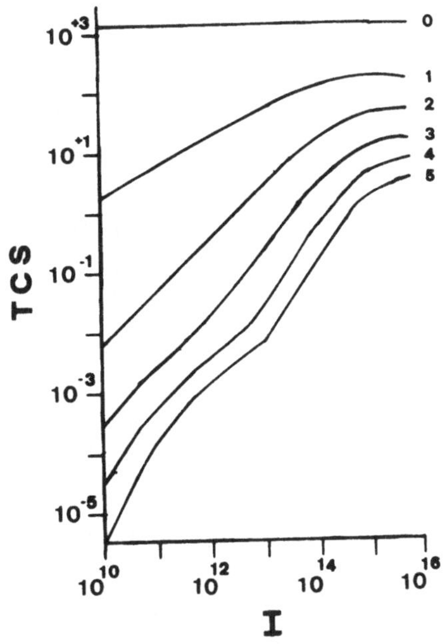

Fig. 5 Total Cross Sections (in units of πa_o^2) vs the field intensity (in W/cm^2). Chaotic (CH) field case, $\Delta\omega = 0$. Caption as in Fig. 4.

realistic lasers, and on the laser properties in intense field domains. Though difficult, multiphoton free-free transitions have been repeatedly demonstrated to be experimentally accessible[19], and very intense lasers too are becoming available in several pure research laboratories. So, it is hoped that some of the predictions contained in the theory outlined here may be tested in a near future.

Acknowledgements

This work has been partially supported by the Italian Ministry of Education, the National Group of Structure of Matter and the Sicilian Committee for Nuclear and Structure of Matter Researches.

References

1. M.H. Mittleman, Introduction to the Theory of Laser-Atom Interactions, Plenum Publishing Co., New York (1982).
2. F. Ehlotzky, Can. J. Phys. 59, 1200 (1981) and ibidem to be published.
3. L. Rosenberg, Adv. At. Mol. Phys. 18, 1 (1982).
4. G. Ferrante, in: Fundamental of Atomic Collisions, Ed.s H. kleinpoppen, H.O. Lutz and J.S. Briggs, Plenum Publishing Co., New York (1985).
5. For reviews and references see, for instance, J.H. Eberly, in: Laser Spectroscopy, Springer-Verlag, Berlin (1979), p.80 and A.T. Georges, P. Lambropoulos and P. Zoller, ibidem, p.368.
6. P. Zoller, J. Phys. B: At. Mol. Phys. 13, L249 (1980).
7. R. Daniele and G. Ferrante, J. Phys. B: At. Mol. Phys. 14, L635 (1981).
8. E.L. Beylin and B.A. Zon, Kvantovaja Elektronika 9, 1692 (1982) (in russian).
9. R. Daniele, F.H.M. Faisal and G. Ferrante, J. Phys. B: At. Mol. Phys. 16, 3831 (1983).
10. F. Trombetta, C.J. Joachain and G. Ferrante, in: Collisions and Half-Collisions with Lasers, Ed.s C. Guidotti and N.K.H. Raham, Harwood, London (1984).
11. F. Trombetta, G. Ferrante, K. Wodkiewicz and P. Zoller, "Field Correlation Effects in Laser Assisted Electron Scattering. The Phase Diffusion Model", J. Phys. B: At. Mol. Phys. (in press) and references therein.
12. W. Becker, M.O. Scully, K. Wodkiewicz and M.S. Zubairy, Phys. Rev. A30, 2245 (1984).
13. P. Zoller, in: Laser Physics, Academic Press, Sidney (1980), p.99.
14. A.T. Georges, Phys. Rev. A21, 2034 (1980).
15. F. Trombetta, G. Ferrante and K. Wodkiewicz, "Electron Scattering in the Presence of Strong Stochastics Field. The Gaussian Amplitude Model", (in preparation).
16. N.G. Van Kampen, Phys. Rep. 24, 171 (1976).
17. G.S. Agarwal, Phys. Rev. A1, 1445 (1970).
18. B.R. Mollow, Phys. Rev. 175, 1555 (1968).
19. A. Weingartshofer, J.K. Holmes, G.Candle, E.M. Clarke and H. Kruger, Phys. Lett. 39, 269 (1977);
A. Weingartshofer, E.M. Clarke, J.K. Holmes and C. Jung, Phys. Rev. A19, 2371 (1979);
A. Weingartshofer, J.K. Holmes, J. Sabbagh and S.L. Chin, J. Phys. B: At. Mol. Phys. 16, 1805 (1983);
A. Weingartshofer and C. Jung, in: Multiphoton Ionization of Atoms, Academic Press, Canada (1984).

LASER-ASSISTED ATOM-ATOM COLLISIONS

F. ROUSSEL
Service de Physique des Atomes et des Surfaces
CEN.SACLAY - 91191 Gif-sur-Yvette Cedex France

The basic laser-assisted atom-atom collision processes are reviewed in order to get a simpler picture of the main physical facts. The processes can be interpreted in terms of photoexcitation of the quasimolecule formed during the collisional process. Last results of our laboratory in this field are also presented.

I. INTRODUCTION

Laser-assisted atom-atom collisions form an exciting field of research which has been actively studied during the last decade. The possibility of switching on inelastic transitions, such as energy transfer from an atom to another, charge exchange, association, or ionization, which would be adiabatically impossible in the absence of the laser field has stimulated a considerable amount of theoretical contributions and some experimental results. The crucial characteristic of all the processes studied here is that their occurence is entirely dependent on, or very sensitive to the presence of the laser field during the collision. The first to study this field were Gudzenko and Yakovlenko /1/ in 1972. They used the term "radiative collisions of atoms" to describe collisions in which there is occurence of two simultaneous processes : The collision of an atom A with an atom B, and during this collision, absorption or emission of a photon.

In the case of a long-range interaction, this situation can be described with an atomic representation, as theoretically studied by Geltman /2,3/ for bound-bound transitions. We consider that the absorption of a photon is possible if the energy $h\nu$ of this photon is equal to the energy defect between $A_i - A_f$ and $B_{f'} - B_{i'}$ as shown in figure 1a, so that collision and absorption of light are unseparable. In a molecular representation of a laser-assisted process, as shown in figure 1b, we can see that for R=Rc where the energy of separation is equal to the energy $h\nu$ of the photon, a resonant absorption is possible. If the curve $V_i + h\nu'$ (dashed curve) crosses V_f, radiative collision process is possible. This molecular approach of radiative collisions has been studied by Kroll and Watson /4/, Lau /5-7/, Gallagher and Holstein /8/, and by Gudzenko and Yakovlenko /1, 9/.

This process can be viewed as a radiative transition in a quasi-molecule. This is fundamentally a three-body interaction as two atoms and a photon are simultaneously involved.

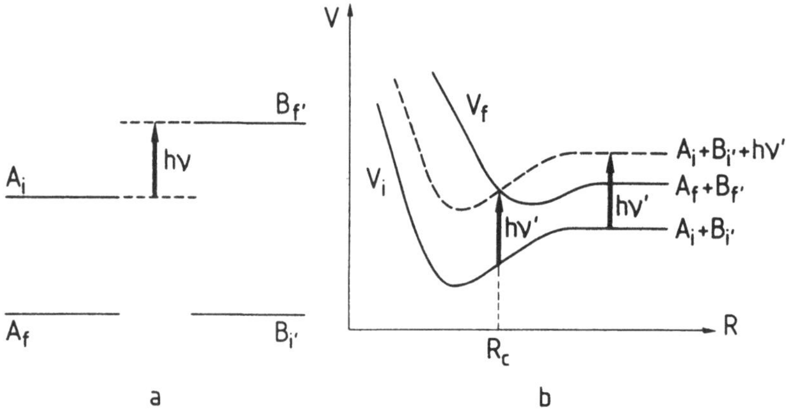

Figure 1 - Atomic and molecular representation of laser-assisted atom-atom collisions : (a) atomic picture; (b) molecular picture. The dashed curve represents the potential curve V_i translated by $h\nu'$.

We will distinguish two types of laser-assisted atom-atom collisions : optical collisions, where only one atom is changing its state during the collision, while the other acts as a spectator atom; and radiative collisions where the states of both atoms are changing during the collision.

II. OPTICAL COLLISIONS

Optical collisions have been the subject of many

theoretical and experimental investigations. This is due to the fact that measurements of collisional broadening of atomic transitions can lead to determination of the potential curves of quasi molecules involving noble gas atoms.

These investigations have been performed by using different methods /10-17/, such as resonant ionization spectroscopy, absorption, or fluorescence techniques for example.

Let us study the following process :

$$A(nl) + B + h\nu \longrightarrow A(n'l') + B$$

the transition from (nl) to (n'l') may be dipole-allowed or dipole forbidden for A alone. Sayer et al /15/ have studied a case where this transition is forbidden for A alone. They have investigated the processes

$$Cs(6S_{1/2}) + A + h\nu \longrightarrow Cs(5\,D_{5/2}\,;\,7\,S_{1/2}) + A$$

where A is a noble gas, by a fluorescence method. The formation of the two excited states of Cs was detected via the cascade decay to the $6S_{1/2}$ fundamental state of cesium. These experiments have led to determination of the potential curves for cesium-noble gas quasi molecules that have been compared with theoretical calculations of Pascale and Pascale and Vandeplanque /18/. In the case of helium, the experiment has shown a well-marked avoided crossing between the $7\,S_{1/2}$ and $5D_{5/2}$ states of cesium, which was not predicted by theory.

Recently, Pradel et al. /19/ have performed a laser assisted optical ionization. They have investigated the process

$$He\,(2^1S,\,2^3S) + He\,(1^1S) + h\nu \longrightarrow He^+ + He(1^1S) + e^-$$

where the energy $h\nu$ of the photon (3,49 eV) is not sufficient to ionize the metastable He atom without the presence of the $He(1^1S)$ perturber. In fact the photon is absorbed by the quasimolecule formed during the collision, giving rise to associative ionization. As the kinetic energy of the He metastable atom is high (50 eV) the quasimolecular ion dissociates and it is finally an atomic ion He^+ which is dectected, the state of the $He(1^1S)$ perturber remaining unchanged.

Optical collisions have stimulated many theoretical contributions as for example those of Nayfeh /20/, Yeh and Berman /21/, Light and Szöke /22/, Julienne /23/, Nienhuis et al. /24/, Gallagher and Holstein /8/, and recently Kulander and Rebentrost /25/.

III. RADIATIVE COLLISIONS

Figure 2 shows four basic and simple radiative collisions. The two upper processes, the Laser-Induced Collisional Energy Transfer (LICET) and the pair excitation and its inverse process, the Radiative Collisional Fluorescence (RCF), are bound-bound processes. The two lower processes, the Laser-Induced Collisional Ionization (LICI) and the Laser-Induced Charge Transfer (LICT) are bound-free transitions. We will not consider the latter case, which does not belong to the atom-atom collisions field, but we shall give experimental examples of the three others situations and also of the laser photoassociation, in order to get a somewhat simple picture of the main physical facts.

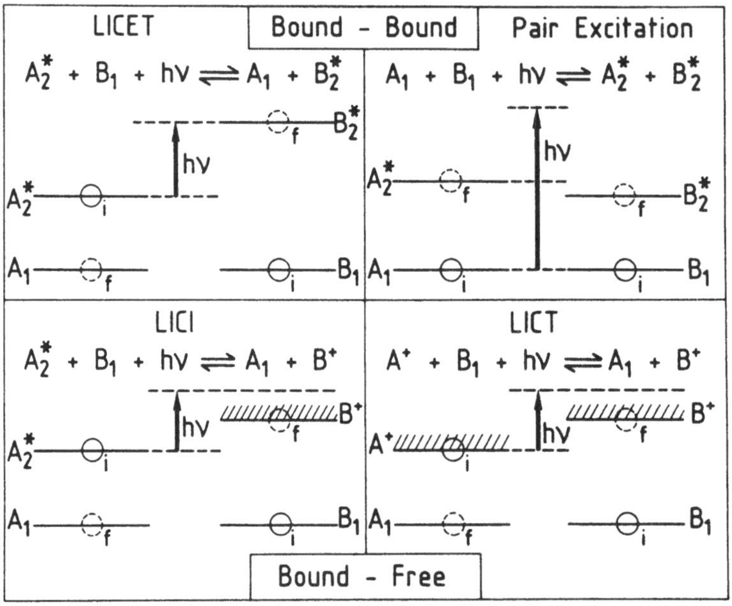

Figure 2 - The four basic radiative collision processes.

III. 1 Laser-Induced Collisonal Energy Transfer

In this process, the energy of an excited atom A is transferred to an atom B. The energy defect of the reaction is compensated by the absorption of a photon during the collision. The first experimental evidence of this process was reported by Harris et al./26/ with Calcium and Strontium. In a further experiment /27/, they have measured the cross section for the process versus the laser intensity, and they have shown the linear dependance of the cross section with the laser power in the weak field regime as theoretically predicted by Gudzenko and Yakovlenko /1/.

Quantitative studies of the LICET process have been performed by Brechignac et al. /28/ with Europium and Strontium. They conclude from their studies that a universal, strongly asymetrical profile characterizes this type of interaction. The assymetry of the Line shape can be explained by considering the molecular potential curves V_i and V_f for the initial and final states of the quasimolecule formed during the collision.

The comparison between the experimental line shape and theoretical calculations from Gallagher and Holstein /8/, Harris and White /29/, Crance and Stenholm /30/ as well as the experiment of Cheron and Lemery /31/, have suggested that higher order interactions such as dipole-quadrupole may play a role in the process. A theoretical calculation by Berman /32/ assumed that coherences can be produced in such a process. Experimental evidence of this fact was given by Débarre who performed also an experiment in a Na-Ca mixture in which the virtual level is optically excited /33/.

Gallagher et al. /34/ have performed recently an experiment of Microwave-assisted collisional energy transfer (MACET) with Sodium atoms excited in Rydberg states. The large dipole moments of the Rydberg states reduce the power requirement to a few watts per square centimeter, so that it is possible to reach a strong field regime where a decrease of the cross section was observed. This decrease is interpreted by the authors in terms of destructive interferences between the sidebands of the considered levels.

All these processes are well interpreted when considering the optical transition between the initial and the final states of the quasimolecule formed during the collision.

III. 2 - Pair excitation - Radiative collisional fluorescence.

The pair excitation process can be described as the absorption of a photon during the collision, resulting in the simultaneous excitation of both atoms to excited states. The collisional absorption is maximized when the photon energy is equal to the sum energy of the excited atoms for infinite separation. The inverse process is called radiative collisional fluorescence.

The first observation of this latter process was performed in 1978 by White et al. /34/ with Ba atoms. They have studied the process Ba(6p) + Ba(5d) $\longrightarrow$ 2Ba(6s) + hν (339,4 nm). A fluorescence spectrum was observed in which the radiative collisional fluorescence is seen to be much larger than the two Ba atomic lines at 333.74 and 341.38 nm. The maximum emission occurs at the sum energy of the excited levels for infinite separation as predicted by theoretical calculations /1,9/ .

The same authors /35/ observed, in 1979, for the first time pair excitation processes for bound states in Ba-Ba and Ba-Tl systems with an incoherent source. Pair excitation processes have been observed for bound states and also for the continuum in Cesium and Rubidium systems by Hotop and Niemax /36/ . They found an enhancement of the cross section near dipole-allowed transitions; vice versa, the transition is weak when the energy separation from any characteristic atomic energy is large.

In 1981, White /37/ observed pair excitation and radiative collisional fluorescence in laser-induced dipole-quadrupole collisions with two Ba atoms. This process permits collisional energy transfer to states inacessible to dipole-dipole collisions.

In pair excitation processes, it is not an evidence to consider the role played by the quasimolecule formed during the collision, because this process is taking place at large internuclear distance. This is more evident in the following processes of photoassociation, as the final state of the reaction is a molecule! Scheingraber and Vidal /38/ have reported laser induced photoassociation of Mg_2 molecules. A similar process has been observed in Hg_2 by Ehrlich and Osgood /39/. Grieneisen et al /40/ have reported formation of Xe_2^* molecules by collision pair excitation of two Xe atoms. Inoue et al. /41/ have observed the same process for XeCl* molecules. These photoassociation processes are of great interest as they contribute to formation of molecules used in exciplexes lasers. Recently, Ku et al./42/ have reported the laser photoassociative reaction of Xe and Br atoms : Xe + Br $(^2P_{3/2}) + h\nu \rightarrow$ Xe Br (B, v') following a pulsed dc discharge in Xe/Br_2 mixtures. This process can be viewed as a pair excitation, occuring at short internuclear distance.

III. 3 - Laser Induced Collisional Ionization

The first experimental observation of such a process was made in 1978 by Weiner et al. /43/ with lithium. They have observed formation of Li^+ as theoretically predicted by Geltman /3/ and of Li_2^+. Li^+ is interpreted as resulting from a laser-assisted Penning ionization and Li_2^+ from a laser-assisted associative ionization. Weiner and coworkers have made several other experimental observations for Na-Na /44/, Na-Li /45/ and Na-Ba /46/ collisions. This latter process was a two-step process in which Na*(4d) is first populated by a radiative collision between Na(3s) and Ba*(6p), and then photoionized into Na^+ by absorption of a second photon. In 1984, Keller and Weiner /47/ reported a similar two-step process in which Ba*(6p) is first populated by a radiative collision between Ba(6s) and Na*(3p) and then photoionized into Ba^+ by absorption of a second photon. This set of two experiments clearly demonstrate that we are in presence of laser-induced collisional energy transfer, followed by photoionization of the excited state.

In 1980, Brechignac, Cahuzac and Debarre /48/ have observed a two photon laser-induced collisional ionization of

Cs in a Cs-Sr mixture. If this process can be viewed as a three step process in the atomic picture, in the molecular picture it can be more simply seen as a two-photon absorption in the transient Sr-Cs quasi-molecule. This process has stimulated theoretical calculations by Crance and Feneuille /49/.

Laser-modified Penning ionization has been experimentally observed by Goble, Hollingsworth, and Winn /50/ in the following system :

$$Ar^*(3P_2^o) + Ca + h\nu \longrightarrow Ar + Ca^+ (5p\ ^2P) + e^- ,$$

where Ca^+ is created in an excited state. The process can be viewed as the photoionization of the autoionizing quasimolecule Ca Ar*. It is a laser-modified Penning ionization, which has been theoretically studied by Bellum, Lam and George /51/ and recently by Saha, Dahler and Nielsen /52/.

III. 4 - Laser Ionization Based on Resonance Saturation

A special case of LICI is the laser ionization based on resonance saturation (LIBORS). The first experimental observation was made in a dense Na vapor by Lucatorto and McIlrath /53/ in 1976. Exciting the 3p level at resonance, they have obtained total ionization of the vapor. This surprising result can be explained in terms of electron seeding, followed by an avalanche. At low atomic densities, indeed, only the seeding process is observed. There is no avalanche. The avalanche process has been predicted by Measures /54/. It is due to superelastic collisions of electrons with Na*(3p) atoms. After two superelastic collisions, the electron has a sufficient energy to ionize a Na*(3p) atom during a collision. Another electron is liberated, it is the avalanche process.

Experimental evidence of superelastic collisions has been reported by our group /55/, measuring the mean electronic temperature Te versus the vapor density, and by Le Gouët et al. /56/ by electron spectroscopy. If there is evidence of superelastic collisions, what is the origin of the seed electrons ? Several mechanisms have been proposed, such as Multiphoton ionization of the 3p state, associative ionization, Energy pooling collisions followed by photoionization of the excited state, or laser-modified associative ionization.

This latter process was first supposed to be consistent with the experimental observations. The cross sections previously reported for associative ionization /57/ and for energy pooling collisions /58/ were indeed to small to explain the large ion yields obtained. Thus, we have concluded that Na_2^+ ions were obtained by laser-modified associative ionization, while Na^+ ions were mostly coming from the photodissociation of Na_2^+ /59/.

However, further experimental results have led us to reexamine this conclusion. The photodissociation of Na_2^+ / 6 0 / cannot explain the Na^+ ion yield. There is a direct channel to obtain Na^+ ions. Moreover, recent experiments /61,62/ have shown that energy pooling collisions, followed by photoionization /63/ are an important process, and finally, measurements of the cross section for associative ionization recently obtained /64,65/ are ten times larger than those obtained by de Jong and van der Valk /57/.

In fact, there is a possibility for a temperature effect on such a cross section and non-assisted associative ionization seems to be responsible of the large Na_2^+ ion yield. Collisions involving excited states seems to be the major process for producing Na^+ ions. Moreover it is now seen that molecules play a role /66/.

Structure has been observed in the Na_2^+ ion yield, off resonance, in a two-laser experiment /44/ and by our group in a single laser experiment /67/. While Boulmer and Weiner /68/ interpret the spectrum obtained in terms of laser assisted atomic collisions, we considered that resonant multiphoton ionization of the molecules is the dominant process /69/. Our last results /70/ show that the off-resonance Na_2^+ spectrum quite disappears when the Na_2 density is decreased by a factor of ten, the Na^+ density remaining unchanged. This is the signature of a molecular process.

IV - CONCLUSION

The basic processes of laser-assisted atom-atom collisions are demonstrating the quasimolecular behavior of the colliding partners. Some other possibilities, which does not belong to the atom-atom collision field may be considered, such as laser-induced charge transfer or chemical radiative collisions, giving rise to a change in the chemical structure of colliding molecules. Theoretically, the study of the strong field regime, where the perturbative method does not apply is now of interest. But experimental evidence of such laser-assisted processes is not allways easy to obtain and in particular, one has to be careful when speaking of laser assisted processes in alkali-alkali collisions.

ACKNOWLEDGMENT

The author would like to thank Dr. C.Manus for helpful discussions.

REFERENCES

/1/ L.I. Gudzenko and S.Y. Yakovlenko, Zh. Eksp. Teor. Fiz. **62**, 1686 (1972) [Sov. Phys. JETP **35**, 877 (1972)].

/2/ S. Geltman, J. Phys. B **9**, L569 (1976).

/3/ S. Geltman, J. Phys. B **10**, 3057 (1977).

/4/ N.M. Kroll and K.M. Watson, Phys. Rev. A **13**, 1018 (1976).

/5/ A.M.F. Lau, Phys. Rev. A **13**, 139 (1976).

/6/ A.M.F. Lau, Phys. Rev. A **14**, 279 (1976).

/7/ A.M.F. Lau and C.K. Rhodes, Phys. Rev. A **15**, 1570 (1977).

/8/ A. Gallagher and T. Holstein, Phys. Rev. A **16**, 2413 (1977).

/9/ L.I. Gudzenko and S.Y. Yakovlenko, Phys. Lett. **46**A, 475 (1974).

/10/ M.H. Nayfeh, G.S. Hurst, M.G. Payne and J.P. Young, Phys. Rev. Lett. **39**, 604 (1977). M.H. Nayfeh, G.S. Hurst, M.G. Payne and J.P. Young, Phys. Rev. Lett. **41**, 302 (1978). M.H. Nayfeh, W. Glab, A. Mc.Cown, Phys. Rev. A **24**, 1142 (1981). W. Glab, G.B. Hillard, and M.H. Nayfeh, Phys. Rev. A **25**, 3431 (1982).

/11/ G. Moe, A.C. Tam and W. Happer, Phys. Rev. A**14**, 349 (1976). A.C. Tam, G. Moe, W. Park and W. Happer, Phys. Rev. Lett. **35**, 85 (1975).

/12/ K.H. Weber and K. Niemax, Opt. Comm. **28**, 317 (1979).

/13/ J.L. Carlsten and A. Szöke, J. Phys. B **9**, L321 (1976).

/14/ R.E.M. Hedges, D.L. Drummond and A. Gallagher, Phys. Rev. A **6**, 1519 (1972).

/15/ B. Sayer, M. Ferray and J. Lozingot, J. Phys. B **12**, 227 (1979). B. Sayer, M. Ferray, J.P. Visticot and J. Lozingot, J. Phys. B **13**, 177 (1980), M. Ferray, J.P. Visticot, J. Lozingot and B. Sayer, J. Phys. B **13**, 2571 (1980).

/16/ P. Thomann, K. Burnett and J. Cooper, Phys. Rev. Lett. **45**, 1325 (1980).

/17/ R. Kachru, T.W. Mossberg and S.R. Hartmann, Phys.Rev. A **21**, 1124 (1980).

/18/ J. Pascale, J. Chem. Phys. **67**, 204 (1977). J. Pascale and J. Vandeplanque, J. Chem. Phys. **60**, 2278 (1974).

/19/ P. Pradel, D. Dubreuil, J. Heuzé, J.J. Laucagne, P. Monchicourt and G.Spiess, to be published.

/20/ M.H. Nayfeh, Phys. Rev. A **16**, 927 (1977). M.H. Nayfeh, Phys. Rev. A **20**, 1927 (1979). M.H. Nayfeh and W. Glab, Phys. Rev. A **25**, 1619 (1982).

/21/ S. Yeh and P.R.Berman, Phys. Rev. Lett. **43**, 848 (1979). S. Yeh and P.R. Berman, Phys. Rev. A **19**, 1106 (1979).

/22/ J. Light and A. Szöke, Phys. Rev. A **18**. 1363 (1978).

/23/ P.S. Julienne, J. Appl. Phys. **48**, 4140 (1977). P.S. Julienne, J. Chem. Phys. **68**, 32 (1978). P.S. Julienne , Phys. Rev. A **26**, 3299 (1982).

/24/ G. Nienhuis, J. Phys. B **15**, 535 (1982). F. Schuller and G. Nienhuis, Can. J. Phys. **62**, 183 (1984).

/25/ K.C. Kulander and F. Rebentrost, J. Chem. Phys. **80**, 5623 (1984).

/26/ R.W. Falcone, W.R. Green, J.C. White, J.F. Young and S.E. Harris, Phys. Rev..A **15**, 1333 (1977).

/27/ W.R. Green, J. Lukasik, J.R. Willison, M.D. Wright, J.F. Young and S.E. Harris, Phys. Rev. Lett **42**, 970 (1979).

/28/ C. Brechignac, Ph. Cahuzac and P.E. Toscheck, Phys. Rev A **21**, 1969 (1980).

/29/ S.E. Harris and J.C. White, IEEE J. Quant. Electron.Q.E.**13**, 972 (1977).
/30/ M. Crance and S. Stenholm, J. Phys. B **13**, 1563 (1980).
/31/ B. Cheron and H. Lemery, Opt. Comm. **42**, 109 (1982).
/32/ P.R. Berman, Phys. Rev. A **22**, 1838 (1980), P.R. Berman, Phys. Rev. A **22**, 1848 (1980), P.R. Berman and E.Giacobino, Phys. Rev. A **28**, 2900 (1983).
/33/ A. Debarre, J. Phys. B **15**, 1693 (1982), A. Debarre, J. Phys. B **16**, 431 (1983).
/34/ J.C. White, G.A. Zdasiuk, J.F. Young and S.E. Harris, Phys. Rev. Lett. **41**, 1709 (1978), Phys. Rev. Lett **12**, 480 (1979).
/35/ J.C. White, G.A. Zdasiuk, J.F. Young and S.E. Harris, Opt. Lett. **4**, 137 (1979).
/36/ R. Hotop and K. Niemax, J. Phys. B **13**, L93 (1980).
/37/ J.C. White, Phys. Rev. A **23**, 1698 (1981), J.C. White, Opt. Lett. **6**, 242 (1981).
/38/ H. Scheingraber and C.R. Vidal, J. Chem. Phys. **66**, 3694 (1977).
/39/ D.J. Ehrlich and R.M. Osgood, Jr. Phys. Rev. Lett. **41**, 547 (1978).
/40/ H.P. Grieneisen, K. Hohla and K. L. Kompa, Opt. Comm. **37**, 97 (1981).
/41/ G. Inoue, J.K. Ku and D.W. Setser, J. Chem. Phys. **76**, 733 (1982).
/42/ J.K. Ku, D.W. Setser and D. Oba, Chem. Phys. Lett. **109**, 429 (1984).
/43/ A.V. Hellfeld, J. Caddik and J. Weiner, Phys. Rev. Lett. **40**, 1369 (1978).
/44/ P. Polak-Dingels, J.F. Delpech, and J. Weiner, Phys. Rev. Lett. **44**, 1663 (1980), J.Weiner and P. Polak-Dingels, J. Chem. Phys. **74**, 508 (1981).
/45/ P. Polak-Dingels, J. Keller, J. Weiner, J.C. Gauthier and N. Bras, Phys. Rev. A **24**, 1107 (1981).
/46/ P. Polak-Dingels, R. Bonanno, J. Keller and J. Weiner, J. Phys. B **15**, L41 (1982).
/47/ J. Keller and J. Weiner, Phys. Rev. A **29**, 2230 (1984).
/48/ C. Brechignac, Ph. Cahuzac and A. Debarre, J. Phys. B **13**, L383 (1980).
/49/ M. Crance and S. Feneuille, J. Phys.B **13**, 3165 (1980).
/50/ J.H. Goble, W.E. Hollingsworth and J.S. Winn, Phys. Rev. Lett. **47**, 1888 (1981).
/51/ J.C. Bellum, K.S. Lam and T.F. George, J. Chem. Phys. **69**, 1781 (1978), J.C. Bellum and T.F. George, J. Chem. Phys. **70**, 5059 (1979).
/52/ H.P. Saha and J.S. Dahler, and S.E. Nielsen, Phys. Rev. A **28**, 1487 (1983), H.P. Saha and J.S. Dahler, Phys. Rev. A **28**, 2859 (1983).
/53/ T.B. Lucatorto and T.J. McIlrath, Phys. Rev. Lett. **37**, 428 (1976).
/54/ R.M. Measures, J. Appl. Phys. **48**, 2673 (1977), R.M. Measures, N. Drewell and P. Cardinal, J. Appl. Phys. **50**, 2662 (1979).
/55/ F. Roussel, P. Breger, G. Spiess, C. Manus and S. Geltman, J. Phys.B **13**, L631 (1980), B. Carré, F. Roussel, P. Breger and G. Spiess, J. Phys. B **14**, 4289 (1981).
/56/ J.L.Le Gouët, J.L. Picqué, F. Wuilleumier, J.M. Bizau, P. Dhez, P.M. Koch and D.L. Ederer, Phys. Rev. Lett. **48**, 600 (1982).

/57/ A. De Jong and F. Van der Valk, J. Phys. B **12**, L561 (1979).

/58/ G.H. Bearman and J.J. Leventhal, Phys. Rev. Lett. **41**, 1227 (1978). V.S. Kushawaha and J.J. Leventhal, Phys. Rev. A **22**, 2468 (1980).

/59/ B. Carré, F. Roussel, P. Breger and G. Spiess, J. Phys. B **14**, 4271 (1981).

/60/ F. Roussel, B. Carré, P. Breger and G. Spiess, J. Phys. B **16**, 1749 (1983).

/61/ V.S. Kushawaha and J.J. Leventhal, Phys. Rev. A **25**, 570 (1982).

/62/ J. Huennekens and A. Gallagher, Phys. Rev. A **25**, 771 (1983).

/63/ B. Carré, G. Spiess, J.M. Bizau, P. Dhez, P. Gerard, F. Wuilleumier, J.C. Keller, J.L. Le Gouët, J.L. Picqué, D.L. Ederer and P.M. Koch, Opt. Comm. **52**, 29 (1984), B. Carré, F. Roussel, G. Spiess, J.M. Bizau, P. Gérard and F. Wuilleumier, to be published.

/64/ J. Huennekens and A. Gallagher, Phys. Rev. A **28**, 1276 (1983).

/65/ R. Bonanno, J. Boulmer and J. Weiner, Phys. Rev. A **28**, 604 (1983).

/66/ M.Allegrini, W.P. Garver, V.S. Kushawaha and J.J. Leventhal, Phys. Rev. A **28**, 199 (1983).

/67/ F. Roussel, B. Carré, P. Breger and G. Spiess, J. Phys B **14**, L313 (1981).

/68/ J. Boulmer and J. Weiner, Phys. Rev. A **27**, 2817 (1983).

/69/ C.Y.R. Wu, F. Roussel, B. Carré, P. Breger and G. Spiess, J. Phys.B **18**, 239 (1985).

/70/ F. Roussel, P. Breger and G. Spiess, to be published.

LASER SPECTROSCOPY OF COLLISION COMPLEXES: A CASE STUDY

Nils Andersen*
Physics Laboratory II, H.C. Ørsted Institute
DK-2100 Copenhagen, Denmark

1. INTRODUCTION

This paper adresses the question of what can be learned about atomic interactions by light scattering, and in particular from the redistribution in frequency and polarization of the photons scattered off collision complexes. This is currently a very active area of atomic physics, and the reader is referred to two recent reviews [1,2] which provide excellent overview and perspective. Much of the presentation below is based on the papers [3,4] which may be consulted for further details and extensive lists of references.

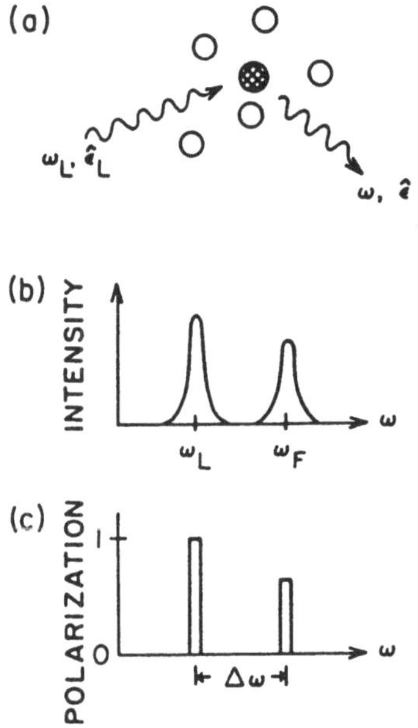

Figure 1

Figure 1(a) shows schematically a typical geometry for a redistribution experiment. An atom surrounded by perturbers is irradiated by photons from e.g. a laser with frequency ω_L and polarization $\hat{\varepsilon}_L$. The

---

* Present adress: Institute of Physics, University of Aarhus, DK-8000 Aarhus C, Denmark.

intensity and polarization $\hat{\varepsilon}$ of the frequency analysed photons scattered in some direction are studied. When the incident frequency ω_L is not far from the frequency ω_0 of a resonance transition af the atom, the scattered light will (for reasonably low laser intensity) consist of two components, see Figure 1(b) and (c). One component has the same frequency as the incident photons and is due to Rayleigh scattering. This component is also present without perturber atoms. The other one has the frequency ω_0 characteristic of the fluorescence of the free atom, and an intensity proportional to the perturber density. This second component has a polarization different from that of the Rayleigh component. We shall now discuss the physics behind this fluorescence emission.

2. THE EXCITATION-DEEXCITATION PROCESS

The redistribution in intensity and polarization may be interpreted as follows, cf. Figure 2. The atom A may undergo a collision with one of

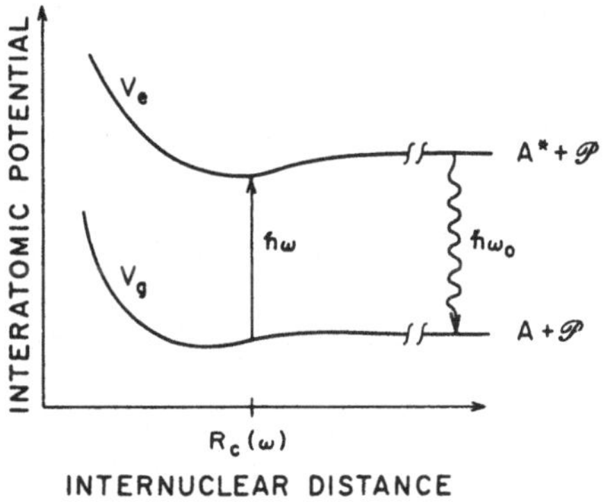

Figure 2

the surrounding perturber atoms, thereby forming a quasi-molecule which at thermal energies has a lifetime τ_c of typically a picosecond. For detunings $\Delta\omega = |\omega_0-\omega_L| >> \tau_c^{-1}$ (the quasi-static region) the collision complex may absorb a photon at a well-localised internuclear distance $R_c=R_c(\omega_L)$ - the Condon point - where the energy difference between the ground state potential curve V_g and that of an excited state V_e matches the photon energy

$$\hbar\omega_L = V_e(R_c) - V_g(R_c)$$

The collision complex now evolves along the excited potential curve V_e.

The quasi-molecule separates into two non-interacting atoms at some distance R_{Dec} - the decoupling radius - and the excited atom decays - typically after some nanoseconds - by emission of a fluorescence photon, thus contributing to the second component of Figure 1. Since this process depends on potential-curve geometry we may thus get information about the atom-perturber molecule from studying the changes in intensity and polarization when ω and thereby the Condon-point, varies. Specifically, the fluorescence intensity $I(\omega)$ corresponding to the frequency ω is proportional to the number of atoms at the corresponding distance $R_C(\omega)$

$$I(\omega)d\omega \propto 4\pi R_C^2 dR$$

or, using $\Delta V = V_e - V_g = \hbar\omega$

$$I(\omega) \propto 4\pi R_C^2 \left|\frac{d}{dR}\Delta V\right|_{R_C}^{-1}$$

Thus the fluorescence intensity depends on the Condon point and the slope of the potential curve difference at this distance.

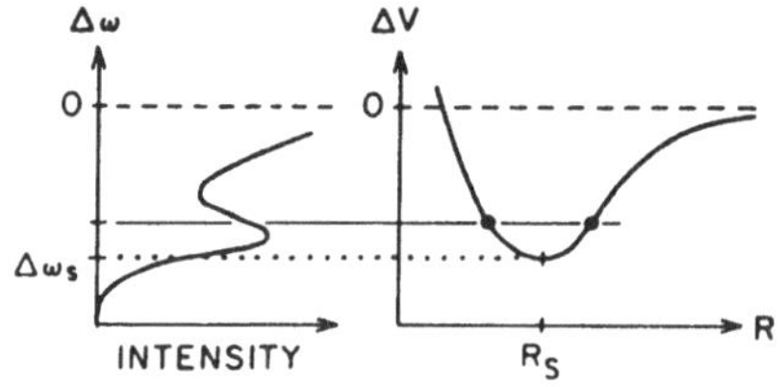

Figure 3

As indicated in Figure 3 this picture is modified in regions where several (or no) Condon-points contribute to the fluorescence. This is especially so near a maximum or minimum of ΔV, giving rise to a so-called satellite, followed by an exponential drop at larger detunings.

How the polarization depends on excitation geometry and molecular quantum numbers is illustrated on Figure 4, where the simple case of Σ-excitation is shown in (a): a Σ-orbital is excited at the Condon-point (i). At short internuclear distances the orbital is locked to the internuclear axis and rotates with it, but will eventually decouple at some radius R_{Dec}. From here on the orbital stays fixed in space until it decays at point (iii) (We here neglect the effect of subsequent, depolarizing collisions, which may be corrected for by investigating the pressure dependence). Excitation to a Π-orbital follows the same scheme,

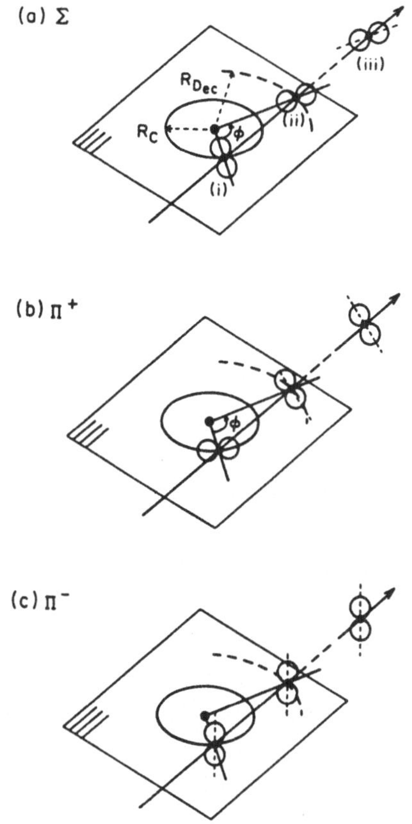

Figure 4

but with an important modification: A Π-orbital may either be (b) a Π^+ (located in the scattering plane) or (c) a Π^- (perpendicular to the scattering plane), or some mixture hereof. The Π^+-orbital has, apart from a 90°-rotation in the scattering plane, the same fate as the Σ just discussed, while the Π^- on the contrary stays fixed in space during the whole collision. Assuming straight line trajectories (SLT) and averaging over all orientations of the scattering plane with respect to the polarization-direction of the incident photons, one gets the following expression for the polarizations when observing the linear polarization $P_L = (I_{\parallel}-I_{\perp})/(I_{\parallel}+I_{\perp})$ with linearly polarized photons

Σ-excitation: $P_L = 9x^2/(25+3x^2)$

Π-excitation: $P_L = (15+15x+9x^2)/(55+5x+3x^2)$

Here $x = R_C/R_{Dec}$ $(0 < x < 1)$ is the ratio between the Condon-radius and the decoupling radius. Σ-polarization is thus located in the range 0 − 32 %, while Π-polarization is restricted to 27 − 62 %. One may show that trajectory effects always tend to increase the polarization compared to the SLT-value. If one measures the circular polarization $P_c = (I_+-I_-)/(I_++I_-)$ when exciting with photons with + helicity one gets

Σ-excitation: $P_c = 0$

Π-excitation: $P_c = 25x/(55+5x+3x^2)$

The polarization thus contains information about the collision dynamics, and in particular about the distance at which the break-up of the quasimolecule to separated-atoms takes place. We also notice that (under the assumptions above) one can predict the outcome of a circular polarization experiment if the relevant molecular quantum number and the linear polarization (at the same detuning) are known.

3. THE BARIUM-RARE GAS COLLISION COMPLEXES

When selecting proper systems for a redistribution experiment several practical criteria are of importance. For example, $^1P \rightarrow {}^1S$ fluorescence is particularly convenient, since the molecular curve structure is simple when no fine structure or hyperfine structure is present. Also, one wants a convenient wavelength for tunable dye lasers. Finally one might select a heavy rare gas as perturber since they are easily polarized, and therefore pronounced structures in the potential curves may be looked for. Based on these and other criteria the Ba-Xe system is a convenient choice since the green BaI 6^1S-6^1P resonance transition has a wavelength of 5535 Å (Notice, however, that from the point of view of the theoretician this choice might be quite far from the favourite one !).

Figure 5(a) and (b) shows the light intensity (lower panels) and linear polarizations (upper panels) for (a) the red and (b) the blue wings for the Ba-Xe system. The intensity (or the redistribution coefficient k_r) is measured per Barium and per perturber atom, and has been multiplied by $\Delta\omega^2$ in order to better display the structures. Since a Lorentzian shape corresponds to $(\Gamma^2+\Delta\omega^2)^{-1}$, a Lorentz-curve would in this plot yield a horizontal line in the far-wing region. The absolute scale is determined by normalization to known Rayleigh cross sections.

Starting with the red wing we notice a satellite around 40 cm^{-1} at which value the polarization drops to about 10 %, indicating predomi-

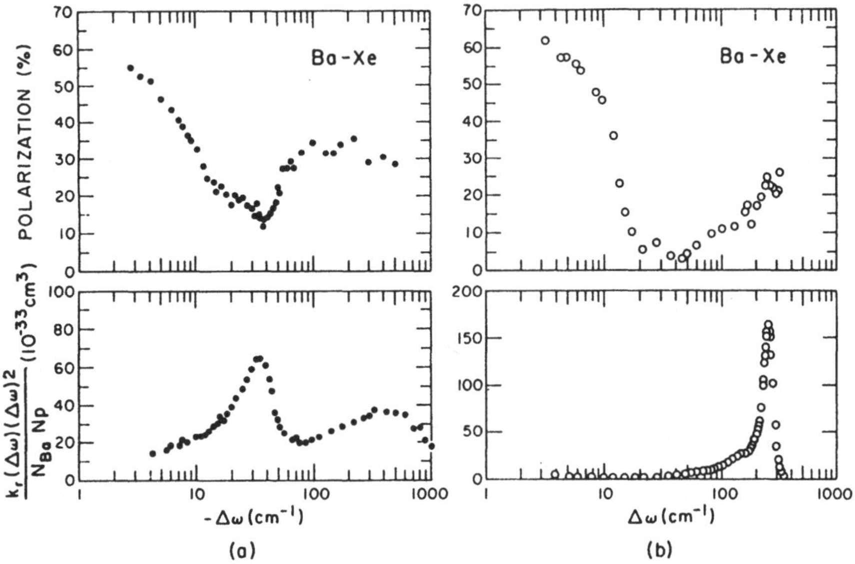

Figure 5

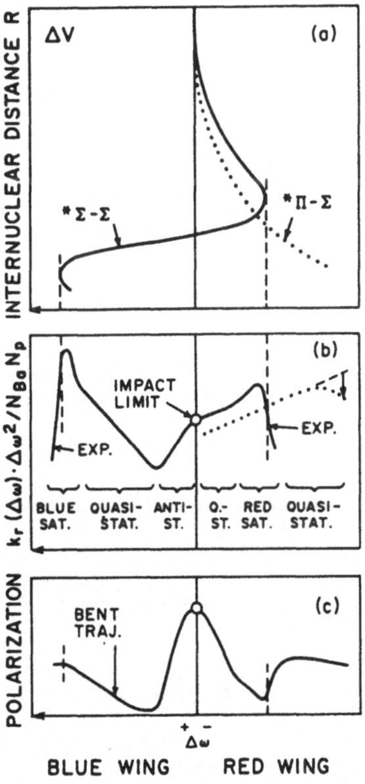

Figure 6

nantly Σ-excitation. Beyond the satellite the polarization goes up to about 30 %, apparently due to Π-excitation. The blue wing polarization drops to almost zero near 40 cm^{-1}, i.e. Σ-excitation. There is very little blue wing intensity until we reach the satellite at about 250 cm^{-1}. The rise in the polarization out to this value is ascribed to trajectory effects, i.e. deviations from SLT.

A schematic interpretation of the findings of Figure 5 is given in Figure 6 which indicates, as function of detuning, the qualitative behaviour of (a) the potential curve difference ΔV, and the corresponding intensities (b) and linear polarizations (c) resulting from this. A more detailed discussion is given in [3].

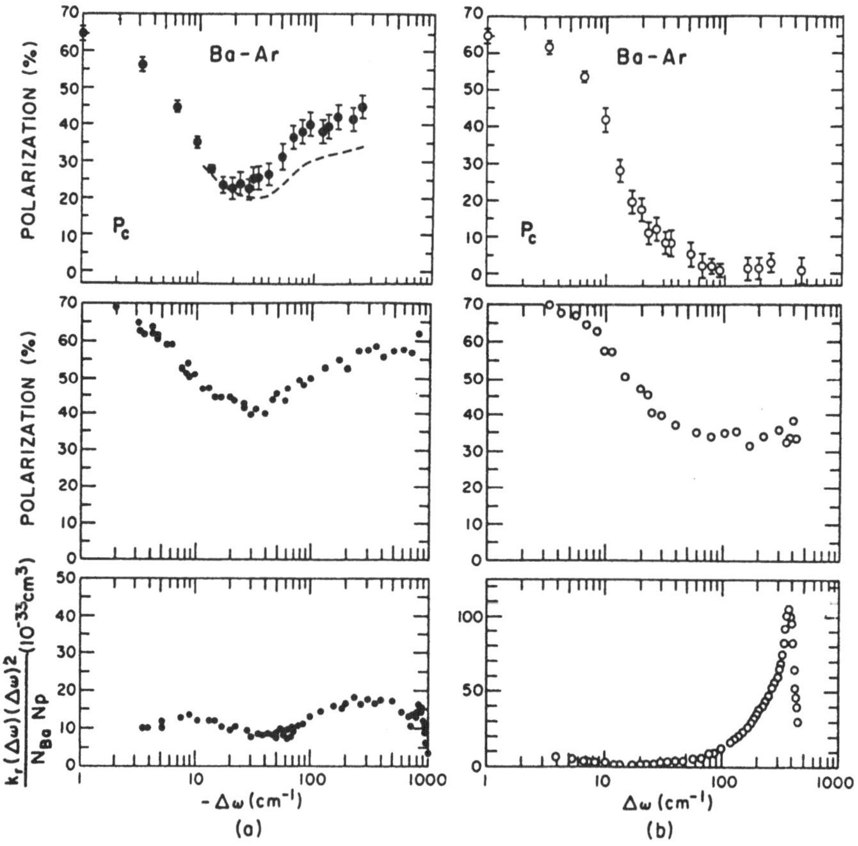

Figure 7

Turning finally to the circular polarization P_c, Figure 7 shows P_c (upper panels) measured for the (a) red wing and (b) blue wing of the Ba-Ar system, together with the linear polarizations (middle panels) and intensities (lower panels). The Ba-Ar curves look similar to the corresponding curves for Ba-Xe though most structures are less pronounced, and the potential curves may show a behaviour similar to that given in Figure 6(a). For the blue wing P_c goes to zero as expected for a Σ-curve. For the red wing the dashed curve shows P_c predicted by the reorientation model from the linear polarization, assuming SLT. The curve follows nicely the experimental points, but deviates gradually at large detunings, indicative of increasingly important trajectory effects. Further details may be found in [4].

4. CONCLUSIONS

It is hoped that the presentation above, though brief and somewhat simplified, has served to illustrate how careful analysis of the redistribution in intensity and polarization of photons scattered off collision complexes may serve as a rich source of information about interatomic forces and molecular quantum numbers and guide the development of simple models that allow us to visualize the development of the excited electron charge cloud during the collision.

5. ACKNOWLEDGMENTS

My involvement in the experiments described above took place during stays in 1983 and 1984 as Visiting Fellow at the Joint Institute for Laboratory Astrophysics of the National Bureau of Standards and the University of Colorado in Boulder, Colorado. I am much indebted to my collegues at JILA, and in particular Joe Alford, Keith Burnett, and Jinx Cooper, for introducing me to this fascinating topic.

REFERENCES

[1] N. Allard and J. Kielkopf, Rev. Mod. Phys. 54 (1982) 1103.
[2] K. Burnett, Physics Reports (1985), in press.
[3] W.J. Alford, N. Andersen, K. Burnett, and J. Cooper, Phys. Rev. A 31 (1984) 2366.
[4] W.J. Alford, N. Andersen, M. Belsley, J. Cooper, D.M. Warrington, and K. Burnett, Phys. Rev. A 31 (1985), in press.

TWO-PHOTON COLLISIONAL REDISTRIBUTION OF RADIATION

G. Alber
Joint Institute for Laboratory Astrophysics, University of Colorado and National Bureau of Standards, Boulder, Colorado 80309 USA

I. INTRODUCTION

Scattering of light by atoms undergoing collisions with a gas of foreign perturbers is an important method for studying interatomic interactions. There has been considerable activity in recent years to obtain this collisional information from one-photon scattering experiments (see, e.g., Refs. 1-3). Typically in such an experiment a laser with frequency ω_L excites an atom (radiator) from its ground state to an excited state and the spectrum of the spontaneously emitted radiation is then investigated. For low perturber densities, such that different collisions do not overlap in time, and weak laser fields the spectrum consists of two peaks, namely (1) a Rayleigh peak centered at the laser frequency ω_L, and (2) a redistributed peak centered around the atomic transition frequency, which vanishes in the absence of collisions as long as the lower state is a ground state.[16]

Depending on the value of the detuning of the laser from the atomic transition frequency $|\omega_o-\omega_L|$ in comparison with the inverse duration of a collision τ_c the redistributed peak contains qualitatively different kinds of collisional information.[4-6,11] In the impact region, where $|\omega_o-\omega_L|\tau_c \ll 1$, the redisributed intensity is determined by collisional S-matrix elements, because collisions are instantaneous in comparison with the time evolution of the radiator in the laser field (Markov limit). In the non-Markovian limit, where $|\omega_o-\omega_L|\tau_c \gg 1$, the laser photon is absorbed instantaneously in comparison with the duration of a collision τ_c and details of the intra-collisional time evolution manifest themselves in the redistributed intensity. In particular, effects due to the degeneracy of the radiator states are important[5,6] and determine the degree of polarization of the total redistributed intensity for example.[2,3]

In this discussion we want to generalize these investigations to a two-photon process, where a first laser excites the radiator from the ground state to an excited state and a second laser induces a transition from there to a final state. We shall study the spectrum of the radiation corresponding to the spontaneous transition from the final to an intermediate state. Of particular interest is the redistributed peak of this spectrum, which occurs at the transition frequency between the final and the excited state. The degeneracy of radiator states will be properly taken into account in a $J=0 \rightarrow J=1 \rightarrow J=0$ type transition. Our main interest will focus on two aspects of this problem: (a) What can we learn about collisions from such an experiment? (b) How does the degeneracy of radiator states affect redistribution?

Though most of our discussion will focus on weak laser fields, we shall also consider a situation of current experimental interest,[3] where this is no longer the case.

We represent the model problem, which we study here, in Sec. II. Section III deals with nondegenerate radiator states and explains the basic physics of two-photon collisional redistribution. In Sec. III we discuss the modifications brought about by the degeneracy of radiator states and consider laser intensity effects in a special case.

II. BASIC EQUATIONS

We study a system schematically shown in Fig. 1. A neutral atom (radiator), which is surrounded by N perturber atoms interacts with a classical electromagnetic field $\vec{E}(t)$. In addition, it is coupled to the vacuum modes of the electromagnetic field giving rise to spontaneous emission of photons. The perturbers, usually noble gas atoms, which collide with the radiator, are assumed to be structureless. This is certainly a good approximation as long as the excitation energies of a perturber atom are much larger than typical excitation energies of the radiator. The classical electromagnetic field

$$\vec{E}(t) = \sum_{j=1}^{2} \vec{e}_j \, \varepsilon_j \, e^{-i\omega t} + c.c. \qquad (1)$$

consists of two monochromatic laser fields of frequencies ω_1, ω_2 and polarizations $\vec{e}_1$, $\vec{e}_2$. In particular, we study the following excitation process: The first laser field $(\omega_1, \vec{e}_1)$ excites the radiator (e.g. an alkaline earth atom) from its ground state $|g\rangle$ with total angular momentum J=0 to an excited state $|e_i\rangle$ with J=1 and the second laser photon $(\omega_2, \vec{e}_2)$ induces a transition from this manifold to the final state $|f\rangle$ with J=0.

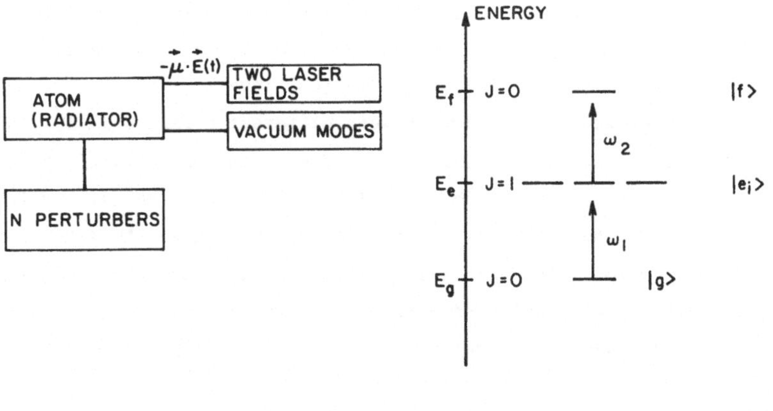

Fig. 1. Schematic representation of physical system and excitation process.

In the further treatment of this problem, we shall assume:

1) The interaction between the radiator and the two laser fields may be described by an effective Hamiltonian for the degenerate three-level system $|g\rangle$, $\{|e_i\rangle\}$, $|f\rangle$

$$H_{eff} = E_g|g\rangle\langle g| + \sum_{i=1,2,3} (E_g-\hbar\Delta_1)\,|\bar{e}_i\rangle\langle\bar{e}_i| + (E_g-\hbar\Delta_1-\hbar\Delta_2)\,|\bar{f}\rangle\langle\bar{f}| -$$

$$-\frac{\hbar}{2}\sum_{i=1,2,3}\{\Omega^*_{fe_i}|\bar{e}_i\rangle\langle\bar{f}| + \Omega_{fe_i}|\bar{f}\rangle\langle\bar{e}_i| + \Omega^*_{e_ig}|g\rangle\langle\bar{e}_i| + \Omega_{e_ig}|\bar{e}_i\rangle\langle g|\} \quad (2)$$

with detunings from resonance $\Delta_1 = (E_g+\hbar\omega_1-E_e)/\hbar$, $\Delta_2 = (E_e+\hbar\omega_2-E_f)/\hbar$ and Rabi frequencies $\Omega_{e_ig} = (2/\hbar)\cdot\langle e_i|\vec{\mu}\cdot\vec{e}_1|g\rangle\cdot\varepsilon_1$, $\Omega_{fe_i} = (2/\hbar)\cdot\langle f|\vec{\mu}\cdot\vec{e}_2|e_i\rangle\cdot\varepsilon_2$. Here $|\bar{e}_i\rangle = |e_i\rangle\exp(-i\omega_1 t)$ and $|\bar{f}\rangle = |f\rangle\cdot\exp[-i(\omega_1+\omega_2)\cdot t]$ are rotating atomic states. Thereby we make the dipole and the rotating wave approximation. Quadratic Stark shifts and the direct coupling between $|g\rangle$ and $|f\rangle$ via all other (nonresonant) atomic states have been neglected, which is certainly valid for sufficiently low laser intensities. We also assume that ionization from the final state $|f\rangle$ is unimportant.

2) The influence of the vacuum modes of the electromagnetic field on the dynamics of the system may be described within the Markov approximation by (time independent) spontaneous decay rates. This assumes that the correlation time associated with these modes, which is typically of order 10^{-16}-10^{-18} s, is much smaller than all other time scales in the radiator-N perturbers-laser fields-system,[4] which is usually the case.

3) The neutral perturbers are assumed to be statistically independent. Correlations between perturbers can be included in our formulation to some extent by considering independent "quasi particles." We treat the coupling between radiator and perturbers in the binary collision approximation (BCA),[4] which amounts to a single collision occurring at a time. This allows us to disentangle different collisions and reduces the complicated (N+1)-body problem to a two-body problem, namely the collision of a single perturber with the radiator in the presence of the laser fields. The BCA is valid as long as the duration of a collision τ_c is much smaller than a typical inverse (impact) collision rate γ_c, which is a measure for the time between different collisions.[4,7] In addition, we neglect for simplicity all effects due to the velocity of the radiator and assume a spherically symmetric collision environment. Velocity effects could be included in our formulation in an approximative way by a Doppler convolution. Inelastic collisions are also neglected. The effective collisional interaction V therefore couples only degenerate radiator states.

According to these approximations the reduced density operator of the radiator $\sigma(t)$ obeys the non-Markovian equation of motion[5,8]

$$\frac{d}{dt}\,\sigma(t) = [L_{eff}+\Gamma+N\ Tr_P(\rho_P V_1)]\ \sigma(t) + \int_0^t dt'\ M(t-t')\ \sigma(t') \qquad t \geq 0 \qquad (3a)$$

with $M(\tau) = N\ Tr_P\{V_1 G_1(\tau) V_1 \rho_P\}$ and

$$\frac{d}{d\tau}\,G_1(\tau) = (L_{eff} + L_P + V_1 + \Gamma)\ G_1(\tau) \qquad \tau > 0 \quad , \qquad (3b)$$

$$G_1(\tau=0) = 1 \quad .$$

Thereby we define $L_{eff}\cdot = \frac{1}{i\hbar}[H_{eff},\cdot]$, $V_1\cdot = \frac{1}{i\hbar}[V,\cdot]$, $L_P\cdot = \frac{1}{i\hbar}[(\hat{p}^2/2M),\cdot]$, with the radiator-perturber interaction potential V and the kinetic momentum $\hat{p}$ and mass M of a single perturber. ρ_P is the density operator characterizing one perturber. The tetradic operator Γ describes the effects due to spontaneous emission of photons within the Markov approximation. The collision ("memory") kernel $M(\tau)$ is determined by the dynamics of the radiator-perturber collision in the presence of the laser fields. In a Markov-type treatment of collisions we would describe collisions also by a time-independent collision rate, setting $\int_0^t dt'\ M(t-t')\ \sigma(t') \approx \Gamma_{coll}\sigma(t)$. This would require the duration of a single collision τ_c to be much smaller than all other times, which determine the time evolution of $\sigma(t)$. However, it is precisely the breakdown of the Markov approximation as far as collisions are concerned, which is our main interest. Initial radiator-perturber correlations, which decay on a time scale of order τ_c,[4] are neglected in Eq. (3a), which restricts its validity to times t much larger than τ_c.

We are particularly interested in the spectrum of the light $I(\omega,\vec{\varepsilon})$ corresponding to the spontaneous transition of the radiator from $|f\rangle$ to $\{|e_i\rangle\}$, which is determined by the dipole autocorrelation function. We find[8]

$$I(\omega,\vec{\varepsilon}) \propto 2\ \mathrm{Re}\ \{\sum_i \langle e_i|\vec{\mu}\cdot\vec{\varepsilon}^{\,*}|f\rangle\ \langle\bar{f}|G(z=\omega-\omega_2)|\bar{e}_i\rangle\} \qquad (4a)$$

with

$$\{z - i[L_{eff} + \Gamma + M(z)]\}\ G(z) = i\sigma(t\to\infty)\ d^+ \quad . \qquad (4b)$$

$\vec{\varepsilon}$ is the polarization of the emitted photon and $M(z)=\int_0^\infty d\tau\ e^{iz\tau}\ M(\tau)+N\ Tr_P(\rho_P V_1)$. Consistent with the BCA we have thereby neglected contributions to the spectrum due to fluorescence during a collision, because they are negligible in the situation of interest here.[8]

III. NONDEGENERATE THREE-LEVEL SYSTEM

To demonstrate some basic features of two-photon collisional redistribution we discuss next the (somewhat unrealistic) case where the intermediate radiator state, denoted $|e\rangle$, is nondegenerate. In particular, we consider weak fields in the sense that

$$|\Omega_{fe}| \quad , \quad |\Omega_{eg}| \ll 1/\tau_c \quad , \tag{5a}$$

$$|\Omega_{fe}| \quad , \quad |\Omega_{eg}| \ll |\Delta_1|,|\Delta_2|,|\Delta_1+\Delta_2| \tag{5b}$$

and large detunings such that the secular approximation[9] holds, i.e.,

$$|\Delta_1|,|\Delta_2|,|\Delta_1+\Delta_2| \gg \gamma_c,\gamma_e,\gamma_f \quad , \tag{5c}$$

where γ_e and γ_f are the spontaneous decay rates of $|e\rangle$ and $|f\rangle$. Equation (5a) allows us to calculate the influence of the laser fields on the collision kernel $M(\tau)$ perturbatively. Equation (5b) implies that the ground state of the radiator is effectively undepleted during the interaction with the laser fields and $\sigma(t)$ can be calculated perturbatively.

1. Dressed States and the Weak Field Spectrum

In the weak field limit [Eq.(5b)] the dressed states, which diagonalize the effective Hamiltonian H_{eff}, i.e., $H_{eff}|j\rangle = E_j|j\rangle$, $j = I,II,III$, are approximately given by (see also Fig. 2)

$$\begin{aligned} |I\rangle &= |g\rangle - \frac{1}{2}\frac{\Omega_{eg}}{\Delta_1}|\bar{e}\rangle + \frac{1}{4}\frac{\Omega_{fe}\cdot\Omega_{eg}}{\Delta_1\cdot(\Delta_1+\Delta_2)}|\bar{f}\rangle \quad , \\ |II\rangle &= \frac{1}{2}\frac{\Omega_{eg}}{\Delta_1}|g\rangle + |\bar{e}\rangle - \frac{1}{2}\frac{\Omega_{fe}}{\Delta_2}|\bar{f}\rangle \quad , \\ |III\rangle &= \frac{1}{4}\frac{\Omega_{fe}\cdot\Omega_{eg}}{\Delta_2(\Delta_1+\Delta_2)}|g\rangle + \frac{1}{2}\frac{\Omega_{fe}}{\Delta_2}|\bar{e}\rangle + |\bar{f}\rangle \quad . \end{aligned} \tag{6}$$

Within the secular approximation [Eq. (5c)] the spectrum of the light $I(\omega,\vec{\varepsilon})$ corresponding to the transition $|f\rangle \to |e\rangle$ in lowest (fourth) order in the fields is then given by

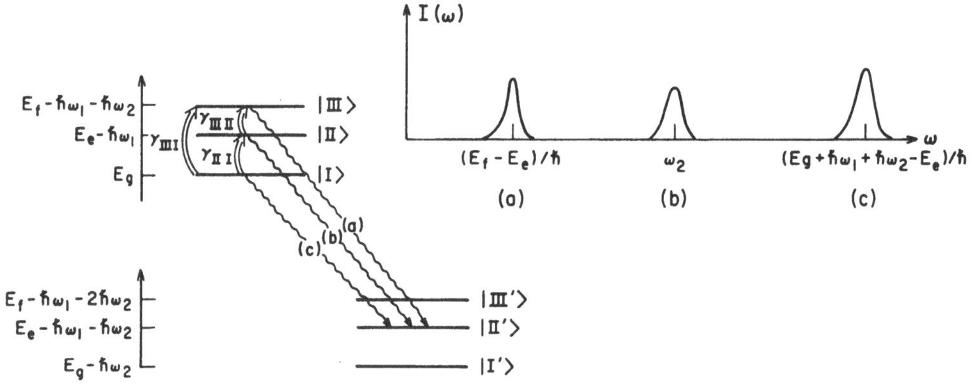

Fig. 2. Dressed states and weak field spectrum.

$$I(\omega,\vec{\varepsilon}) \propto \left|\langle f|\vec{\mu}\cdot\vec{\varepsilon}|e\rangle\right|^2 \cdot \sum_{i=I,II,III} \left|\langle f|i\rangle\right|^2 \cdot \sigma_{ii}(t\to\infty) \cdot \frac{2\Gamma_i}{[\omega-\omega_2-(E_i-E_{II})/\hbar]^2 + \Gamma_i^2} \quad , \tag{7}$$

which is obtained directly from Eqs. (4a) and (4b) by inserting the complete set of dressed states of Eq. (6). This corresponds to three well-separated peaks as shown in Fig. 2. Their widths Γ_i are determined by the corresponding spontaneous decay and collisional dephasing rates. The stationary dressed states density matrix elements are calculated from Eq. (3a). In particular we find within the secular approximation in the weak field limit[8]

$$\begin{aligned} \gamma_f\, \sigma_{III\,III}(t\to\infty) &= \gamma_{III\,II}\, \sigma_{II\,II}(t\to\infty) + \gamma_{III\,I} \quad , \\ \gamma_e\, \sigma_{II\,II}(t\to\infty) &= \gamma_{II\,I}\, \sigma_{I\,I}(t\to\infty) \quad , \quad \sigma_{I\,I}(t\to\infty) = 1 \quad . \end{aligned} \tag{8a}$$

Note that the coupling due to spontaneous decay is negligible under the conditions of Eqs. (5b) and (5c). In our further treatment we shall for convenience describe collisions within the classical path approximation with straight line trajectories, which is essentially valid for detunings less than $kT/\hbar$[10] (T = temperature of perturber bath). Then the collision rates in Eq. (8a) are related to dressed states collisional S-matrix elements by[8]

$$\gamma_{ij} = \{|\langle i|U(\infty,-\infty)|j\rangle|^2\}_{av} \qquad i,j = I,II,III \quad , \tag{8b}$$

where the time evolution operator $U(t,-\infty)$ in the presence of collisions and the laser fields is determined by

$$i\hbar \frac{d}{dt} U(t,-\infty) = \left(H_{eff} + V^{(b,v)}(t)\right) U(t,-\infty) \tag{8c}$$

with the initial condition $U(-\infty,-\infty) = 1$. $\{\ldots\}_{av} = (N/V)\cdot\int_0^\infty dvf(v)\cdot v\cdot 2\pi\int_0^\infty dbb\ldots$ represents the averaging over all impact parameters b and velocities v of the perturber with $\int_0^\infty dvf(v) = 1$. The effective collisional interaction is given by $V^{(b,v)}(t) \equiv V(R(t))$ with $R^2(t) = b^2 + v^2\cdot t^2$. The matrix elements of $V(R(t))$ are, of course, those appropriate for dressed states and are evaluated assuming no collisional coupling between different (bare) radiator states. Solving Eq. (8c) under the weak field conditions of Eqs. (5a) and (5b), we find

$$\begin{aligned} \gamma_{II\,I} &= \frac{1}{4} \cdot \frac{\Omega_{eg}^2}{\Delta_1^2} \cdot \left\{\left|\int_{-\infty}^{\infty} dt\, e^{i\Delta_1 t} \cdot \frac{1}{\hbar} V_{eg}^{(b,v)}(t)\, \langle e|U_I(t,0)|e\rangle\langle g|U_I(t,0)|g\rangle^* \right|^2\right\}_{av} \\ &\equiv \frac{1}{4} \cdot \frac{\Omega_{eg}^2}{\Delta_1^2} \cdot 2\gamma_{eg}(\Delta_1) \quad , \end{aligned}$$

$$\gamma_{III\ II} = \frac{1}{4} \cdot \frac{\Omega_{fe}^2}{\Delta_2^2} \cdot \left\{ \left| \int_{-\infty}^{\infty} dt\ e^{i\Delta_2 \cdot t} \cdot \frac{1}{\hbar} V_{fe}^{(b,v)}(t) \langle f|U_I(t,0)|f\rangle \langle e|U_I(t,0)|e\rangle^* \right|^2 \right\}_{av}$$

$$\equiv \frac{1}{4} \cdot \frac{\Omega_{fe}^2}{\Delta_2^2} \cdot 2\gamma_{fe}(\Delta_2) \quad , \tag{8d}$$

$$\gamma_{III\ I} = \frac{1}{16} \cdot \Omega_{fe}^2 \cdot \Omega_{eg}^2 \cdot \left\{ \left| \int_{-\infty}^{\infty} dt \frac{1}{\Delta_1 + \Delta_2} \left(\frac{V_{fe}^{(b,v)}(t)}{\hbar\Delta_1} - \frac{V_{eg}^{(b,v)}(t)}{\hbar\Delta_2} \right) \right. \right.$$

$$\cdot \langle f|U_I(t,0)|f\rangle \langle g|U_I(t,0)|g\rangle^* e^{i(\Delta_1+\Delta_2)t} +$$

$$+ i \int_{-\infty}^{\infty} dt_1 \frac{V_{fe}^{(b,v)}(t_1)}{\hbar\Delta_2} \cdot \langle f|U_I(t_1,0)|f\rangle \langle e|U_I(t_1,0)|e\rangle^* e^{i\Delta_2 \cdot t_1} \cdot$$

$$\left. \left. \cdot \int_{-\infty}^{t_1} dt_2 \frac{V_{eg}^{(b,v)}(t_2)}{\hbar\Delta_1} \langle e|U_I(t_2,0)|e\rangle \langle g|U_I(t_2,0)|g\rangle^* e^{i\Delta_1 \cdot t_2} \right|^2 \right\}_{av}$$

with the difference potentials $V_{fe}^{(b,v)}(t) = \langle f|V^{(b,v)}(t)|f\rangle - \langle e|V^{(b,v)}(t)|e\rangle$, etc. The first two of these equations indicate the relation between the dressed states collision rates and the collisional coherence decay rates $\gamma_{eg}(\Delta_1)$ and $\gamma_{fe}(\Delta_2)$, which can also be obtained directly from the appropriate matrix elements of the collision operator $M(z=0)$. These can be written in various equivalent forms,[4,5,8] e.g.

$$\gamma_{eg}(\Delta_1) = \mathrm{Re}\left\{ \int_{-\infty}^{\infty} dt_o \int_0^{\infty} dt \left(\Delta_1 + i\frac{\gamma_e}{2}\right) \cdot \exp\left[i\left(\Delta_1 + i\frac{\gamma_e}{2}\right)t\right] \right.$$

$$\left. \cdot \langle e|U_I(t+t_o,t_o)|e\rangle \langle g|U_I(t+t_o,t_o)|g\rangle^* \cdot \frac{1}{\hbar} V_{eg}^{(b,v)}(t_o) \right\}_{av} \quad , \tag{8e}$$

where t_o is the time of closest approach. The time evolution operator in the interaction picture is defined by

$$U_I(t_1,t_2) = \exp\left[-\frac{i}{\hbar} \int_{t_1}^{t_2} dt\ V^{(b,v)}(t)\right] \quad .$$

The physical significance of the collisional rates in Eq. (8d) is discussed in Sec. III.2.

In the absence of collisions we therefore have $\sigma_{I\ I}(t\to\infty) = 1$, $\sigma_{II\ II}(t\to\infty) = \sigma_{III\ III}(t\to\infty) = 0$. The weak field spectrum consists of a single peak around frequency $\omega \approx (E_g + \hbar\omega_1 + \hbar\omega_2 - E_e)/\hbar$, which corresponds to a Raman process (denoted (c) in Fig. 2). The integrated intensity of this peak is given by

$$I_{Raman} \propto 2\pi \; |\langle f|\vec{\mu}\cdot\vec{\varepsilon}|e\rangle|^2 \cdot \frac{1}{16} \frac{|\Omega_{fe}|^2 \; |\Omega_{eg}|^2}{\Delta_1^2(\Delta_1+\Delta_2)^2} \quad . \tag{9}$$

In the presence of collisions two additional peaks appear in the spectrum due to the fact that now also $\sigma_{II\ II}(t\to\infty) \neq 0$ and $\sigma_{III\ III}(t\to\infty) \neq 0$. One peak is centered on frequency $\omega \approx \omega_2$ (denoted (b) in Fig. 2) and corresponds to "collisional-induced Rayleigh" scattering of the laser photon $(\omega_2,\vec{e}_2)$. Its integrated intensity is given by

$$I_{Rayleigh} \propto 2\pi \; |\langle f|\vec{\mu}\cdot\vec{\varepsilon}|e\rangle|^2 \cdot \frac{1}{4} \cdot \frac{|\Omega_{fe}|^2}{\Delta_2^2} \cdot \sigma_{II\ II}(t\to\infty) \quad . \tag{10}$$

The second peak is centered on the radiator transition frequency $\omega \approx (E_f-E_e)/\hbar$ and represents the process of two-photon collisional redistribution. Its integrated intensity

$$I_{red} \propto 2\pi \; |\langle f|\vec{\mu}\cdot\vec{\varepsilon}|e\rangle|^2 \cdot \sigma_{III\ III}(t\to\infty) \tag{11a}$$

is a direct measure for the stationary dressed state population $\sigma_{III\ III}(t\to\infty)$, which involves the absorption of the two laser photons $(\omega_1,\vec{e}_1)$ and $(\omega_2,\vec{e}_2)$. From Eq. (8a) we find

$$\sigma_{III\ III}(t\to\infty) = \frac{1}{\gamma_f} \{\gamma_{III\ II} \cdot \frac{\gamma_{II\ I}}{\gamma_e} + \gamma_{III\ I}\} \quad . \tag{11b}$$

Thus the first term represents a process in which the laser photons are absorbed in subsequent collisions; the second term describes absorption of both photons during a single collision.

2. Total Redistributed Intensity

In the following we study the total redistributed intensity as given by Eq. (11a) in three different limiting situations in order to demonstrate the physical significance of the various dressed states collision rates.

a) $|\Delta_1|\tau_c,\ |\Delta_2|\tau_c,\ |\Delta_1+\Delta_2|\tau_c \ll 1$

The time scales associated with the absorption of both laser photons, i.e., $1/|\Delta_1|,\ 1/|\Delta_2|,\ 1/|\Delta_1+\Delta_2|$, are much larger than the duration of a collision τ_c. Any collision in Eq. (8d) can therefore be completed and the dressed states collision rates reduce to collisional (tetradic) S-matrix elements between bare atomic states and are independent of the laser frequencies ω_1 and ω_2 (impact or Markovian limit). Therefore we find for example from Eq. (8e)[17]

$$\gamma_{eg}(\Delta_1) \xrightarrow[|\Delta_1|\tau_c\to 0]{} -\mathrm{Re}\ \{\langle e|U_I(\infty,-\infty)|e\rangle\langle g|U_I(\infty,-\infty)|g\rangle^* - 1\}_{av} \quad . \tag{12}$$

In Ref. 8 it is shown that in the impact limit

$$\gamma_{III\ I} \to \frac{1}{16} \cdot \Omega_{fe}^2 \cdot \Omega_{eg}^2 \cdot \left\{\frac{2\cdot\gamma_{fe}(\Delta_2=0)}{\Delta_1^2\cdot\Delta_2\cdot(\Delta_1+\Delta_2)} + \frac{2\cdot\gamma_{eg}(\Delta_1=0)}{\Delta_1\cdot\Delta_2^2\cdot(\Delta_1+\Delta_2)} - \frac{2\cdot\gamma_{fg}(\Delta_1+\Delta_2=0)}{\Delta_1\cdot\Delta_2\cdot(\Delta_1+\Delta_2)^2}\right\} , \tag{13}$$

which may be obtained directly by evaluating the dressed states matrix elements of the collision operator M(z=0) in the impact limit. Equation (13) inserted in Eq. (11a) yields the total redistributed intensity. In particular we see that for low perturber densities, i.e., $\gamma_e \gg \gamma_c$, the single collision contribution [Eq. (13)] dominates, whereas for high densities, i.e., $1/\tau_c \gg \gamma_c \gg \gamma_e$, the subsequent collision part determines the redistributed intensity.

b) $|\Delta_1\tau_c| \gg 1$, $|\Delta_2|\tau_c \ll 1$

The time for absorbing the photon ω_2 is much larger than the duration of a collision τ_c so that $\gamma_{fe}(\Delta_2)$ again has to be evaluated in the impact limit and contains therefore only global information about the collision process in form of S-matrix elements. However, the situation for the laser photon ω_1 is quite different. The duration of a collision is much larger than the associated absorption time $1/|\Delta_1|$. The photon ω_1 is therefore absorbed instantaneously during the collision and we become able to investigate details of the intra-collisional evolution process (quasi-static limit). $\gamma_{eg}(\Delta_1)$ is now frequency dependent thereby indicating the breakdown of a Markov-type treatment of the influence of the perturbers on the radiator, which would lead to frequency independent collision rates. For the large detunings Δ_1 we are considering at the moment $\gamma_{eg}(\Delta_1)$ can be evaluated from Eq. (8d) by stationary phase methods. In such a treatment the dominant contribution to the integral in Eq. (8d) comes from points of stationary phase R_S, where the laser detuning Δ_1 can match the corresponding interatomic difference potential, i.e., $\hbar\Delta_1 = V_{eg}(R_S)$. R_S indicates the interatomic distances, where the laser photon is absorbed instantaneously (see Fig. 3a). In particular, if two of these internuclear separations (R_1 and R_2) come close together, $\gamma_{eg}(\Delta_1)$

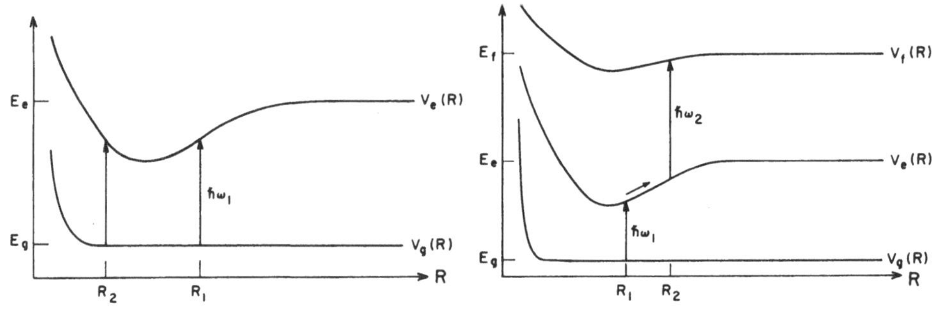

Fig. 3. Schematic representation of absorption during a collision for a one-photon process (a) and a molecular two-step process (b).

shows an oscillatory dependence on the detuning due to interferences between both stationary phase contributions. However, as long as they are well separated their contributions add up independently. In particular, the contribution of one such stationary phase point to the collision rate is asymptotically given by[8]

$$\gamma_{eg}(\Delta_1) = \frac{N}{V} \cdot \Delta_1^2 \cdot \frac{4}{3}\pi^2 \cdot \frac{R_S^2}{\frac{1}{\hbar}\left|(dV_{eg}/dR)(R)\right|_{R=R_S}} \quad . \tag{14a}$$

This explicitly shows the detailed information about the interatomic difference potential contained in $\gamma_{eg}(\Delta_1)$. In the absence of a stationary phase point (anti-static limit), $\gamma_{eg}(\Delta_1)$ goes to zero. It is also instructive to compare the quasi-static limit with the impact limit. A calculation by Cooper[11] shows that for an internuclear difference potential of the form $V_{eg}(R) = C_n/R^n$

$$\gamma_{eg}(\Delta_1) = C\,\gamma_{eg}(\Delta_1=0) \cdot (|\Delta_1|\tau_c)^{1-3/n} \tag{14b}$$

as long as $|\Delta_1|\tau_c > 1$. C is a constant of order unity (~1.7 for n = 4,6).

From the results of Ref. 8 we further see that

$$\gamma_{III\ I} \rightarrow \frac{1}{16} \cdot (\Omega_{fe}^2/\Delta_2^2) \cdot (\Omega_{eg}^2/\Delta_1^2) \cdot 2\gamma_{eg}(\Delta_1) \quad . \tag{15}$$

Inserting Eqs. (15) and (8d) into Eq. (11b) we obtain the total redistributed intensity. Whereas for low perturber densities the single collision contribution dominates, for high perturber densities the total redistributed intensity is completely determined by the subsequent collision contribution.

c) $|\Delta_1|\tau_c$, $|\Delta_2|\tau_c$, $|\Delta_1+\Delta_2|\tau_c \gg 1$

Now both laser photons are absorbed on time scales, i.e., $1/|\Delta_1|$, $1/|\Delta_2|$, $1/|\Delta_1+\Delta_2|$, much smaller than the duration of a collision τ_c. In this extreme non-Markovian limit, information about all three difference potentials is contained in the total redistributed intensity. $\gamma_{eg}(\Delta_1)$ and $\gamma_{fe}(\Delta_2)$, which determine the dressed states collision rates $\gamma_{II\ I}$ and $\gamma_{III\ II}$, are now both frequency dependent and are determined by the local properties of the corresponding interatomic difference potentials around the points of stationary phase.

A general asymptotic evaluation of the quantity $\gamma_{III\ I}$, given by Eq. (8d), is quite complicated due to the fact that different stationary phase points may be close together and end point contributions must also to be considered. However, from Eq. (8d) we see that it is determined by the squared sum of the amplitudes representing a direct instantaneous absorption of both laser photons (first term) and a two-step process (second term). The single collision contribution to the total redistributed intensity may therefore be written as $\gamma_{III\ I} = \gamma_{TS} + \gamma'_{III\ I}$, where[15]

$$\gamma_{TS} = \frac{1}{16} \cdot \Omega_{fe}^2 \cdot \Omega_{eg}^2 \cdot 4 \cdot \{\int_{-\infty}^{\infty} dt_o \int_0^{\infty} d\tau \, \langle\langle f,e|\pi \cdot \delta(\Delta_2 - iV_1^{(b,v)}(\tau+t_o))|f,e\rangle\rangle \cdot$$

$$\cdot \langle\langle e,g|\pi\delta(\Delta_1 - iV_1^{(b,v)}(t_o))|e,g\rangle\rangle\}_{av} \tag{16a}$$

describes a molecular (two-photon) two-step process (see Fig. 3b): The first laser photon $(\omega_1,\vec{e}_1)$ is absorbed instantaneously at the internuclear separation R_1, the collision complex then propagates (with the radiator in the excited state) to the internuclear distance R_2, where the second laser photon $(\omega_2,\vec{e}_2)$ is absorbed. $\gamma'_{III\ I}$ is determined by the direct instantaneous excitation process $|g\rangle \to |f\rangle$ during a collision associated with a stationary phase of $\Delta_1+\Delta_2$, interferences between this process and the molecular two-step excitation and end point contributions of the second term in the expression for $\gamma_{III\ I}$ in Eq. (8d). A simple estimate shows that[8]

$$\gamma_{TS}/\gamma'_{III\ I} = C' \cdot (|\Delta_2|\tau_c)^{(n-1)/2n} \tag{17}$$

for internuclear difference potentials of the form C_n/R^n and $R_1 \ll R_2$. C' is a constant of order unity. For sufficiently large detunings the molecular two-step process therefore gives the dominant single collision contribution to the total redistributed intensity. In particular for a situation as shown in Fig. 3b we find for $R_1 \ll R_2$

$$\gamma_{TS} = \frac{1}{16} \cdot \Omega_{fe}^2 \cdot \Omega_{eg}^2 \cdot \frac{N}{V} \cdot \int_0^{\infty} dv \, f(v) \, \frac{1}{v} \cdot \frac{(2\pi)^3 \cdot 2 \cdot R_1^2}{\frac{1}{\hbar}\left|\frac{dV_{fe}}{dR}(R)\right|_{R=R_2} \cdot \frac{1}{\hbar}\left|\frac{dV_{eg}}{dR}(R)\right|_{R=R_1}} , \tag{16b}$$

which can be directly obtained from Eq. (16a). If Δ_1 or Δ_2 is anti-statically detuned, γ_{TS} goes to zero.

For low perturber densities, i.e., $\gamma_e \gg \gamma_c$, the subsequent collision contribution is negligible compared with γ_{TS} and the total redistributed intensity is determined by γ_{TS} provided $\gamma'_{III\ I} \ll \gamma_{TS}$. According to Eqs. (14b) and (17) for high perturber densities, i.e., $1/\tau_c \gg \gamma_c \gg \gamma_e$, the ratio between the subsequent collision and the molecular two-step contribution varies as $(\gamma_c/\gamma_e) \cdot (|\Delta_2|\tau_c)^{-2/n}$; thus the subsequent collision contribution can give a dominant contribution to the total redistributed intensity even for $\gamma_{TS} \gg \gamma'_{III\ I}$.

IV. DEGENERATE THREE-LEVEL SYSTEM

We study now the total redistributed intensity for the J=0 → J=1 → J=0 transition shown in Fig. 1.

1. Weak Fields

Let us first of all consider weak fields in the sense of Eqs. (5a) and (5b). We further make the secular approximation and therefore assume the validity of Eq. (5c). The total redistributed intensity, which has been calculated for this case in Ref. 8, can be written in the form

$$I_{red} \propto 2\pi \cdot \frac{1}{3} \left|\langle f \| \mu \| e\rangle\right|^2 \cdot \sum_K D(\vec{e}_1,\vec{e}_2;K) \frac{1}{\gamma_f} \left\{\gamma^{(K)}_{III\ II} \frac{\gamma^{(K)}_{II\ I}}{\gamma_e + \gamma^K} + \gamma^{(K)}_{III\ I}\right\} \quad , \tag{18}$$

where $D(\vec{e}_1,\vec{e}_2;K)$ determines the dependence on the polarizations of the laser fields. Due to the degeneracy of the intermediate state the dressed states collision rates $\gamma^{(K)}_{III\ II}$, $\gamma^{(K)}_{II\ I}$ and $\gamma^{(K)}_{III\ I}$ now depend on a multipole index $K = 0,1,2$. The first term in Eq. (18) is due to subsequent collisions. $\gamma^{(K)}$ is the collisional destruction rate of the orientation (K=1) and alignment (K=2) of the excited state manifold. Neglecting inelastic collisions implies $\gamma^{(K=0)} = 0$. The second term, i.e., $\gamma^{(K)}_{III\ I}$, describes the single collision contribution to the total redistributed intensity.

We investigate now the physical significance of these dressed states collision rates by considering two limiting cases. This will also help us to understand the physical difference between degeneracy and nondegeneracy in two-photon redistribution experiments.

a) $|\Delta_1|\tau_c \gg 1$, $|\Delta_2|\tau_c \ll 1$

The second laser photon $(\omega_2,\vec{e}_2)$ is absorbed on a time scale that is long compared to the duration of a collision, i.e., $1/|\Delta_2| \gg \tau_c$. $\gamma^{(K)}_{III\ II}$ therefore contains information about the collision process in the form of S-matrix elements and we find[8]

$$\gamma^{(K)}_{III\ II} = \frac{1}{4} \cdot \frac{\Omega^2_{fe}}{\Delta^2_2} \cdot \left[2\gamma_{fe}(\Delta_2=0) - \gamma^{(K)}\right] \quad . \tag{19a}$$

(Rabi frequencies are now defined in terms of reduced dipole matrix elements, i.e., $\Omega_{fe} = (2/\hbar)\cdot(1/\sqrt{3})\cdot\langle f\|\mu\|e\rangle\cdot\varepsilon_2$, etc.) As the photon $(\omega_1,\vec{e}_1)$ is absorbed instantaneously during the collision, i.e., $1/|\Delta_1| \ll \tau_c$, $\gamma^{(K)}_{II\ I}$ as well as $\gamma^{(K)}_{III\ I}$ contain detailed information about the intra-collisional evolution. From the results of Ref. 8 we can obtain

$$\gamma^{(K)}_{II\ I} = \frac{1}{4} \cdot \frac{\Omega^2_{eg}}{\Delta^2_1} \left(\Gamma^{(K)}_{eg}(\Delta_1) - \gamma^{(K)}\right) \tag{19b}$$

and

$$\gamma^{(K)}_{III\ I} = \frac{1}{16} \frac{\Omega^2_{fe}}{\Delta^2_2} \cdot \frac{\Omega^2_{eg}}{\Delta^2_1} \cdot \left(\Gamma^{(K)}_{eg}(\Delta_1) - \gamma^{(K)}\right) \tag{19c}$$

with

$$\Gamma_{eg}^{(K)}(\Delta_1) = - \sum_{\mu_1 \cdots \mu_5, Q} (-)^{\mu_2+\mu_4} \begin{pmatrix} 1 & 1 & K \\ \mu_2 & -\mu_1 & Q \end{pmatrix} \begin{pmatrix} 1 & 1 & K \\ \mu_4 & -\mu_3 & Q \end{pmatrix} \cdot$$

$$\cdot\ 2\Delta_1 \cdot \mathrm{Im}\{\int_{-\infty}^{\infty} dt_o \langle\langle e\mu_1, e\mu_2 | U_I(\infty, t_o) | e\mu_5, e\mu_4 \rangle\rangle \cdot$$

$$\cdot \int_0^{\infty} d\tau \exp[i(\Delta_1 + i \frac{\gamma_e}{2})\tau] \langle\langle e\mu_5, g0 | U_I(t_o, -\tau+t_o) V_1^{(b,v)}(-\tau+t_o) | e\mu_3, g0 \rangle\rangle\}_{av} \cdot \tag{19d}$$

$U_I(t_2,t_1) = T \exp\,[\int_{t_1}^{t_2} dt\ V_1^{(b,v)}(t)]$ is the tetradic interaction picture time development operator.

Burnett and Cooper[5] have shown that $\Gamma_{eg}^{(K=0)}(\Delta_1) = 2 \cdot \gamma_{eg}(\Delta_1)$. This implies that the K = 0 component of the total redistributed intensity is identical to the result obtained for the nondegenerate case in Sec. III.2b. In general, $\Gamma_{eg}^{(K)}(\Delta_1)$ describes a process[12] in which the laser photon $(\omega_1, \vec{e}_1)$ is absorbed instantaneously at a certain internuclear distance R_1, thereby preparing the collision complex in a particular molecular state (in Fig. 4 a Π state). In the subsequent completion of the collision, characterized by the "half collision" quantity $U_I(\infty, t_o)$, the collision complex is reoriented (due to the mixing in of the molecular Σ state at large internuclear distances) and finally ends up in a certain linear combination of excited radiator states. If this reorientation during the completion of the collision were unimportant ($U_I(\infty, t_o) = 1$), $\Gamma_{eg}^{(K)}(\Delta_1)$ would reduce to the coherence decay rate $2\gamma_{eg}(\Delta_1)$, which is the case for the K = 0 component. The reorientation has been measured in one-photon redistribution experiments.[2] It can also be studied in the two-photon redistribution situation described in this section.[3] In particular, for linear polarization of both laser fields we find

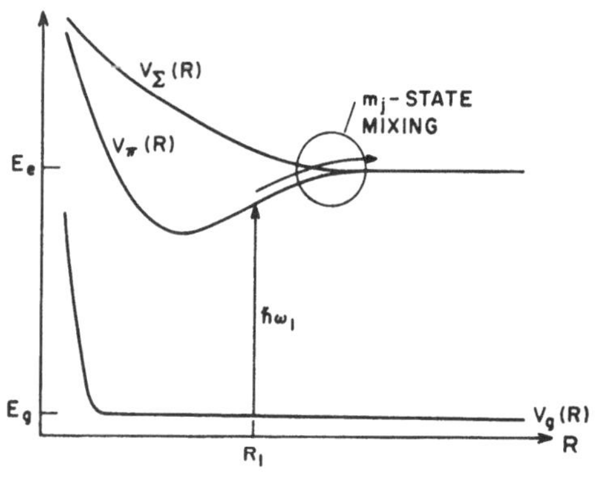

Fig. 4. Representation of the quasi-static contribution to $\Gamma_{eg}^{(K)}(\Delta_1)$.

$$\text{pol} = \frac{I_{red}(\vec{e}_1 \parallel \vec{e}_2) - I_{red}(\vec{e}_1 \perp \vec{e}_2)}{I_{red}(\vec{e}_1 \parallel \vec{e}_2) + I_{red}(\vec{e}_1 \perp \vec{e}_2)} = \frac{3\alpha^{(2)}(\Delta_1)}{2 \cdot \left(1 + \frac{\gamma^{(2)}}{\gamma_e}\right) + \alpha^{(2)}(\Delta_1)} \tag{20}$$

with $\Gamma_{eg}^{(K)}(\Delta_1) - \gamma^{(K)} = 2\gamma_{eg}(\Delta_1)\cdot\alpha^{(K)}(\Delta_1)$. In the absence of reorientation we have $\alpha^{(K)}(\Delta_1) = 1$ and pol $\to \{3/[3+2(\gamma^{(2)}/\gamma_e)]\}$ reflecting the fact that all depolarization effects are due to subsequent collisions described by $\gamma^{(2)}$. In the extreme opposite limit of complete reorientation, i.e., $\alpha^{(K)} = 0$ for $K = 1,2$ (note that always $\alpha^{(K=0)}(\Delta_1) = 1$), pol $\to 0$.

b) $|\Delta_1|\tau_c,\ |\Delta_2|\tau_c,\ |\Delta_1+\Delta_2|\tau_c \gg 1$

When the absorption of both photons is instantaneous compared to the duration of a collision, the situation is quite complicated. Qualitatively most of the discussion of Sec. III.2c for the nondegenerate case still applies. However, in addition to possible interference effects between the stationary phase points R_1 and R_2 in the molecular two-step process shown in Fig. 3b there might now also be interferences between the amplitudes describing, for example, a $|g\rangle \xrightarrow{\omega_1} |\Pi\rangle$ and a $|\Sigma\rangle \xrightarrow{\omega_2} |f\rangle$ transition. Such an interference between these two transition amplitudes can be important for small times τ of intermediate propagation [see Eq. (16a)] in the sense that the associated phase difference in the time development operator is small, i.e.,

$$|\Delta\phi| = \left| \int_{t_o}^{\tau+t_o} dt\ [V_{\Pi}^{(b,v)}(t) - V_{\Sigma}^{(b,v)}(t)] \right| < 1 \quad .$$

2. Intermediate Fields

As soon as the weak field conditions [Eqs. (5a) and (5b)] are violated, things become more difficult. The most complicated situation certainly arises for strong fields so that the Rabi frequencies exceed the inverse duration of a collision. In this case, the collision dynamics described by Eq. (3b) is modified by the laser fields and $G_1(\tau)$ can no longer be calculated perturbatively. For a realistic calculation the degeneracy of atomic states should also be taken into account as it may give rise to significant effects.[13] However, for intermediate fields such that only Eq. (5b) [but not Eq. (5a)] is violated, the collision dynamics may still be evaluated perturbatively but intensity effects become important in the coupling of the various density matrix elements in Eq. (3a). We shall discuss now these types of effects for the case where $|\Delta_1|\tau_c \gg 1$ but $|\Delta_2|\tau_c \ll 1$, which is of interest for some recent experiments.[3]

In particular, we consider a case where $|\Omega_{fe_1}|, |\Omega_{e_1g}| \ll |\Delta_1|$. Within this constraint and Eq. (5a) the second laser field $|\Omega_{fe_1}|$ may be arbitrarily strong compared to $|\Delta_2|$. If $|\Omega_{fe_1}| \gtrsim |\Delta_2| \gtrsim \gamma_e, \gamma_f, \gamma_c$ the spectrum of the radiation corresponding to the spontaneous transition $|f\rangle \to \{|e_1\rangle\}$ becomes quite complicated.

In addition to Raman-, "collisional-induced Rayleigh" -- and redistributed peak also "strong field" peaks appear due to the ac-Stark splitting associated with the manifolds $\{|e_i\rangle, |f\rangle\}$. We define the total redistributed intensity in this case by

$$I_{red} = \int_{(E_f-E_e)/\hbar-\eta}^{(E_f-E_e)/\hbar+\eta} d\omega\, I(\omega,\vec{\varepsilon}) \qquad (21)$$

with $\max\{|\Omega_{fe_i}|, |\Delta_2|, \gamma_e, \gamma_f, \gamma_c\} \ll \eta \ll |\Delta_1|$. Neglecting terms of order $|\Delta_2/\Delta_1|^2 \ll 1$ in comparison with unity, this quantity is proportional to $\sigma_{ff}(t\to\infty)$, because the contribution due to the Raman peak is negligible [see Eq. (9)].

For linear polarizations of both laser fields with $\vec{e}_1 \parallel \vec{e}_2$ the stationary final state population is obtained from the set of equations[14]

$$\begin{aligned}
&\left(\gamma_e + \tfrac{2}{3}\gamma^{(2)} + W\right)\sigma_{oo}(t\to\infty) = \tfrac{1}{3}\gamma^{(2)}\left(\sigma_{++}(t\to\infty) + \sigma_{--}(t\to\infty)\right) + \\
&\qquad + \left(W + \frac{\gamma_{f\to e}}{3}\right)\sigma_{ff}(t\to\infty) + R_o + R \quad , \\
&\left(\gamma_e + \tfrac{1}{3}\gamma^{(2)}\right)\left(\sigma_{++}(t\to\infty) + \sigma_{--}(t\to\infty)\right) = 2\cdot\tfrac{1}{3}\cdot\gamma^{(2)}\sigma_{oo}(t\to\infty) + \\
&\qquad + 2\cdot\frac{\gamma_{f\to e}}{3}\sigma_{ff}(t\to\infty) + 2\cdot R_1 \quad , \\
&(\gamma_f + W)\,\sigma_{ff}(t\to\infty) = W\,\sigma_{oo}(t\to\infty) - R
\end{aligned} \qquad (22)$$

with the rates

$$R_o = \frac{1}{4}\cdot\Omega_{eg}^2\cdot\frac{\gamma_e + \frac{2}{3}\cdot\Gamma_{eg}^{(2)}(\Delta_1) + \frac{2}{3}\gamma_{eg}(\Delta_1)}{\Delta_1^2} \quad , \quad R = W\cdot\frac{1}{4}\cdot\frac{\Omega_{eg}^2}{\Delta_1^2} \quad ,$$

$$W = \frac{1}{4}\cdot\Omega_{fe}^2\cdot\frac{\gamma_e + \gamma_f + 2\gamma_{fe}(\Delta_2=0)}{\Delta_2^2 + \left(\frac{\gamma_e+\gamma_f}{2} + \gamma_{fe}(\Delta_2=0)\right)^2} \quad , \quad R_1 = \frac{1}{4}\cdot\Omega_{eg}^2\cdot\frac{-\frac{1}{3}\Gamma_{eg}^{(2)}(\Delta_1) + \frac{2}{3}\gamma_{eg}(\Delta_1)}{\Delta_1^2} \quad ,$$

and $\sigma_{oo} \equiv \langle em_j=0|\sigma|em_j=0\rangle$, $\sigma_{++} \equiv \langle em_j=+1|\sigma|em_j=+1\rangle$, etc. R_o and R_1 are effective rates populating the excited states from the ground state. R is a stimulated effective rate from $|f\rangle$ to $|em_j=0\rangle$. It is due to a two-photon Raman process, which involves the ground state. W is the stimulated transition rate between the final and the excited state. This set and a similar set of equations, which determines the stationary final state population for linear polarizations and $\vec{e}_1 \perp \vec{e}_2$ are graphically represented in Fig. 5. Quadratic Stark shifts and ionization have thereby been neglected. It is interesting to note that the quantity

$$\mathrm{pol} = \frac{I_{red}(\vec{e}_1 \parallel \vec{e}_2) - I_{red}(\vec{e}_1 \perp \vec{e}_2)}{I_{red}(\vec{e}_1 \parallel \vec{e}_2) + I_{red}(\vec{e}_1 \perp \vec{e}_2)} = \frac{3\cdot\alpha^{(2)}(\Delta_1)}{2\cdot\left(1 + \frac{\gamma^{(2)}}{\gamma_e}\right) + \alpha^{(2)}(\Delta_1)} \qquad (23)$$

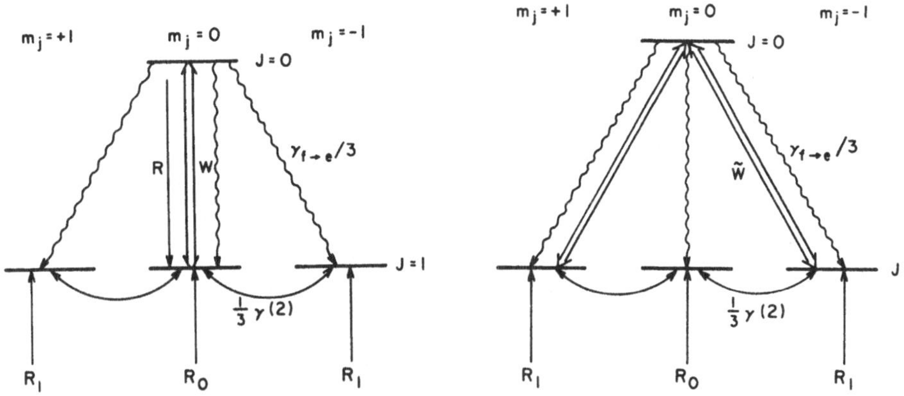

Fig. 5. Graphical representation of the set of rate equations for linear polarizations ((a): $\vec{e}_1 \parallel \vec{e}_2$, (b): $\vec{e}_1 \perp \vec{e}_2$) with $\tilde{W} = (1/2)\cdot W\cdot\{[\gamma_e+\gamma^{(2)}]/[\gamma_e+\gamma^{(2)}+(1/2)\cdot W]\}$.

is completely independent of the transition rate W and consequently of the intensity of the second laser. A similar result applies for circular polarizations, i.e.,

$$\text{pol} = \frac{I_{red}(\vec{e}_2=\vec{e}_1^{\,*}) - I_{red}(\vec{e}_2=\vec{e}_1)}{I_{red}(\vec{e}_2=\vec{e}_1^{\,*}) + I_{red}(\vec{e}_2=\vec{e}_1)} = \frac{3\cdot\alpha^{(1)}(\Delta_1)\cdot\left(\gamma_e/[\gamma_e+\gamma^{(1)}]\right)}{2+\alpha^{(2)}(\Delta_1)\cdot\left(\gamma_e/[\gamma_e+\gamma^{(2)}]\right)} \quad . \qquad (24)$$

This surprising result is consistent with recent experiments.[3] It is interesting that this result would not be obtained without the inclusion of the two-photon Raman rate R.

V. CONCLUSIONS

We have discussed two-photon collisional redistribution, focusing on the kind of collisional quantities that can be measured in such experiments. We found that in the extreme non-Markovian case for sufficiently large detunings the total (two-photon) redistributed intensity is mainly determined by a molecular two-step absorption process. We also considered an intermediate case, where the first laser is strongly detuned and the second almost on resonance, and found that the total redistributed intensity is determined by the same type of collisional quantities, which have previously been measured by one-photon scattering experiments. In particular, we emphasized the importance of the degeneracy of the radiator states and pointed out that for intermediate intensities of the second laser field the polarization dependent quantity pol is intensity independent, as indicated by recent experiments.[3]

I am indebted to John Cooper for numerous valuable discussions and for his continuous support. This work was supported by National Science Foundation grant PHY82-00805 to the University of Colorado.

LITERATURE

1. J. L. Carlsten, A. Szöke and M. G. Raymer, Phys. Rev. A 15, 1029 (1977).
2. W. J. Alford, N. Andersen, K. Burnett and J. Cooper, Phys. Rev. A 30, 2366 (1984).
3. W. J. Alford, N. Andersen, M. Belsley, J. Cooper, D. M. Warrington and K. Burnett, Phys. Rev. A (in press).
4. K. Burnett, J. Cooper, R. J. Ballagh and E. W. Smith, Phys. Rev. A 22, 2005 (1980).
5. K. Burnett and J. Cooper, Phys. Rev. A 22, 2027 (1980).
6. G. Nienhuis, J. Phys. B 16, 1 (1983).
7. E. W. Smith, J. Cooper and L. J. Roszman, J.Q.S.R.T. 13, 1523 (1973).
8. G. Alber and J. Cooper, Phys. Rev. A (in press).
9. K. Burnett, J. Cooper and P. D. Kleiber, Phys. Rev. A 25, 1345 (1982).
10. E. W. Smith, C. R. Vidal and J. Cooper, J. Res. Natl. Bur. Std. 73A, 389 (1969).
11. J. Cooper, Astrophys. J. 228, 339 (1979) [in particular Eq. (A6)].
12. J. Cooper, in Spectral Line Shapes, Vol. 2, edited by K. Burnett (de Gruyter, Berlin, 1983), p. 737.
13. J. Light and A. Szöke, Phys. Rev. A 18, 1363 (1978).
14. G. Alber and J. Cooper (in preparation).
15. This can be obtained directly from Eq. (8d) by stationary phase evaluation for Δ_1 and Δ_2. See also the discussion in Ref. 8.
16. G. G. Lombardi, D. E. Kelleher and J. Cooper, Astrophys. J. (in press).
17. E. W. Smith, J. Cooper and C. R. Vidal, Phys. Rev. 185, 140 (1969) [in particular their Eq. (56)].

PART II: Multiphoton Ionization

STUDIES OF MULTIQUANTUM PROCESSES IN ATOMS

Charles K. Rhodes
Department of Physics, University of Illinois at Chicago
P.O. Box 4348, Chicago, Illinois 60680

ABSTRACT

From an analysis of multiquantum processes in atoms, which includes information derived from ion charge state distributions, photoelectron spectra, and photon spectra, and involving the atomic number (Z), intensity, frequency, and polarization dependencies, an approximate description of the electronic motions involved in these processes has emerged. The data strongly indicate that an organized motion of an entire shell, or a major fraction thereof, is directly involved in the nonlinear coupling. With this picture, the outer atomic subshells are envisaged as being driven in coherent oscillation by the intense ultraviolet wave. An immediate consequence of this motion is an increase in multiphoton coupling resulting directly from the larger magnitude of the effective charge involved in the interaction. In this way, a multielectron atom undergoing a nonlinear interaction responds in a fundamentally different fashion from that of a single electron counterpart. The strong highly nonlinear coupling which develops between the radiation field and the atom can result in the transfer of energy by a direct intra-atomic process to inner-shell excitations. These coherent atomic motions, which can be related to fast atom-atom encounters ($\gtrsim$ 10 MeV/amu), should enable the selective excitation of atomic inner-shell states in the KeV range to be produced by intense irradiation of atoms at ultraviolet wavelengths.

I. INTRODUCTION

Studies of atomic and molecular ionization, conducted under collision-free conditions at ultraviolet wavelengths, have manifested unexpected characteristics. The initial studies[1,2] of the Z-dependence of collision-free multiphoton ionization of atoms at 193 nm clearly exhibited anomalous behavior in terms of the gross rate of energy transfer. The general class of physical processes studied in those experiments was

$$N\gamma + X \rightarrow X^{q+} + qe^- \ . \qquad (1)$$

A prominent feature of those studies was the unusually strong nonlinear coupling found characteristic of certain heavy materials.[1] These experiments clearly demonstrated that standard theoretical techniques were incapable, by a discrepancy as great as several orders of magnitude, of describing those results. Subsequent work,[3] as well as other studies[4,5] conducted at wavelengths of 1.06 μm and 0.53 μm, has confirmed the anomalous nature of the coupling strength.

Data have been obtained for ion charge state spectra and electron energy distributions.[6] A salient feature of the Z-dependence of the ion charge state data is the clear influence of atomic shell structure[1] on the observed ion spectra. For Ar, Kr, and Xe, the maximum charge states observed would correspond to the complete removal of certain atomic sub-shells. Indeed, for these materials they are the 3p, the 4p, and both the 5s and 5p shells, respectively.

From an analysis of the data obtained to date, which includes information on the atomic number (Z), intensity, frequency, and polarization dependencies, the following approximate description of the electronic motions involved in these processes has emerged.[7] The data strongly indicate that an organized motion of an entire shell, or a major fraction thereof, is directly involved in the nonlinear coupling. With this picture, the outer atomic subshells are envisaged as being driven in coherent oscillation by the intense ultraviolet wave. In this way, a multielectron atom undergoing a nonlinear interaction responds in a fundamentally different fashion from that of a single electron atom.

Although all standard theoretical approaches[8-13] fail to provide a description of the observed phenomena, a relatively simple model,[14] valid at sufficiently high intensity, can be contemplated. In this case, we imagine an atom composed of two parts, (a) an outer shell of electrons driven in coherent oscillation by the radiative field and (b) a remaining atomic core for which direct coupling to the radiation field is neglected. Coupling between these two systems can occur, since the outer electrons could, through inelastic "collisions," transfer energy to the core. Simple estimates[14] indicate that, for intensities corresponding to an electric field $E \gg e/a_o^2$, that oscillating atomic current densities j on the order of $10^{14} - 10^{15}$ amps/cm^2 could be established.

II. MECHANISM OF COUPLING

A. Simple Classical Estimate

In the limit of high intensity, it is possible to formulate an estimate of the coupling of the coherently driven outer electrons with the remaining atomic core by relatively simple procedures. This is now done at two levels of approximation, initially with the neglect of the influence of the coherence characterizing the motion of the outer electrons and, subsequently, with its inclusion.

We now furnish an estimate based simply on the magnitude of the ambient current density j. Since the electron kinetic energies are considerably above their corresponding binding energies, it is possible to use a first order Born approximation[15] in a manner similar to that used to the study of electron collisions for K- and L-shell ionization[16] and shell specific ionization processes in highly charged ions.[17,18] Indeed, in the case of xenon ions, cross sections for electron impact ionization are available.[19]

In this elementary classical picture,[14] the transition rate R can be written as

$$R \simeq \frac{j}{e}\,\sigma_e \qquad (2)$$

in which e is the electronic charge and σ_c is the cross section characterizing the excitation of the atomic core by inelastic electron collisions arising from the current density j. If $j \sim 10^{14}$ amps/cm^2 and $\sigma_e \sim 10^{-19}$ cm^2, then $R \sim 6 \times 10^{13}$ sec^{-1}. Furthermore, if the radiatively driven current density j is damped by electron emission in a time τ on the order of $\sim 10^{-15}$ sec, an approximate time scale characterizing autoionization, the overall transition probability $P \sim R\tau \sim 6 \times 10^{-2}$ indicating a significant probability of energy transfer.

B. Role of Coherence

The coherence associated with the motions of the outer-shell electrons induced by intense irradiation at ultraviolet frequencies has important consequences[20] for the coupling of energy to atomic inner-shells which are ignored in the estimate given above. As described below, the influence of this type of coherent atomic motion can be related to certain properties[21] of energetic ($\gtrsim$ 10 MeV/amu) atom-atom collisions.

The role of coherence in the motion of the outer electrons in the excitation of the core is readily described by appeal to descriptions of energetic atom-atom (A/B) collisions. In this comparison,[20] a correspondence is established between the scattering of the coherently driven outer electrons (a) from the atomic core (b) and the respective interaction of the electrons in the projectile atom A with the target atom B. Consider the process

$$A + B(o) \xrightarrow{\sigma_{no}} A + B^*(n) \tag{3}$$

in which A is a ground state neutral atom with atomic number Z_A and $B^*(n)$ represents an electronically excited configuration of the target system with quantum numbers collectively represented by (n). In the plane-wave Born approximation (PWBA), the cross section σ_{no} can be written in the form presented by Briggs and Taulbjerg[21] as

$$\sigma_{no} = \frac{8\pi e^4}{v^2} \int_{K_{min}}^{K_{max}} |\varepsilon_{no}^{B}(\vec{K})|^2 [|Z_A - \sum_j \omega_j \langle\phi_j^A|\exp(i\vec{K}\cdot\vec{s}_A)\phi_j^A\rangle|]^2 \frac{dK}{K^3} \tag{4}$$

in which

$$\varepsilon_{no}^{B}(\vec{K}) = \int dr_B^3 \psi_{nB}^*(\vec{r}_B)\exp(i\vec{K}\cdot\vec{r}_B)\psi_{oB}(\vec{r}_B) \quad . \tag{5}$$

In expressions (4) and (5), v is the relative atom-atom velocity, the ϕ_j^A are orthnormal spin orbitals representing the electrons on the projectile atom, ω_j is the statistical weight of the shell, $\vec{K}$ is the momentum transfer in the collision, and ψ_{oB} and ψ_{nB} represent the electron wavefunctions of the target system. The summation over the index j appearing in equation (4) extends over all occupied orbitals so that in the limit $\vec{K} \to 0$, the summation tends to the number of electrons N_A associated with the projectile atom.[21] In the low momentum transfer limit, in which complete screening occurs, the amplitudes of the electrons combine coherently and the contribution to the cross section σ_{on} arising from the motion of the electrons in atom A is increased by a factor of N_A^2 over that of a single electron at the same collision velocity v. Alternatively, for sufficiently low momentum transfer such that $Ka_o \ll \hbar$, the electron cloud acts as a coherent scattering center with a mass $N_A m_e$, a charge $N_A e = Z_A e$, a velocity v, and a kinetic energy $N_A(1/2\ m_e v^2)$. Significantly, on account of the coherence, the single particle energies $(1/2\ m_e v^2)$ add so that, in principle, this value could be below the magnitude required to produce the excitation of the target atom B.

In sufficiently high field strengths, coherently accelerated electrons in outer atomic shells (a) can interact with the remaining atomic core system (b) in a manner closely analogous to the atom-atom scattering described above. If a PWBA description is used, the cross section representing energy transfer can be derived directly from expression (4) with $Z_A = 0$. The basic physical concepts are simply represented in the high field limit ($E \gg e/a_o^2$), a regime in which the driven electronic velocities correspond approximately to those characteristic of atom-atom collisions at a collision energy of $\sim$ 10 MeV/amu. Therefore, the motion of these electrons simulates the electronic collisional environment that would occur in fast atom-atom encounters, but with the important absence of the nuclear contribution arising from the Z_A term in expression (4).

It is now possible to estimate the contribution to σ_{on} for inner-shell excitation arising from coherently excited atomic shells. For this we take expression (4) with $Z_A = 0$ and restrict K_{max} to $\lesssim \hbar/a_o$, to fulfill the condition for full shielding. We further take Z_1 to denote the number of electrons in the outer shells and expand equation (5) for $\varepsilon^B_{no}(\vec{K})$ so that only the leading dipole term x_{on} is retained. Finally, for a core excitation energy ΔE we put $K_{min} \simeq \Delta E/v$, the condition that holds for ΔE much less than the collision energy. With these modifications, the coherent piece σ^c_{on} can be written as

$$\sigma^c_{on} \simeq \frac{8\pi e^4 x^2_{on} Z^2_1}{v^2 \hbar^2} \int_{\Delta E/v}^{\hbar/a_o} \frac{dK}{K} , \tag{6}$$

a result which, with the exception of the restriction on K_{max} and the Z_1^2 factor, is exactly the form of the well known result for inelastic scattering of electrons on atoms developed by Bethe.[22] The final result, valid for

$$\alpha\left(\frac{v}{c}\right)\left(\frac{m_e c^2}{\Delta E}\right) > 1 , \tag{7}$$

is

$$\sigma^c_{on} \simeq 8\pi\alpha^2 \left(\frac{c}{v}\right)^2 Z_1^2 x^2_{on} \ln\left[\alpha\left(\frac{v}{c}\right)\left(\frac{m_e c^2}{\Delta E}\right)\right] \tag{8}$$

in which α is the fine structure constant.

Obviously, all types of possible excited configurations cannot fully benefit from this type of coherent motion regardless of the field strengths used. Indeed, the limitation can be estimated from equation (7). At sufficiently high intensity in the limit $v \to c$, the maximum value of ΔE_{max} is given by

$$\Delta E_{max} \sim \alpha m_e c^2 = 3.73 \text{ KeV} . \tag{9}$$

C. Frequency of Irradiation

The physical picture presented above also enables a statement concerning the frequency of irradiation ω to be formulated. For the excitation of inner-shell states in the kilovolt range by the quasi-free coherently driven motion of outer-shell electrons, two basic assumptions are involved. The first, as noted above, concerns the field strength E such that the condition

$$E \gg E_o = e/a_o^2 \tag{10}$$

holds, enabling the electrons to be regarded as approximately free. The second consideration involves the energy scale of the motion ε_e, in this case, taken to be sufficiently great to readily excite inner-shell states in the desired kilovolt range. With the neglect of relativistic corrections, we can express the electron energy as

$$\varepsilon_e = 1/2\ m_e v_e^2 \tag{11}$$

with the quantity v_e representing the velocity of induced electronic motion. For a free electron, the maximum value of v_e, commonly known as the quiver velocity, is given by[23]

$$v_e = \frac{eE}{m_e \omega} \tag{12}$$

for a field with angular frequency ω.

For stated values of E and ε_e which fulfill the assumptions of the model, a frequency scale generally characteristic of those physical conditions is now defined by combination of expressions (10), (11), and (12). If we take $E \cong 3\ E_o$ to satisfy equation (10) and $\varepsilon_e \sim 10^3$ eV as reasonable values, then

$$\omega \simeq (30)\alpha^3 \frac{c}{\lambda\!\!\!^{-}_c} , \tag{13}$$

a frequency which corresponds to an ultraviolet wavelength of $\sim$ 200 nm. We conclude that ultraviolet wavelengths naturally match the physical conditions characteristic of the coherent atomic motions envisaged in this description.

III. EXPERIMENTAL RESULTS

At bottom, there are three fundamental categories of measurement which can be used to unravel the nature of the physical processes involved. They are (1) ion charge state spectra, (2) photoelectron measurements, and (3) measurements of radiation.

A. Ion Spectra

The determination of ion charge state spectra under collision-free conditions is a simple unambiguous experimental method that provides direct information on the gross scale of the energy transfer rate. We recall that it was from measurements[1,2] of this nature that the anomalously strong coupling was seen and the original clues concerning the importance of the shell properties of the atoms in the nonlinear coupling were first obtained. The evidence currently available strongly suggests, at least in rough approximation, that the greater the number of electrons in the outer shell, as designated solely by the principal quantum number, the greater the strength of the nonlinear coupling. Since a rather thorough discussion of the ion results is available in the literature, we direct the interested reader to those sources.[1,2,7,14]

B. Photoelectron Spectra

Photoelectron energy distributions provide extremely valuable information on the detailed nature of the electronic motions occurring in reactions such as that shown in equation (1). For example, we currently see very substantial differences in the electron distributions produced by Ar and Kr even though the ion spectra for these materials show that the outer p-shell is completely stripped in both cases.[14]

The most significant results, however, appear in connection with the behavior of xenon. Indeed, in contrast to argon and krypton, the xenon electron energy spectrum exhibits a dramatic change with increasing 193 nm intensity in the 10^{14} - 10^{15} W/cm^2 range. The first ionization line, which corresponds to two-photon absorption with a corresponding electron energy of 0.7 eV, nearly disappears, while the three-photon process, arising from continuum-continuum transitions,[24,25] becomes dominant, generating a final state distribution of the ions having approximately 80 percent in the excited $5s^2 5p^5\ {}^2P_{1/2}$ state with the remaining 20 percent in the $5s^2 5p^5\ {}^2P_{3/2}$ ground level.

In addition to the ladder of continuum-continuum lines, new sharp electron features appear in the range from 8 eV to 20 eV at an intensity of $\sim 10^{15}$ W/cm^2. These lines have been tentatively assigned to

$N_{45}OO$-Auger lines following excitation of the 4d inner-shell. The most prominent lines are those associated with $N_{45}O_1O_1$ transitions which terminate on the $4d^{10}5s^05p^6$ double hole state. Although the lines are shifted to a higher energy ($\sim$ 1 eV) by an amount comparable to that observed for the continuum-continuum transitions, their relative spacing, number and, to a lesser extent, relative intensities fit well to values previously reported in the literature[26,27] for such Auger transitions. Moreover, a total of six electron lines is observed representing a triplet of pairs of transitions, significantly, all of which exhibit the known[26,27] $4d_{3/2} - 4d_{5/2}$ splitting in xenon of $\sim$ 2 eV.

The general trend[28] of the electron spectrum for xenon as a function of intensity is illustrated in Fig. (1). Note the appearance of a group of lines at an intensity of $\sim 10^{15}$ W/cm^2 which are attributed to Auger decay of 4d-vacancies in the atom. This spectral region,[6] which is believed to represent $N_{45}OO$ Auger processes, is shown in higher resolution in Fig. (2).

Although these preliminary results do not constitute a proof of the mechanism involved, we simply observe that energy transfer from coherently driven valence shell electrons could produce such inner-shell excitation. Furthermore, with the model presented in Section II along with consideration of the known[19] inelastic electron scattering cross sections for xenon ions, an estimate can be made of the intensity at which such Auger lines should appear. Inelastic scattering studies[19] show that the 4d excitation in xenon has a threshold at $\sim$ 67.6 eV, closely followed by a broad maximum at $\sim$ 100 eV. If we approximate the motion of the N_A outer electrons in xenon as that of a free electron, the maximum electron quiver energy ε_e can be written, in a form which reexpresses equation (11), as

$$\varepsilon_e = (1.79 \times 10^{-13})\lambda^2 I \tag{14}$$

with units expressed as ε_e(eV), λ(μm), and I(W/cm^2). The 4d-threshold at $\sim$ 67.6 eV corresponds to an intensity for single electron motion of $I \sim 10^{16}$ W/cm^2, a value somewhat above that used in the experimental studies. However, if the picture of the coherent motion is valid, the single particle energy can be reduced, for a fixed threshold requirement, by a factor of N_A, the number of electrons participating in the coherent outer-shell motion. For xenon, previous ion studies[1,7,14] indicated that $N_A = 8$ is a reasonable value, the total number of electrons in the n = 5 shell ($5s^25p^6$). This reduces the threshold intensity for 4d-vacancy production to $\sim 1.2 \times 10^{15}$ W/cm^2, a value quite

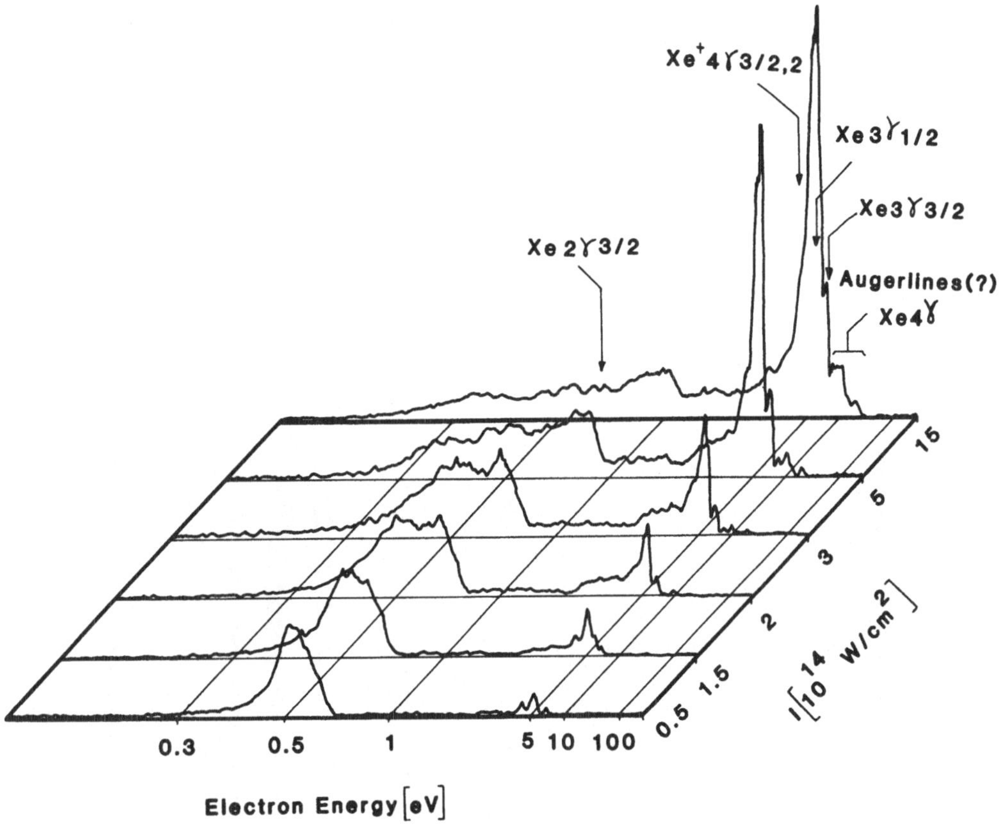

Fig. (1): Overall xenon time-of-flight photoelectron spectrum from ∿ 0.1 to ∿ 100 eV. The uncertainty in the intensity scale is approximately a factor of two. Irradiation was at 193 nm with a pulse duration of ∿ 5 ps with a lens with a focal distance of 20.5 cm.

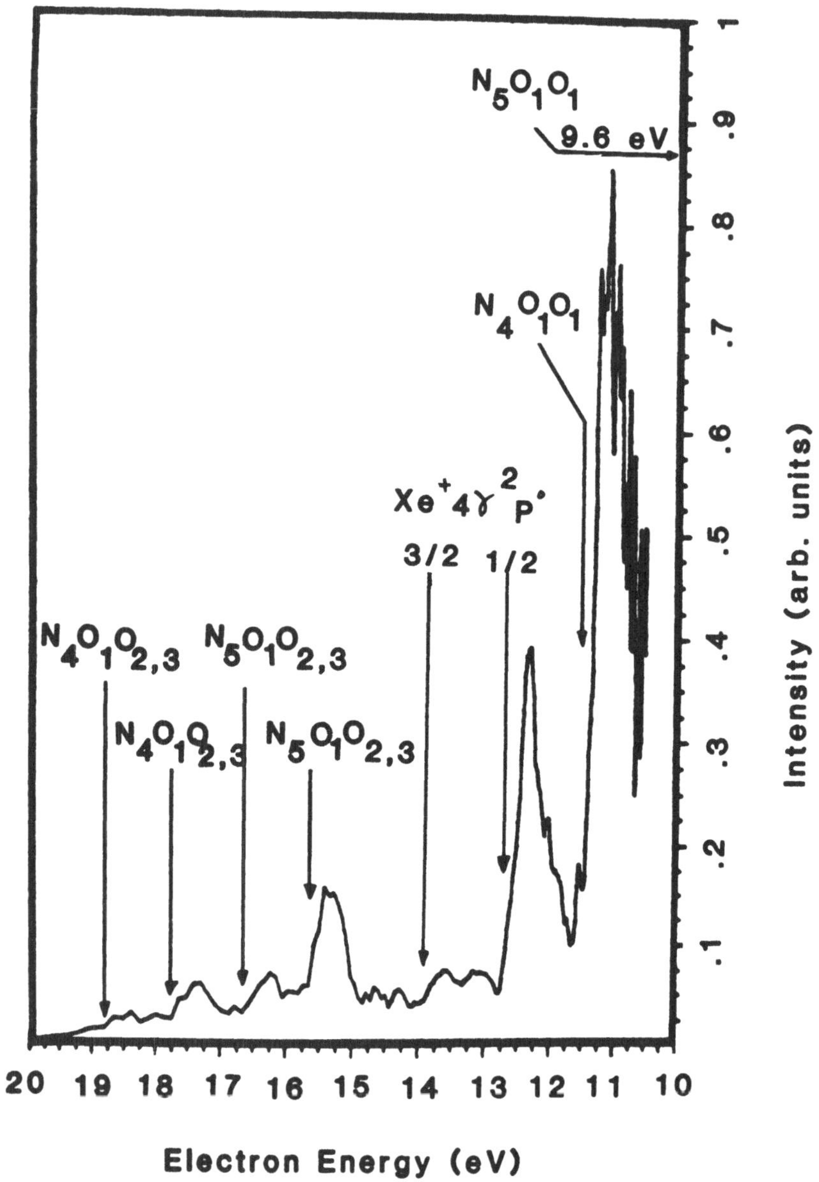

Fig. (2): Prominent transitions observed in the electron spectrum of xenon irradiated with 193 nm at $\sim 10^{15}$ W/cm^2 are shown. Both Auger ($N_{45}OO$) and continuum-continuum features ($4\gamma \rightarrow Xe^+_{3/2}$, $Xe^+_{1/2}$) are apparent. The splitting between the three N_4-N_5 pairs has the common value of $\sim$ 2 eV, the known $4d_{5/2}$ - $4d_{3/2}$ separation in xenon. The arrows indicate the high energy edges of the observed features which represent the true energies of the lines.

close to that ($\sim 10^{15}$ W/cm^2) corresponding to the appearance of the electron lines presumed to arise from Auger decay. On the basis of the very approximate nature of this analysis, the closeness of this agreement should not be accorded great significance. It is sufficient that a model formulated on the basis of a general statement concerning the nature of the fundamental electronic motions, and otherwise essentially free of arbitrary adjustable parameters, produces results in qualitative rapport with experiment.

C. Stimulated Emission Spectra

We note that further evidence supporting this general physical picture is present in the characteristics of certain stimulated spectra that have been observed in krypton.[29-32] In this case, the states believed to be involved are those having multiple excitations and inner-shell excitations[33-37] in closely coupled subshells, such as $4s4p^6n\ell$ and $4s^24p^4n\ell n'\ell'$, a class of levels which is exactly of the type expected to be strongly excited on the basis of the physical model presented in Section II.

IV. CONCLUSIONS

Basic physical studies of collision-free nonlinear atomic processes, through an analysis involving combined measurements of ion charge state distributions, photoelectron energy spectra, and photon spectra arising from intense ultraviolet irradiation, have produced data which strongly indicate that multielectron atoms respond in a manner fundamentally distinguished from single electron counterparts. The confluence of the evidence suggests that, under appropriate circumstances, the outer atomic subshells can be driven in coherent oscillation, and this ordered electronic motion can, by direct intra-atomic coupling, lead to the rapid excitation of atomic inner-shell states.

An atom, in a radiative field whose amplitude is comparable to or greater than an atomic unit, experiences a violent perturbation that has important features in common with certain well studied collisional phenomena such as ion-atom collisions, electron-ion collisions, and beam-foil interactions.[38,39] Indeed, in the case of beam-foil collisions, a radiative environment at an intensity of 3×10^{18} W/cm^2 at an ultraviolet wavelength approximates, in several important respects, the conditions associated with the passage of an argon ion through a carbon foil with a kinetic energy of ~ 1 GeV. This rough similarity leads to the consideration of the concept of an "optical solid" in which stationary atoms in a sufficiently intense radiative field will experience an interaction comparable to that of energetic ions traversing solid matter. A natural expectation is an extreme level of excitation comparable to that required to establish the conditions needed to produce stimulated emission in the kilovolt range. In addition, the coherence of the radiative environment can act to introduce a measure of control on the energy transfer that will enable considerable selectivity in the energy flow to be achieved.

V. ACKNOWLEDGMENTS

The author wishes to acknowledge fruitful discussion with T. S. Luk, H. Egger, U. Johann, A. P. Schwarzenbach, K. Boyer, and A. Szöke. This work was supported by the Office of Naval Research, the Air Force Office of Scientific Research under contract number F49630-83-K-0014, the Department of Energy under grant number DeAS08-81DP40142, the Lawrence Livermore National Laboratory under contract number 5765705, the National Science Foundation under grant number PHY 81-16626, the Defense Advanced Research Projects Agency, and the Avionics Laboratory, Air Force Wright Aeronautical Laboratories, Wright Patterson Air Force Base, Ohio.

VI. REFERENCES

1. T. S. Luk, H. Pummer, K. Boyer, M. Shahidi, H. Egger, and C. K. Rhodes, Phys. Rev. Lett. 51, 110 (1983).

2. T. S. Luk, H. Pummer, K. Boyer, M. Shahidi, H. Egger, and C. K. Rhodes, Excimer Lasers - 1983, AIP Conference Proceedings No. 100, edited by C. K. Rhodes, H. Egger, and H. Pummer (American Institute of Physics, New York, 1983) p. 341.

3. A. L'Huillier, L-A. Lompré, G. Mainfray, and C. Manus, in Laser Techniques in the Extreme Ultraviolet, AIP Conference Proceedings No. 119, edited by S. E. Harris and T. B. Lucatorto (AIP, New York, 1984) p. 79.

4. A. L'Huillier, L-A. Lompré, G. Mainfray, and C. Manus, Phys. Rev. Lett. 48, 1814 (1982).

5. A. L'Huillier, L-A. Lompré, G. Mainfray, and C. Manus, Phys. Rev. A27, 2503 (1983).

6. U. Johann, T. S. Luk, H. Egger, H. Pummer, and C. K. Rhodes, "Evidence for Atomic Inner-Shell Excitation in Xenon from Electron Spectra Produced by Collision-Free Multiphoton Processes at 193 nm," Conference on Lasers and Electro-Optics '85, Baltimore, Maryland, to be published.

7. T. S. Luk, U. Johann, H. Egger, H. Pummer, and C. K. Rhodes, "Collision-Free Multiple Ionization of Atoms and Molecules at 193 nm," to be published.

8. P. Lambropoulos in Advances in Atomic and Molecular Physics, Vol. 12, edited by D. R. Bates and B. Bederson (Academic Press, New York, 1976) p. 87.

9. H. B. Bebb and A. Gold, Phys. Rev. 143, 1 (1966).

10. Y. Gontier and M. Trahin, J. Phys. B13, 4384 (1980); Y. Gontier and M. Trahin, Phys. Rev. 172, 83 (1968).

11. H. R. Reiss, Phys. Rev. A1, 803 (1970); ibid., Phys. Rev. Lett. 25, 1149 (1970); ibid., Phys. Rev. D4, 3533 (1971); ibid., Phys. Rev. A6, 817 (1972).

12. H. S. Brandi and L. Davidovich, J. Phys. B12, L615 (1979).

13. L. V. Keldysh, Zh. Eksp. Teor. Fiz. 47, 1945 (1964)[Sov. Phys. - JETP 20, 1307 (1965)].

14. C. K. Rhodes, "Studies of Collision-Free Nonlinear Processes in the Ultraviolet Range," in Proceedings of the International Conference on Multiphoton Processes III, Crete, Greece, September 5-12, 1984, edited by P. Lambropoulos and S. J. Smith (Springer-Verlag, Berlin, in press).

15. N. F. Mott and H. S. W. Massey, The Theory of Atomic Collisions (Oxford University Press, London, 1965); H. S. W. Massey and E. H. S. Burhop, Electronic and Ionic Impact Phenomena, Vol. 1 (Oxford University Press, London, 1969); N. M. Kroll and K. M. Watson, Phys. Rev. A8, 804 (1973).

16. James H. Scofield, Phys. Rev. A18, 963 (1978).

17. L. B. Golden, P. H. Sampson, and K. Omidvar, J. Phys. B11, 3235 (1978).

18. L. B. Golden, P. H. Sampson, and K. Omidvar, J. Phys. B13, 2645 (1980).

19. D. C. Griffen, C. Bottcher, M. S. Pindzola, S. M. Younger, D. C. Gregory, and P. H. Crandall, Phys. Rev. A29, 1729 (1984); D. C. Gregory and P. H. Crandall, Phys. Rev. A27, 2338 (1983); C. Achenbach, A. Müller, E. Salzborn, and R. Becker, J. Phys. B17, 1405 (1984).

20. K. Boyer and C. K. Rhodes, "Atomic Inner-Shell Excitation Induced by Coherent Motion of Outer-Shell Electrons," to be published.

21. J. S. Briggs and K. Taulbjerg in Structure and Collisions of Ions and Atoms, edited by I. A. Sellin (Springer-Verlag, Berlin, 1978) p. 105.

22. H. A. Bethe, Ann. Phys. 5, 325 (1930).

23. T. P. Hughes, Plasmas and Laser Light (John Wiley and Sons, New York, 1975).

24. P. Kruit and F. H. Read, J. Phys. E16, 373 (1983).

25. H. G. Muller and A. Tip, Phys. Rev. A30, 3039 (1984).

26. S. Southworth, U. Becker, C. M. Truesdale, P. H. Kobrin, D. W. Lindle, S. Owaki, and D. A. Shirley, Phys. Rev. A28, 261 (1983).

27. L. O. Werme, T. Bergmark, and K. Siegbahn, Phys. Scr. 6, 141 (1972).

28. U. Johann, private communication.

29. T. Srinivasan, H. Egger, T. S. Luk, H. Pummer, and C. K. Rhodes in Laser Spectroscopy VI, edited by H. P. Weber and W. Lüthy (Springer-Verlag, Berlin, 1983) p. 385.

30. H. Egger, T. S. Luk, W. Müller, H. Pummer, and C. K. Rhodes in Laser Techniques in the Extreme Ultraviolet, AIP Conference Proceeding No. 119, edited by S. E. Harris and T. B. Lucatorto (AIP, New York, 1984) p. 64; H. Pummer, H. Egger, T. S. Luk, and C. K. Rhodes, SPIE 461, 53 (1984).

31. K. Boyer, H. Egger, T. S. Luk, H. Pummer, and C. K. Rhodes, J. Opt. Soc. Am. B1, 3 (1984).

32. W. Müller, T. Srinivasan, H. Egger, T. S. Luk, H. Pummer, and C. K. Rhodes, "Properties of Stimulated Emission Below 100 nm in Krypton," to be published.

33. K. Codling and R. P. Madden, J. Res. Natl. Bur. Std. 76A, 1 (1972).

34. K. Codling and R. P. Madden, Phys. Rev. A4, 2261 (1971).

35. D. L. Ederer, Phys. Rev. A4, 2263 (1971).

36. M. Boulay and P. Marchand, Can. J. Phys. 60, 855 (1982).

37. J. A. Baxter, P. Mitchell, and J. Comer, J. Phys. B15, 1105 (1982).

38. Beam-Foil Spectroscopy, Vol. 1 and Vol. 2, edited by I. A. Sellin and D. J. Pegg (Plenum Press, New York, 1976).

39. U. Littmark and J. F. Ziegler, Handbook of Range Distributions for Energetic Ions in All Elements, Vol. 6 (Pergamon Press, New York, 1980).

INTERACTION OF AN INTENSE LASER PULSE WITH A MANY-ELECTRON ATOM : FUNDAMENTAL PROCESSES

L.A. Lompré and G. Mainfray
Service de Physique des Atomes et des Surfaces
Centre d'Etudes Nucléaires de Saclay
F-91191 Gif-sur-Yvette Cédex, FRANCE

1 Introduction

Multiphoton ionization of atoms is a typical example of one of the new field of investigation in atomic physics that the lasers have opened up. The different aspects of the multiphoton ionization of one-electron atoms have been well understood these last few years. They can be correctly described by rigourous theoritical models in the framework of perturbation theory when only one electron is assumed to be involved in the ionization /1-5/. Alkaline atoms which have only one valence electron, and of course atomic hydrogen, are the best examples that satisfy this condition. For example the different aspects of multiphoton ionization of cesium atoms now form a well developed field.

The present paper will be devoted to multiphoton ionization of many electron atoms in the outer shell, as rare gases. It induces removal of several electrons and the production of multiply charged ions. The production of doubly charged ions has been investigated in detail. Doubly charged ions can be produced, either by simultaneous excitation of at least two electrons, or by stepwise process via singly charged ions. This depends mainly on the laser intensity and the photon energy. The basic interaction processes involved are considerably more complicated than for one-electron atoms. A new theoritical model will have to be developed to take into account electron correlation effects.

2 Experimental Results

Recent experiments have emphasized the production of multiply charged ions by multiphoton absorption in rare gas atoms /6-9/. A mode-locked Nd-YAG laser is used to produce a 50 ps pulse which is amplified up to 5 GW at 1064 nm. The second harmonic can be generated at 532 nm up to 1.5 GW when needed. The laser pulse is focused into a vacuum chamber by an aspheric lens corrected for spherical aberrations. The vacuum chamber is pumped down to 10^{-8} Torr and then filled with spectroscopically pure rare gas at a static pressure of 5×10^{-5} Torr. At this pressure, no collisional ionization occurs, and no complications from charge exchange reactions are expected. Only collisionless multiphoton ionization occurs. The ions resulting from the laser interaction with the atoms in the focal volume are extracted with a transverse electric field of 1 $kV.cm^{-1}$, separated by a 20 cm length time-of-flight spectrometer, and then detected in an electron multiplier. The laser intensity is adjusted in order to produce 1 to 10^5 ions. The experiment consists of the measurement of the number of ions corresponding to different charges formed as a function of the laser intensity.

2.1 Multiphoton Ionization of Xe at 532 nm.

Figure 1 is a typical result of the multiphoton ionization of Xe at 532 nm /8/. Up to Xe^{5+} ions are formed. Let us analyse the different processes which occur when the laser intensity I is increased. Figure 1 can be divided into two parts. The first part ($I < 1.5 \times 10^{12}$ $W.cm^{-2}$) is characterized by a laser-neutral atom interaction, while in the second part ($I > 1.5 \times 10^{12}$ $W.cm^{-2}$) a laser-ion interaction occurs. In the first part, the absorption of 6 photons by an atom leads to the removal of one electron and the formation of a Xe^+ ion. This process appears in Fig. 1 through experimental points joined by a straight line with a slope 6 because, out of resonance, a 6-photon ionization rate varies as I^6. When the laser intensity is increased further, approaching the I_S value, the absorption of 15 photons by an atom induces the simultaneous removal of two electrons and the production of a Xe^{2+} ion. This process appears in Fig.1 through experimental points joined by a straight line with a slope 15. The 6-photon and 15-photon ionization processes deplete the number of atoms contained in the interaction volume. A marked change appears in the slope of the curves for both Xe^+ and Xe^{2+} ions beyond the laser intensity I_S. This saturation is a typical effect which occurs in multiphoton ionization experiments when all the atoms in the interaction volume are ionized. The intensity dependence of both curves of Xe^+ and Xe^{2+} ions just beyong I_S arises from ions produced in the expanding interaction volume when the laser intensity is increased further.

The second part of Fig.1, for $I > 1.5 \times 10^{12}$ $W.cm^{-2}$, describes the interaction of the laser radiation with ions, because the interaction volume is filled up with Xe^+ ions in place of atoms. A sudden increase in the number of Xe^{2+} ions occurs when the laser intensity is increased further. This comes from the absorption of 10 photons by a Xe^+ ion. This removes one electron from the Xe^+ and produces a Xe^{2+} ion. This appears in Fig.1 through experimental points joined by a straight line with a slope 10. When the laser intensity is increased further, the 10-photon ionization of Xe^+ ions also saturates and Xe^{3+}, Xe^{4+} and Xe^{5+} ions are formed most

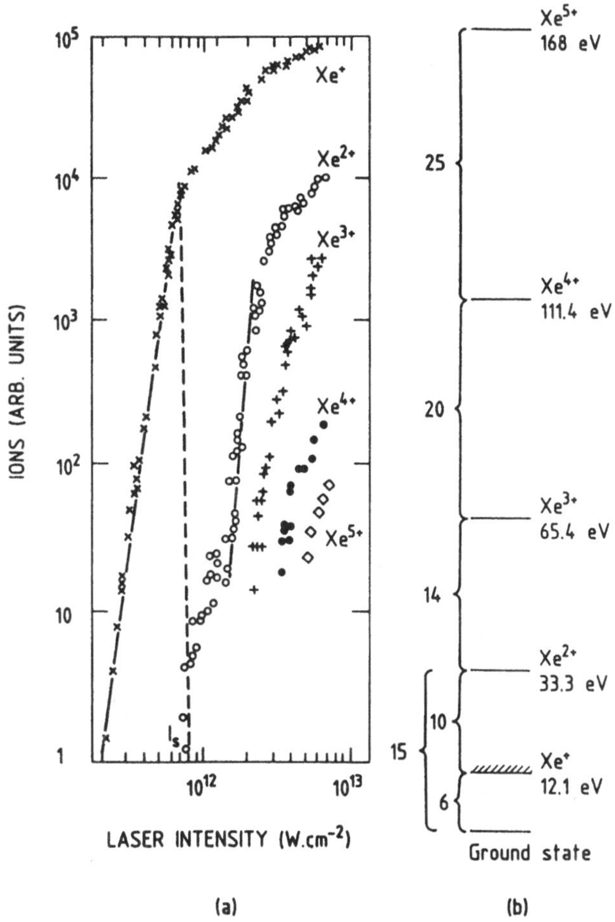

Fig.1 - (a) A log-log plot of the variation of the number of Xe ions formed at 532 nm as a function of the laser intensity.
(b) Schematic representation of the number of multiply charged ions.

likely through stepwise processes. This means Xe^{3+} ions are produced from Xe^{2+} ions by absorbing 14 photons. In the same way, Xe^{4+} ions are produced from Xe^{3+} ions by absorbing 20 photons, and likewise for Xe^{5+} ions. To summ up, Fig.1 is a clear picture of the response of the electrons of Xe atoms to a high laser intensity. Each step of increased intensity gives rise to the removal of an additional electron.

2.2 Multiphoton Ionization of Ne at 532 nm.

Figure 2 shows in a log-log plot the variation of the number of Ne^{+} and Ne^{2+} ions produced as a function of the laser intensity /8/. The Ne^{+} ion curve has a slope of ten which is characteristic of a non-resonant 10-photon ionization of Ne atoms. Ne^{2+} ions are produced in a laser intensity range far beyond the saturation intensity value I_S, that is when the interaction volume is filled up with Ne^{+} ions

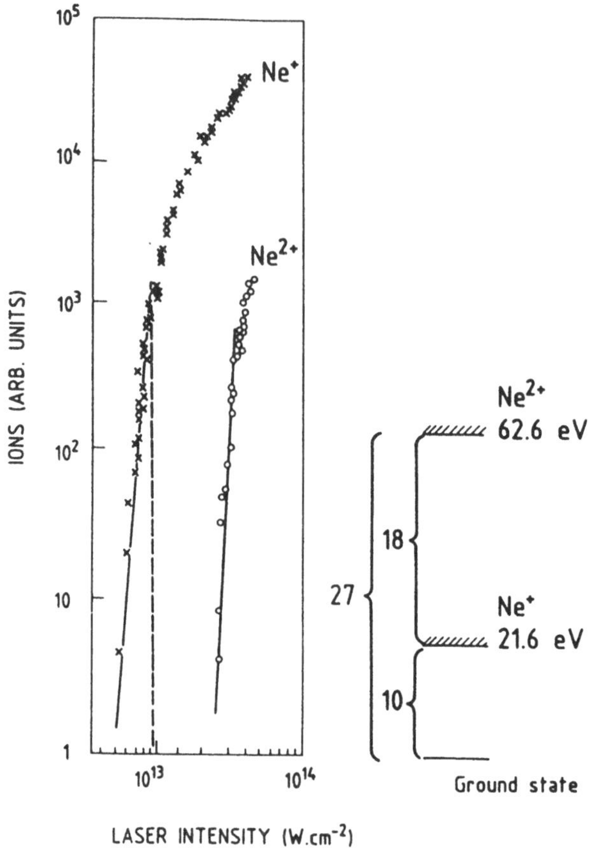

Fig.2 - The laser intensity dependence of Ne^+ and Ne^{2+} ions formed at 532 nm.

and no longer with any Ne atoms. This requires that Ne^{2+} ions are produced through an 18-photon ionization of Ne^+ ions. This is confirmed by the slope 17 ± 2 measured on the Ne^{2+} ion curve. Here, the probability of production of Ne^{2+} ions by a simultaneous excitation of two electrons is much too low to be measured.

2.3 Multiply Charged Ions Produced in Rare Gases at 1064 nm.

The production of multiply charged ions have also been investigated in the five rare gases at 1064 nm /7/. Let us consider here the two most different examples : Xe and He. Figure 3 shows the variation of the number of Xe^+, Xe^{2+}, Xe^{3+}, Xe^{4+}, Xe^{5+} and Xe^{6+} ions as a function of the laser intensity. The general behaviour is similar to that observed at 532 nm, except for two points. First, the two different processes of production of Xe^{2+} ions, namely the simultaneous two-electron removal from Xe atoms, and the one-electron removal from Xe^+ ions, are not so well separated than at 532 nm. Second, the probability of creating

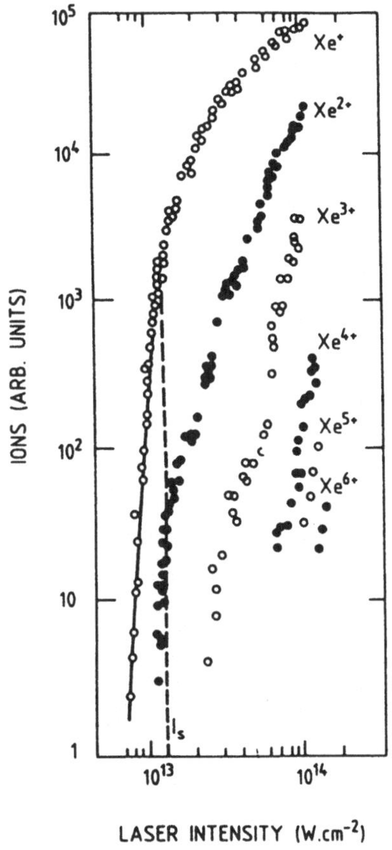

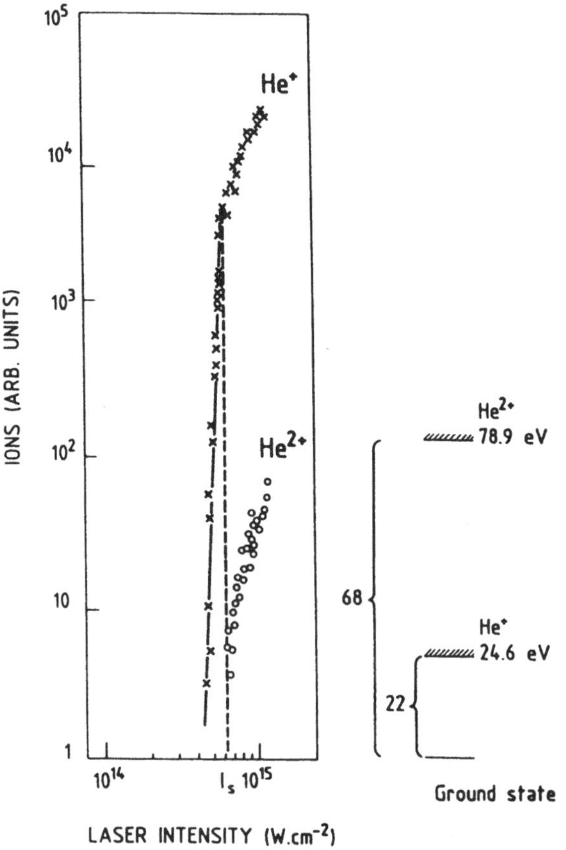

Fig.3 - The laser intensity dependence of Xe ions formed at 1064 nm.

Fig.4 - The laser intensity dependence of He^{+} and He^{2+} ions formed at 1064 nm.

Xe^{2+} ions through a simultaneous two-electron removal from Xe atoms is 30 times larger here than at 532 nm, at the reference intensity I_S. At saturation $I_S = 1.2\times10^{13}$ $W.cm^{-2}$ at 1064 nm the proportion of Xe^{2+} to Xe^{+} ions is 1.5×10^{-2}, whereas it is only 5×10^{-4} at $I_S = 8\times10^{11}$ $W.cm^{-2}$ at 532 nm. It must be pointed out that a large amount of energy can be transmitted to a many-electron atom through multiphoton absorption processes. For example, at least 250 eV have been absorbed by a Xe atom when Xe^{6+} ions are produced.

Figure 4 shows the variation of the number of He^{+} and He^{2+} ions produced as a function of the laser intensity. 68 photons at least have to be absorbed by He atoms to produce He2+ ions which most probably come from a simultaneous excitation of the two electrons. This conclusion is supported by the fact that saturation of both He^+ and He^{2+} ions occurs at the same laser intensity I.

2.4 Electron Energy Measurements

The measurement of only the number of ions does not enable one to distinguish whether ions are in their ground states or in excited states. The further step to get a better understanding of the basic processes involved in the production of multiply charged ions is to analyse the energy of electrons produced in the process of generating singly and doubly charged ions.

As is well known, in the lowest order perturbation theory, an atom is singly ionized by absorbing the minimum number N of photons required to reach the first ionization threshold. However, the absorption of M = N+1, N+2... photons of the same energy can occur if the laser intensity is large enough. For example, using a 50 ps laser pulse, the absorption of one additional photon was observed in cesium atoms at 1064 nm and 10^{11} $W.cm^{-2}$ /10/. The absorption of a number of additional photons was observed in Xe atoms at 532 nm and 10^{12} $W.cm^{-2}$ /11-13/, and the absorption of up to twenty additional photons was observed in Xe at 1064 nm and 10^{13} $W.cm^{-2}$ /11/. The electron energy spectrum consists of a serie of peaks evenly spaced by an amount equal to the photon energy /12,13/.

While the retarding potential method we have used is not the most elegant form of energy analysis, it has a good transmission rate. This is very important because the maximum number of electrons produced at the focal point is only about 10^5 per laser shot. Here, in contrast with ion detection, no extracting electric field is used. About 1% of the electrons produced at the focal point diffuse through the retarding region and can be detected in the electron multiplier. The experiment consists of the measurement of the detected number of electrons as a function of the laser intensity, for different values of the retarding potential V_R.

Figure 5 is a typical result obtained in Xe at 532 nm. The Xe atom has to absorb a minimum number of 6 photons to release one electron. In the 10^{11} - 10^{12} $W.cm^{-2}$ intensity range used here, the Xe atom can ansorb 7, 8, 9...M photons, but with a decreasing absorption rate. Electrons detected at $V_R = 0$ have all the energies of the distribution which consists of a series of peaks which have decreasing amplitudes. Consequently, the laser intensity dependence of electrons measured at $V_R = 0$ is expected to be mainly governed by the intensity dependence of the first peak of the distribution, i.e. I^6, with a possible small contribution of the second or third peak. This is consistent with the slope dlgNe/dlgI = 6.2 ± 0.3 measured at $V_R = 0$. At $V_R = -20$ V, only electrons of the distribution

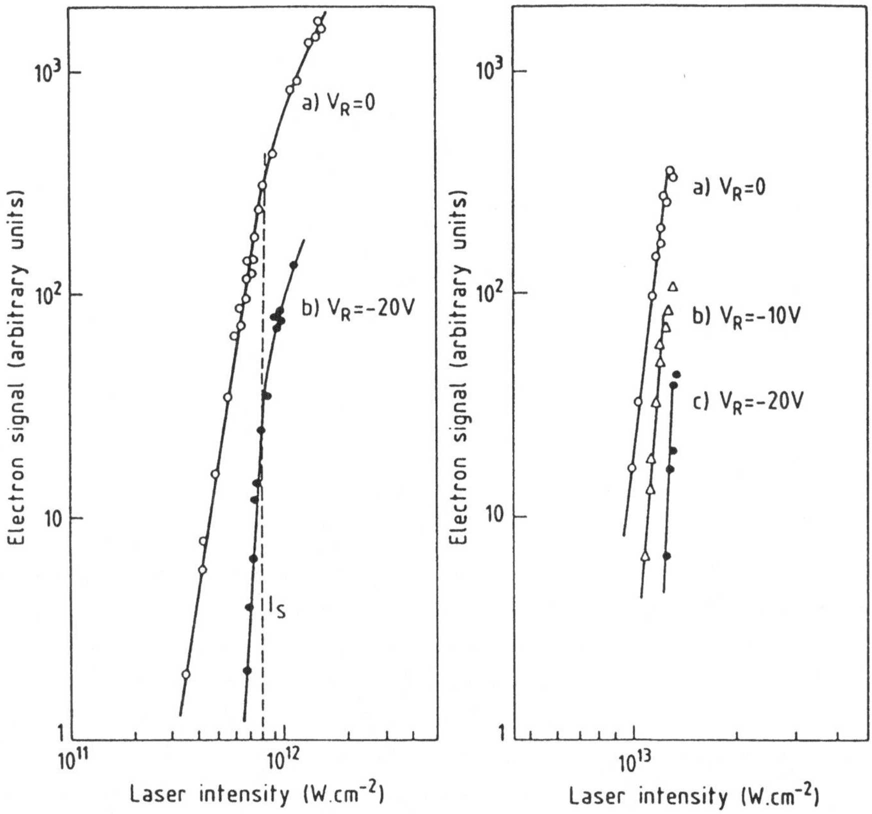

Fig.5 - Multiphoton ionization of Xe at 532 nm. A log-log plot of the variation of the observed number of electrons as a function of the laser intensity at two different values of the retarding potential V_R.

Fig.6 - Multiphoton ionization of Xe at 1064 nm. A log-log plot of the variation of the observed number of electrons as a function of the laser intensity at three different values of the retarding potential V_R.

with energy larger than 20 eV are detected, i.e. electrons released after the absorption of at least 14 or 15 photons by Xe atoms. These two numbers correspond to the $P_{3/2}$ and $P_{1/2}$ ionization limits respectively. The preceding picture is consistent with the slope 16 $\pm$ 1 measured in figure 5(b).Deviation from the I^6 and I^{16} laws observed in figures 5(a) and 5(b) at higher intensities comes from a well known saturation effet which occurs in the production of ions and electrons in all multiphoton ionization experiments. Consequently, the laser intensity dependence of energetic electrons can be measured only in a laser intensity range just below the saturation intensity I_S.

Similar results have been obtained in Xe at 1064 nm and in Ne at 532 nm /11/. Figure 6 shows the variation of the number of electrons detected as a function of the laser intensity for three values of the retarding potential. Electron yield curves 6(a), 6(b) and 6(c) show increasing slopes of 11.3 ± 0.3 at $V_R = 0$, 19 ± 1.5 at $V_R = -10$ V and 30 ± 2 at $V_R = -20$ V. This implies that at least 11, 19 and 28 photons, respectively, have been absorbed in the ionization process.

In conclusion, energetic electrons observed in Xe at 1064 nm imply that up to M = 28 photons have been absorbed in the ionization process, while only 11 photons are necessary to release one electron. The laser intensity dependence of energetic electrons definitively emphasizes the validity of the I^M law. However, the high absorption rate of such large number of photons cannot be easily explained yet.

3 Discussion

The theoretical one-electron model which have been successfully used to describe one electron removal in multiphoton ionization of atomic hydrogen and alkaline atoms in the past few years cannot be applied to explain the production of doubly charged ions induced in many-electron atoms. In this respect, the following example of the production of doubly charged ions is very convincing. The one-electron removal in Xe atoms through 11-photon absorption at 1064 nm requires a laser intensity of 10^{13} W.cm^{-2}. The 29-photon absorption corresponding to the production of Xe^{2+} ions at 1064 nm would require a laser intensity of 10^{15} W.cm^{-2}, a value anticipated from the lowest order pertubation theory in the one-electron model. This is at variance with experimental results (Fig.3) which show that a laser intensity of 1.5×10^{13} W.cm^{-2} is enough to produce Xe^{2+} ions. Figure 3 also shows in other terms that the 29-photon absorption rate giving Xe^{2+} ions is only 100 times less than the 11-photon absorption rate giving Xe^{+} ions at 1.5×10^{13} W.cm^{-2}. In contrast, the 29-photon ionization rate anticipated from the lowest-order pertubation theory in the one-electron model would be about 10^{40} times less than the 11-photon ionization rate.

This suggests that multiphoton ionization of a closed shell atom such as a rare gas atom cannot be described in any circumstances by considering the interaction of the laser field with only one electron of the shell. A collective response of the atomic shell irradiated by the laser pulse should be considered. The absorption of a large number of photons may involve several electrons. The absorbed energy can be redistributed between electrons by means of electron correlations which need to be included in this picture.

The production of multiply charged ions through multiphoton absorption emphasizes both atomic properties and laser characteristics such as intensity, photon energy, pulse duration, etc... The difficult point is that these parameters are not independent. For example, changing the laser photon energy leads to change in the nonlinear order of the interaction, i.e. to change in the number of photons required to reach first and second ionization limits, and consequently the laser intensity required to produce singly and doubly charged ions.

The photon energy seems to play an important role in the production of doubly charged ions through the simultaneous excitation and removal of two electrons. For example, the probability of production of Xe^{2+} ions through a simultaneous two-electron removal from Xe atoms is 30 times less at 532 nm at 10^{12} W.cm^{-2} than at 1064 nm at 10^{13} W.cm^{-2}, as shown by comparison of Figs.1 and 3. This result can more likely be explained in terms of laser wavelength than in terms of laser intensity, as exemplified by Fig.2. This figure shows that at 532 nm and at high laser intensity (10^{13} W.cm^{-2}) no Ne^{2+} ions are produced below the saturation intensity I_S by a simultaneous excitation of two electrons. The longer the laser wavelength, the higher is the simultaneous two-electron removal probability. Such a wavelength dependence in the simultaneous removal of two electrons looks like the well known wavelength dependence in the photoionization cross-section of excited atoms. This wavelength dependence could be amplified further because here, we deal not with a singly excited state, but with a multiply excited atom.

The production of multiply charged ions has been reported in alkaline-earth atoms /14-15/ and in rare gas atoms. All many-electron atoms are expected to behave similarly. There exists no quantitative theoretical model to understand these processes. Going beyond the one-electron model of multiphoton ionization is the first step towards such understanding. This requires taking into account, first, electrons correlation effects, and second doubly excited states for alkaline-earth atoms. For rare gas atoms, the situation is still more complex because this requires the inclusion of multiply excited states. In addition, laser intensity effects on doubly or multiply excited states should be taken into account /16/. Finally, for rare gases with high Z, core excitation could become important.

As there exists no quantitative theoretical model yet, we can attempt to suggest the following tentative picture derived from the preceding experimental observations. Let us consider first the simplest example, i.e. He atom, at 1064 nm. When the laser intensity is low enough so that no ionization occurs, the two electrons can absorb many photons through laser-induced virtual doubly-excited states. Such states could be detected only by fluorescence measurements. When the laser intensity is increased, the two electrons absorb enough energy from the laser field, so that one of the two electrons is released and a He^{+} ion is produced in its ground state or in an excited state. This picture could look like the well known autoionization process. When the laser intensity is increased further, the electrons absorb enough energy so that both are released simultaneously, and a He^{2+} ion is produced. For a many-electron atom, this picture should be extented to all the electrons of the outer shell, and also including possibly inner shells, to explain the production of multiply charged ions observed in rare gases. This picture seems to prevail when a long wavelength laser radiaton is used. On the other hand, this picture could be quite different if a short wavelength laser radiation is used. This is because the minimum number of photons absorbed by the atom to release the first electron is much smaller. As a consequence the laser intensity required to ionize the atom is also much smaller, typically 10^3 times less. This lower intensity would not favor the production of multiply-excited states in the neutral atom. In particular, doubly charged ions would be produced only from multiphoton ionization of singly charged ions, and never through a simultaneous removal of the two electrons of the neutral atom. Obviously, such a picture should be corroborated by further experimental and theoretical data.

4 Conclusion

The interaction of an intense laser field with a many-electron atom is quite an open field because the observation of multiply charged ions produced in rare gas atom raises a number of stimulated questions which cannot be answered at the present time. The production of multiply charged ions is most likely induced by a collective response of the atomic shell. Multiply excited states are expected to play an important role. Furthermore at 1064 nm at least 68 photons, eauivalent to 79 eV, have to be absorbed by a He atom to explain the production of He^{2+} ions. The basic absorption mechanism of such a large number of photons is not yet understood. Further data on electron energy distributions and on fluorescence from excited states would be useful to understand the basic processes involved in the interaction of an intense laser field with a many-electron atom. Finally, an increase in the data concerning atomic spectroscopic behaviour can be expected.

References

1 P. Lambropoulos : Adv. At. Mol. Phys. 12, 87 (1976)

2 J. Eberly and P. Lambropoulos : Multiphoton Processes (Jonn Wiley and Sons, New York 1978)

3 Y. Gontier and M. Trahin : Phys Rev A 19, 264 (1979)

4 J. Morellec, D. Normand and G. Petite : Adv. At. Mol. Phys. 18, 97 (1982)

5 G. Mainfray : J. Physique 43, C2-367 (1982)

6 A. L'Huillier, L-A. Lompré, G. Mainfray and C. Manus : Phys. Rev Lett. 48, 1814 (1982)

7 A. L'Huillier, L-A. Lompré, G. Mainfray and C. Manus : J. Phys. B 16, 1363 (1983)

8 A. L'Huillier, L-A. Lompré, G. Mainfray and C. Manus : Phys. Rev. A 27, 2503 (1983)

9 A. L'Huillier, L-A. Lompré, G. Mainfray and C. Manus : J. Physique 44, 1247 (1983)

10 G. Petite, F. Fabre, P. Agostini, M. Crance and M. Aymar : Phys. Rev. A 29, 2677 (1984)

11 L-A. Lompré, A. L'Huillier, G. Mainfray and J.Y. Fan : J. Phys. B 17, L817 (1984)

12 P. Kruit, J. Kimman and M. van der Wiel : J. Phys. B 14, L597 (1981)

13 F. Fabre, G. Petite, P. Agostini, and M. Clement : J. Phys. B 15, 1353 (1982)

14 I. Aleksakhin, I. Zapesochnyi and V. Suran : J.E.T.P. Lett. 26, 11 (1977)

15 D. Feldman, J. Krautwald, S.L. Chin, A. von Hellfeld and K. Welge : J. Phys. B 15, 1663 (1982)

16 Y.S. Kim and P. Lambropoulos : Phys. Rev. Lett. 49, 1698 (1982)

MULTIPHOTON IONIZATION OF COMPLEX ATOMS

Michèle Crance

Laboratoire Aimé Cotton CNRS II
Bât.505 91405 ORSAY CEDEX FRANCE

1 INTRODUCTION

An atom irradiated by a light field of strong intensity ionises even if the photon energy $\hbar w$ is smaller than the ionisation potential E. In the latter case, at least n photons have been absorbed when an electron is ejected; n is the first integer larger than $E/\hbar w$. For weak intensities, the number of ions created N is proportional to I^n (I is the field intensity) and the interaction time T. The generalised cross section defined by $\sigma^{(n)}=N/(I^nT)$ depends only on the atom and the field frequency. As a function of frequency, $\sigma^{(n)}$ exhibits a resonance structure: $\sigma^{(n)}$ increases when the field frequency or an harmonics is close to a Bohr frequency of the atom. This behaviour is predicted by perturbation theory applied at n-th order and has been observed experimentally. In fact, for frequencies very close to resonance, a perturbative treatment at minimum non vanishing order is no longer valid. By using the projection operator technique, it is possible to describe the atom as a two level system with intensity dependent complex energies. The real part is to be understood as a light shift, the imaginary part corresponds to ionisation probability. In this description, the resonance structure depends on intensity. When the intensity is increased, resonances shift and broaden. For a frequency nearly resonant, the effective order of non linearity $k=d(LnN)/d(LnI)$ differs from n and can reach large values when the shift of the resonance is much larger than its width. The shape of resonances depends on the characteristics of the excitation light pulse -duration, coherence,spatial repartition of intensity. These features, which characterise the behaviour of a resonance in strong field, have been observed in various atoms(1).

As long as only ion yield has been measured, no striking difference has been observed between alkali, alkaline earth or noble gases. Precise studies of resonances have been carried out on alkalis. That is also on one-electron atoms that theoretical works have been successful in interpreting experimental results (2). The problems encountered in interpretation of noble gases ionisation have been attributed to the complexity of their energy spectrum. A noticeable difference between alkalis and noble gas behaviour has been found,at first, when electron energy spectra have been observed. n is the minimum number of photons an atom must absorb to be ionised. In such a case, the electron is ejected with energy $n\hbar w-E$. However, an electron may absorb more than n photons and the electron energy spectrum consists of several peaks at energies $(n+k)\hbar w-E$ (k is a positive integer). For comparable intensities , the electron energy spectra obtained from multiphoton ionisation are rather different for alkalis and noble gases. Several peaks have been observed for noble gases (up to ten in Xenon excited by a Neodyme glas laser) (3). Conversely, after ionisation of Cesium, at most two peaks have been observed and the fast electrons are twenty times less abundant than slow ones (4).

The most striking difference between alkalis and complex atoms has been observed when the formation of multicharged ions has been detected. For an intensity of 10^{14} W cm^{-2}, up to four electrons can be extracted from Xenon by a Neodyme glas laser or its first harmonics. For the same intensity, it is not possible to eject more than one electron of Cesium (5). Experiments carried out at different wavelengths gave similar results (6).In all these experiments, everything happens as if only outershell electrons were involved in multiphoton absorption process. The result is that one outershell electron atoms and complex atoms require completely different treatments. The difference between alkalis and noble gases atoms is not only quantitative (in complexity of calculations) but qualitative: even if only one electron is to be ejected, it is likely that excitation of all outershell electrons has to be taken into account so that it is not realistic to study single ionisation without considering the formation of multicharged ions. A number of questions has to be answered before interpreting multiphoton stripping of atoms. Several mechanisms can

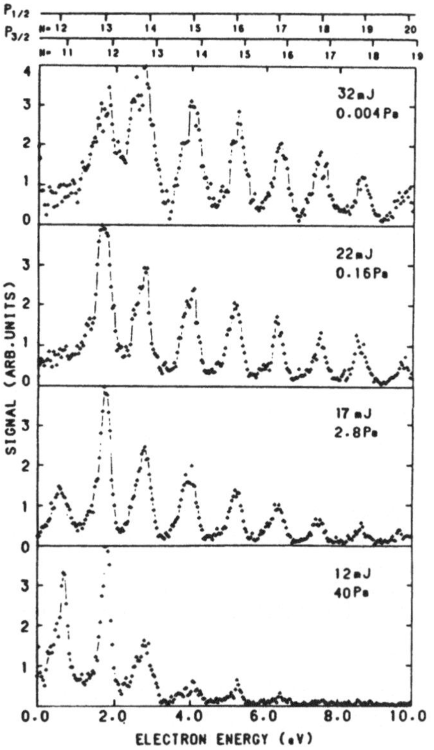

Figure I

Multiphoton ionisation of Xenon at 1.06μ. Electron energy spectrum. from (3).

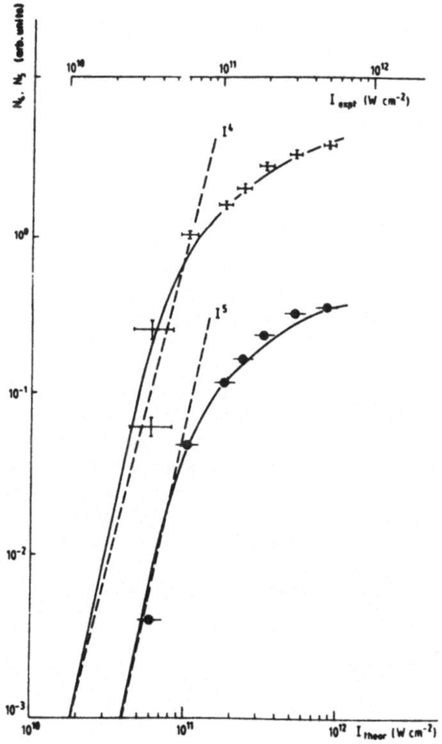

Figure II

Multiphoton ionisation of Cesium at 1.06μ.:

crosses: number of electrons having absorbed 4 photons.

circles: number of electrons having absorbed 5 photons.

from (4)

be invoked to explain the formation of multicharged ions: how can one recognise them and how does their competition depend on the characteristics of the excitation light ? What are the atomic quantities involved and how can one calculate them? The first question concerns the dynamics of multiphoton stripping and shall be discussed in Section 2. The second question concerns the theory of atomic structure and shall be analysed in section 3. In section 4 and 5, we propose two ways to interpret quantitatively multiphoton stripping of complex atoms:

- calculations in two-electron atoms using a compact representation of atomic energy spectrum (square integrable functions and complex dilatation method).
- Statistical heuristic model to describe multiphoton absorption in any atom from simple parameters such as the size and extraction potential of outershell electrons.

2 DYNAMICS OF MULTIPHOTON STRIPPING

The usual way to describe multiphoton absorption is based on perturbation theory at first non vanishing order. In this frame, an energy conservative process such as the removal of k electrons after absorption of n photons consists in n sucessive absorption of a photon leading the atom in n-1 successive intermediate states. This defines a quantum path from the initial state to the final state. Elementary processes are not required to conserve energy so that any state of the atom is to be taken into account as an intermediate state. The quantum paths which contribute most are the ones for which an intermediate state is close to resonance. When several electrons are removed from a complex atom, the energy absorbed is rather large and a number of states involving the excitation of several electrons are good candidate as nearly resonant intermediate state. When studying the dynamics of the process , quantum paths leading from atom A to ion A^{k+} are to be classified according to the number of resonant intermediate states they involve. If there is no resonant intermediate state we shall speak of a direct process. The n photons have to be absorbed coherently and simultaneously, that is in a time short enough for the Heisenberg principle to be satisfied. A transition probability P_{0k} can be defined for the reaction $A + n\hbar w \rightarrow A^{k+} + k\ e^-$. If some of the intermediate states involved in a quantum path are resonant, we speak of a stepwise process. Resonant intermediate states being a , b ,..., Each transition $g \rightarrow a$, $a \rightarrow b$,... corresponds to the simultaneous coherent absorption of several photons and is described by a transition probability P_{ga}, P_{ab},... From the previous considerations, one can deduce immediatly a way to distinguish direct and stepwise process for weak field and short interaction time. As a function of intensity, each probability obeys a power law with an index corresponding to the number of photons involved. The number of atoms in state f after an interaction time T, is the sum of contributions from each type of processes. Contributions are proportional to T, T^2, T^3,... for direct process, two-step process, three-step process...respectively. By atom state we mean any state of the nucleus plus Z electrons system. That is neutral atom in any state, singly charged ion plus one electron, doubly charged ion plus two electrons,...When several processes compete in the formation of ion A^{k+} for a given intensity, one expects that the ratio of stepwise process contribution over direct process contribution increase with the interaction time. For a given interaction time,one expects that the dominant process be the one which requires the smallest number of photons.

Although the previous analysis can be applied to any charge state, we shall restrict further discussion to the formation of A^{++} in ground state. Extraction of two electrons from A requires the absorption of n_3 photons, at least; n_3 is defined as the first integer larger than $[E(A^{++}) - E(A)]/(\hbar w)$; $E(A^{k+})$ is the energy of A^{k+} . Two electrons are ejected which share the energy $e = n_3\hbar w - E(A^{++}) + E(A)$. In a direct process, electrons may have any energy e'_1, e'_2 provided that $e'_1 + e'_2 = e$. If we do not consider coincidental resonances with a bound state of A, a two-step process consists of two transitions:

$$A + n_{1a}\,\hbar w \rightarrow A^+_a + e^- \qquad (1)$$

followed by

$$A^+_a + n_{2a}\,\hbar w \rightarrow A^{++} + e^- \qquad (2)$$

A^+_a stands for singly charged ion in state a . Each process is energy conservative and thus n_{1a} is an integer larger than $[E(A^+_a) - E(A)] / (\hbar w)$. The energy of the first ejected electron is $e_{1a} = n_{1a}\hbar w - E(A^+_a) + E(A)$. The energy of the second ejected electron is $e_{2a} = n_{2a}\hbar w - E(A^{++}) + E(A)$ .If e_{1a} is larger than e, it is not possible to remove a second electron by absorption of $n_3 - n_{1a}$ photons and thus n_{2a} is at least equal to $n_3 - n_{1a} + 1$. The evolution of an atom can be described by rate equations (7) governing the probability for an atom to be in A ground state or in state a of A^+ or in A^{++} ground state. These equations can be integrated formally for any time dependence of the light intensity (7).

In weak field, most singly charged ions will be created in ground state, that is the less energy consuming process to produce A^+. In the formation of A^{++}, the direct process will be more efficient, if it requires less photons ($n_1 + n_2 - 1 = n_3$) than the stepwise process involving the ground state of A^+. For an intensity close to saturation intensity of single ionisation, direct process saturates. A similar mechanism occurs for Above Threshold Ionisation (8) : The depletion of the ground state at saturation prevents further development for all the transitions starting from A as an initial state. The probability per unit time for transition (1) and (2) are respectively P_1, P_2 The probability for direct process

$$A + n_3\hbar w \;—\; A^{++} + 2\,e^- \qquad (3)$$

is P_3 . Saturation intensity for single ionisation is defined by

$$\left[P_1(I_s) + P_3\,(I_s)\right]\,T = 1 \qquad (4)$$

The number of A^{++} created by direct process cannot exceed the value reached for I_s, that is about $P_3(I_s)\,T$,for any intensity higher than I_s. When saturation of single ionisation occurs before the end of the light pulse, only the stepwise process is responsible for further creation of doubly charged ions. For intensity larger than I_s, the stepwise contribution increases approximately as $P_2(I)$, so that the stepwise process contribution is dominant at large intensities. This is in agreement with experiments carried out on Xenon with a frequency doubled Neodýmium glas laser. In this case $n_3 = 15$ and $n_1 = 6$, $n_2 = 10$ when the intermediate state is A^+ ground state. The number of Xe^+ and Xe^{++} created has been studied as a function of intensity for various pulse durations between 5ps and 200ps.

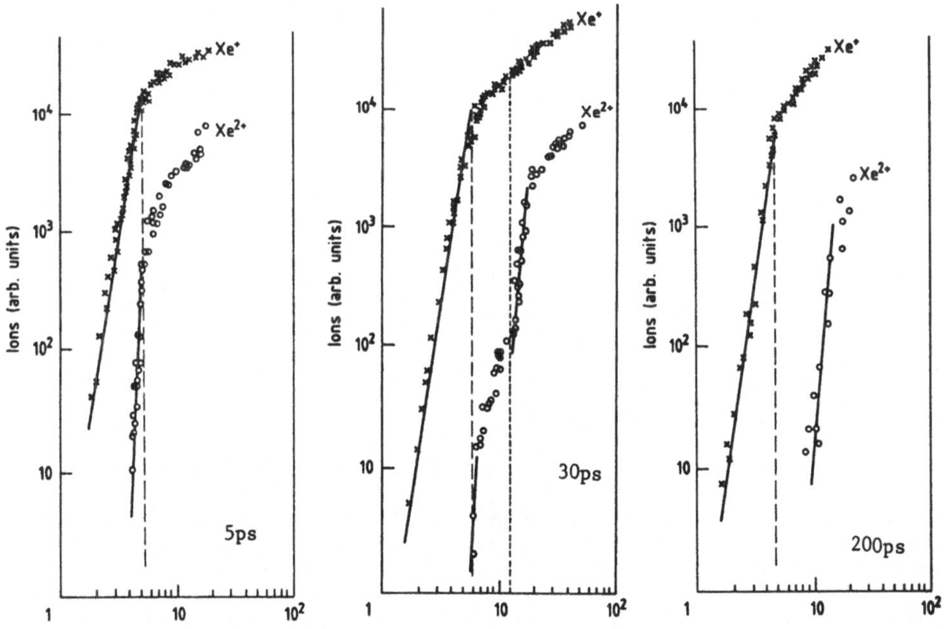

Figure III

The description given above corresponds to experimental curve obtained with T=30ps; when T=200ps, I_s is too low for direct process to be detected and only the stepwise contribution is observed; when T=5ps, intensities higher than I_s have not been reached and only the direct process is observed.

In the case where $n_3 = n_1 + n_2$, for the quantum path involving the ground state of A^+, it is not possible to estimate the role of direct process in weak field on the basis of the previous discussion and we have to examine more closely what type of quantum paths corresponds to direct process. We shall, thus, study the energy spectrum of emitted electrons.

3 ELECTRON ENERGY SPECTRUM IN MULTIPHOTON STRIPPING

An important class of direct processes consists of transitions (1) and (2) in which electrons are ejected with energies $e_1' \neq e_{1a}$ and $e_2' \neq e_{2a}$ but satisfying $e_1'+e_2'=e$ The final states with various electron energies can be distinguished, the probability for atom A to be doubly ionised by absorption of n_3 photons is the integral over e_1' of the probability $P(e_1')$ for A to eject two electrons with energies e_1' and $e_2'=e-e_1'$. In the dressed atom picture, the initial state is $|g,N_o\rangle$: atom A in ground state plus N_o photons, the final state is $|e_1',e_2',N_o-n_3\rangle$ a state of double ionisation continuum dressed by N_o-n_3 photons. The hamiltonian of the system is

$$H = H_o + V \qquad \text{with } H_o = H_a + H_f \tag{5}$$

H_a is the atomic hamiltonian, H_f is the field hamiltonian, V is the atom-field interaction. The resolvent operator corresponding to H_o is

$$G_o(z) = (z - H_o)^{-1} \tag{6}$$

In a perturbative treatment at first non vanishing order, the probability to eject two electrons with energies e_1', e_2' after absorption of n_3 photons is given by:

$$P(e_1') = \frac{1}{\hbar}\langle g,N_o|\ V\ (\ G_o(E_g)\ V\ )^{n_3-1}\ |e_1',e_2'\ ,\ N_o - n_3\rangle \tag{7}$$

E_g is the energy of the initial and final state of dressed atom:

$$E_g = E(A) + N_o\hbar w = e_1' + e_2' + (N_o - n_3)\hbar w$$

The presence of G_o implies a summation over multiplicities of the dressed atom and for each multiplicity, a summation over atomic states. Once n_1 photons have been absorbed, a pole appears in G_o, corresponding to a possible resonance. We have seen that contribution of resonances corresponds to stepwise processes so they have to be removed from the summation. The most important terms in $P(e_1')$ involve some non resonant atomic intermediate state $|a,e_1'\rangle$ of single ionisation continuum of A. a is the state of A^+. An approximate expression for $P(e_1')$ is

$$\overline{P}(a,e_1') = \sum_a \frac{Q(a,e_1')\ R(a,e_2')}{d(a,e_1')^2} \tag{8}$$

$d(a,e_1')$ is the detuning from resonance for state a,e_1' :

$$d(a,e_1') = n_{1a}\hbar w - E(A_a^+) + E(A) - e_1' = e_{1a} - e_1' \tag{9}$$

$$Q(a,e_1') = \langle g,N_o|\ V\ (\ G_o(E_g)\ V\ )^{n_{1a}-1}\ |a,e_1',N_o-n_{1a}\rangle$$

$$R(a,e') = \langle a,N_o-n_{1a}|\ V\ (\ G_o(E_g)\ V\ )^{n_{2a}-1}\ |e_2',N_o-n_{1a}-n_{2a}\rangle \tag{10}$$

Equation 8 allows one to describe the spectrum of electrons ejected in direct double ionisation after absorption of n_3 photons. We show in Figure IV a schematic electron energy spectrum corresponding to a model in which only two states a and b of A^+ areconsidered. State a is such that $n_3 = n_{1a} + n_{2a} - 1$, so that e_{1a} is larger than e, the energy shared by electrons ejected when n_3 photons are absorbed .

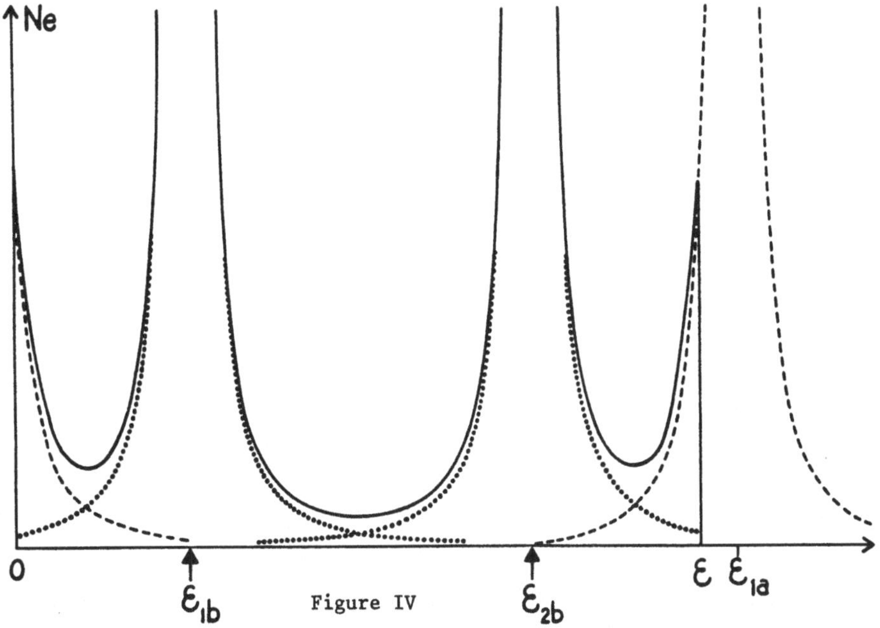

Figure IV

State b is such that $n_3 = n_{1b} + n_{2b}$, so that stepwise process involving b as an intermediate state leads to emission of electrons with energies e_{1b} and e_{2b} lower than e. Solid line shows the contribution of direct process, that is the sum of contributions of transitions involving a as an intermediate state (dashed line) or involving b as an intermediate state (dotted line). Resonances for e_{1b} and e_{2b} are due to stepwise process; the wings of this resonance are to be interpreted as a direct process. We have represented, for energies larger than e, the shape of a resonance at energy e_{1a}, which is forbidden when n_3 photons are absorbed. With the same profile but another scale, it would be the beginning of the electron energy spectrum after absorption of n_3+1 photons.

The number of doubly charged ions is half the area under the solid curve. It is not clear how one can distinguish direct and stepwise processes. In other words, what is the range of energy near e_{1a} or e_{1b} which defines the resonance? From a mathematical point of view, the distinction is easy to draw: the contribution to direct process is given by the integral

$$N_d = ⨍ P(e_1') \, de_1'$$

where symbol $⨍$ means that the divergence at $e_1' = e_{1a}$ has been removed. In the case illustrated in Fig. IV, the contribution of quantum paths involving state a will be included in N_d. For the quantum paths involving state b, the resonance is attributed to stepwise process and the only part of the electron energy spectrum which is counted as direct process is related to the dissymetry of the resonant profile due to the e_1'-dependence of $Q(b,e_1')$ and $R(b,e_1')$. Actually, what is calculated as the direct process is the contribution to double ionisation which is proportional to the interaction time. However, it does not cover the result of all the processes involving a non resonant intermediate state as examined in our analysis of the electron energy spectrum. From a mathematical point of view, direct and stepwise processes can be distinguished by their time dependence in weak field. However, the crucial difference between both processes is related to the coherence in absorption of n_3 photons. In a direct process, two electrons with energies different from resonant ones are to be emitted simultaneously. Only a correlation experiment could draw a meaningful distinction between direct and stepwise processes.

Concerning ion yield measurements, there should be a strong difference in weak field behaviour depending whether $n_3=n_{1a}+n_{2a}$ or $n_3=n_{1a}+n_{2a}-1$ for the quantum paths involving A^+ ground state. For example, an experiment on Argon should give results strikingly different from the ones obtained with Xenon and shown on Fig. IV.

In this section we have identified the dominant processes responsible for multiphoton stripping of atoms and defined the atomic quantities to be calculated. We shall propose two ways to obtain these quantities: one is based on a perturbative treatment of atom-field interaction and is valid for weak field; up to now, this method has been applied only to two-photon double ionisation of Helium (section 4). We present a second method based on an heuristic model of multiphoton absorption which assumes implicitly that the field is strong enough for resonance structure to have been wiped out by shifting and broadening along the light pulse.

4 TWOPHOTON DOUBLE IONISATION OF HELIUM

The method presented here is an extension for two-electron atoms of the method used by Chu and Reinhardt to study two-photon ionisation of Hydrogen (9). We describe Helium states on the basis of two-electron states with angular momenta l_1 and l_2 coupled to give a singlet state with total momentum L. For each orbital momentum l, radial part of wavefunctions is expanded on a basis of complex square integrable functions such as

$$F_{l,n,k,a}(r) = (kr\, e^{ia})^{l+n} \exp(- \frac{kr\, e^{ia}}{2}) \qquad (12)$$

$n \geq l$ is an integer. A finite number of a is introduced which are smaller than $\pi/2$. If an infinite number of n-values were introduced, the F's would span a one dimension Hilbert space. We introduce a finite number of n-values which give a good representation of energy spectrum if the parameters k and a are suitably chosen. The criterium for the choice is the stability of energies for the eigenstates of the atom+field system with respect to basis variations.

In order to describe two-photon absorption, we introduce three multiplicities of the dressed atom: even atomic states ^1S dressed by N_o photons, odd atomic states ^1P dressed by N_o-1 photons, even atomic states ^1S and ^1D dressed by N_o-2 photons. The hamiltonian matrix is calculated on a basis of states $N,(l,n,k,a),(l',n',k',a')$. As a function of the field intensity, each eigenvalue of the hamiltonian matrix can be followed by continuity. One of them, $\hat{E}_g$, goes to $E_g + N_o\hbar\omega$ when the field intensity is decreased. This is the energy of the atomic state g perturbed by interaction with the light field. $\hat{E}_g$ is complex, the imaginary part of $\hat{E}_g$ is proportional to the ionisation probability of state g, or more generally the probability for g-population to vanish by any process to give a continuum state.

For each configuration of each multiplicity, we introduce a finite number of n- and a-values to define the F's. The number of n- and a-values is determined by considering the accuracy of the result. Another criterium is to be followed in choosing the number of a-values. The set of F's for an infinite number of n-values is a basis of a one-dimension Hilbert space for any value of a, in particular for a=0. In the latter case, the F's allow one to define the real energy spectrum. In the description of a resonance, one needs to define an analytic continuation of the hamiltonian in the complex plane. This is the role of the F's with $a \neq 0$. More precisely, this is the presence of complex functions in the basis which gives the contribution to energies corresponding to the poles of the resolvent.

The hamiltonian matrix of the atom + field system is calculated on the basis formed with multiplicities N_o, N_o-1 and $N_o - 2$. In multiplicity N_o, we only need to introduce real F's functions. In multiplicity N_o-1, the presence of complex F's corresponds to the depletion of state g by emission of an electron. This is precisely the process that we have described as the stepwise process. So we define multiplicity N_o-1 by real F's. In multiplicity N_o-2, several configurations are involved: s^2,p^2,d^2,sd. There are essentially two processes responsible for the depletion of state g by absorption of two photons. One is the direct double ionisation which would give two p electrons if correlations were neglected. Another process is the absorption of two photons by a same electron. This process gives rise to He^+ in a bound state and an electron s or d. In order to eliminate this process, we suppress the hamiltonian matrix elements coupling configurations sp of multiplicity N_o-1 to configurations s^2 and sd of multiplicity N_o-2. By studying the role of correlations in continuum states, we estimate the magnitude of neglected terms as less than 6%. We have checked the accuracy of energies by increasing the dimension of the atomic basis. We estimate the accuracy to be of the order of 10%. The probability for double ionisation by direct process has been calculated for frequencies between 1.5 a.u. and 2.15 a.u. (10). Ionisation cross sections of He and He^+ involved in stepwise process are known (11). The cross section for double ionisation by direct process is S_d. The cross sections for transitions (1) and (2) are S_1 and S_2. The ratio $r=S_d/(S_1S_2)$ is a measure of the relative efficiencies of direct and stepwise processes. When both processes involve the same number of photons, this ratio is the interaction time for which their efficiencies would be equal. When stepwise process requires one photon more than direct process, r is the energy per unit area required for both processes to have equal efficiencies. In the latter case, r depends strongly on field frequency since transition (2) may be resonant. Off resonance r is of the order of 50 J cm^{-2} This is qualitative agreement with the results obtained experimentally in the study of Xenon excited by a Neodymium glas laser (5). For an excitation at 0.53μ, the measured ionisation cross sections lead to $r=200$ J cm^{-2}. When double ionisation by stepwise process requires the absorption of two photons, r does not vary strongly with frequency remaining of the order of 10^{-17} s. This result shows that in this case, stepwise process is dominant for any interaction time.

The results obtained on Helium lead to the conclusion that direct double ionisation is to be important only when it involves one photon less than stepwise process through singly charged ion ground state. We do not claim to deduce any definite result when a large number of photons is involved. However, a simple consideration of orders of magnitude gives us some confidence in the generality of our conclusions. In the frame work of perturbation theory, the ratio between stepwise and direct contributions can be written as

$$R = \frac{1}{\hbar T^{-1}} \quad \frac{(e\, a_o\, E)^{2n_1+2n_2-2n_3}}{(2\ \mathrm{Ry})^{2n_1+2n_2-2n_3-1}} \quad \frac{A(n_1)A(n_2)}{A(n_3)} \tag{13}$$

e is the electron charge, a_o is Bohr's radius, 2Ry is one atomic unit of energy. T is the interaction time; $A(n_1)$, $A(n_2)$, $A(n_3)$ are effective hamiltonian matrix elements for processes involving n_1, n_2, n_3 photons expressed in atomic units. E is the field amplitude. Numerous calculations of multiphoton absorption probability for one-electron atoms have shown that the A's, for a given frequency, have an order of magnitude which depends essentially on the number of photons involved. When $n_3=n_1+n_2$, $A(n_1)A(n_2)/A(n_3)$ is expected to be of the order of one. When $n_3=n_1+n_2-1$, $A(n_1)A(n_2)/A(n_3)$ is expected to be one or two orders of magnitude. When we omit the A's in Eq. 13, we obtain a rough order of magnitude for r. Such an approximation leads to $r=2\ 10^{-17}$s, when direct and stepwise processes require the same number of photons, and $r=3$ J cm^{-2} in the opposite situation.

The method described in this section is valid for weak intensities, that is when light shifts are negligible. For higher intensities, one should treat both the electron correlations and the atom-field interaction non perturbatively. For strong field, resonances shift and broaden, so that the number of ions created is described by some averaged cross section which is expected to depend essentially on the characteristics of the initial state and not on the detail of the atomic spectrum. In this perspective we have proposed an heuristic model to describe multiphoton absorption in complex atoms

5 STATISTICAL DESCRIPTION OF MULTIPHOTON STRIPPING

We consider an atom with ionisation potential E_1, irradiated by a light field of frequency w. Ionisation requires the absorption of at least n_1 photons, n_1 being the first integer larger than $E_1/(\hbar w)$. A complex atom has several outershell electrons, say m, which have extraction potential and orbital size of the same order of magnitude. We assume that only outershell electrons are involved in multiphoton ionisation when a few electrons are removed. When p photons are absorbed by the atom, they are shared among outershell electrons. The condition for the atom to be ionised is that one electron has absorbed at least n_1 photons. We assume that all outershell electrons have the same probability to be excited. Therefore, we examine the possible distributions of p photons among m electrons when p n_1 and, for each p we determine the ionisation probability $P_{01}(p)$ as the number of distributions in which one electron has absorbed n_1 photons or more, divided by the number of possible distributions. More generally, the ionisation potential of an atom is E_1, the energy required to remove the k-th electron is E_k. Defining n_k as the integer part of $E_k/(\hbar w)+1$, n_k is the minimum number of photons required for the transition:

$$A^{(k-1)+} + n_k\hbar w \rightarrow A^{k+} + e^- \qquad (14)$$

$A^{(k-1)+}$ and A^{k+} are supposed to be in their ground state. When an atom absorbs p photons, if one electron has absorbed n_1 photons or more, it may escape; if among the remaining electrons, one has absorbed at least n_2 photons it also escape. By examining the distributions of p photons among m electrons, we determine the probabilities $P_{0k}(p)$ for the direct transition $A - A^{k+}$. As a function of p, the P_{0k} are bell-shaped; they vary approximately as the $(n_1+n_2+...+n_k)$-th power of p for small p. When p is increased, the P_{0k} increase until a vamue of a few tenths and then decrease (12).

In the previous analysis, we have treated on the same footing the absorption by an electron of $1, 2, ..., n_1, ...$ photons. Real absorption of less than n_1 photons is forbidden since it is not an energy conserving process. However, virtual absorptions may occur if their duration is short enough for the Heisenberg principle to be satisfied. The energy defect in a virtual absorption is at most E_1. Therefore, $T_1=h/E_1$ is the longest time for which the absorption of any number of photons is allowed. Thus we postulate that the set of $P_{0k}(p)$ can be interpreted as the probabilities for the transitions $A \rightarrow A^{k+}$ for an interaction time T_1 when p photons are available for virtual absorption. We have now to relate p to the light field intensity and the P_{0k} to the quantities introduced in a standard description of multiphoton absorption. Whenan atom is irradiated by a field of intensity I, the photon flux received by an atom is $I r_0^2$, r_0 being the mean radius of an electron orbit in the outershell. In an interaction time T_1, the number of photons available for virtual absorption is taken as $p=I r_0^2 T_1$.

In a standard descriptionof multiphoton absorption, the probability per unit time for the transition $A \rightarrow A^{k+}$ is the product of a generalised cross section and the field intensity at power $n_1+n_2+...+n_k$. For an interaction time T, the probability for the transition $A \rightarrow A^{k+}$, $\mathcal{P}_{0k}(I,T)$, is bell-shaped obeying a power law for weak I. We thus identify $P_{0k}(p)$ with $\mathcal{P}_{0k}(p/(r_0^2T_1),T_1)$. for small p. We apply the same method to calculate the probabilities $P_{kk'}(p)$ for the ion A^{k+} to eject one or more electrons and reach a higher state of charge. By using the latter procedure, we calculate the probability per unit time for any transition $A^{k+} \rightarrow A^{k'+}$ from only the knowledge of extraction potentials and orbitals dimension.

Each multiphoton absorption process can be characterised by the saturation intensity $I_s(T)$: if n photons are involved, the probability for the transition to occur , for weak intensity I, in an interaction time T is given by: $[I/I_s(t)]^n$. The advantage of $I_s(T)$ is to be easily related to experimental data. $I_s(t)$ fixes the limit of the intensity range where power laws are observed. We have used this property to compare experimental data (5) with predictions of the statistical model (13) for excitation at 1.06μ by picosecond pulses (see Table I).

Table I: saturation intensities in unit of $W\ cm^{-2}$

	Xenon	Krypton	Argon	Neon
experimental determination	10^{13}	$2.1\ 10^{13}$	$2.55\ 10^{13}$	$3.5\ 10^{14}$
statistical model prediction	$1.24\ 10^{13}$	$2.15\ 10^{13}$	$4.7\ 10^{13}$	$3\ 10^{14}$

When a large number of photons is required to ionise an atom (several tens), a log-log plot of the number of ions N created as a function f excitation intensity I consists, for weak field, of straight line almost parallel to the N-axis, before it bend, when saturation occurs. We show on Figure V the experimental results obtained for Xenon excited by a CO_2 laser. Xe^+, Xe^{++} and Xe^{+++} have been observed . For each ion yield, saturation intensity is approximately the intensity for which the curve begins. We have marked with an arrow on I-axis the saturation intensity predicted by the statistical model described above.

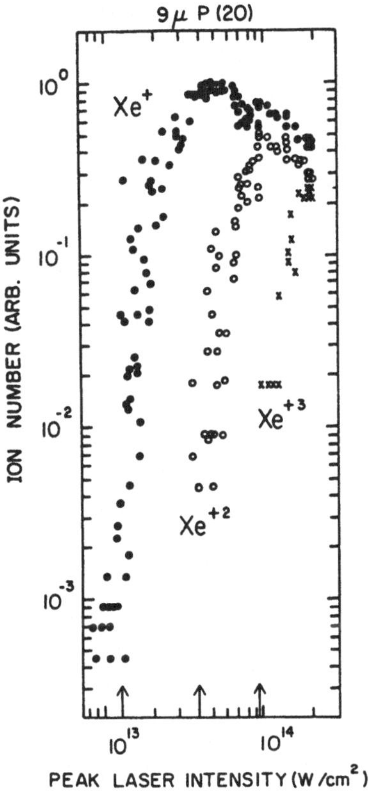

Figure V

Multiphoton ionisation of Xenon irradiated by a CO_2 laser (14)

A similar situation is encountered when atoms in Rydberg states are irradiated by microwaves. Multiphoton ionisation of Hydrogen Rydberg states (principal quantum numbers 63 to 67)has been compared to field ionisation. It has been observed that the field amplitude required for ionisation was the same for microwave (30 MHz to 9.9 GHz) and for static field (15). Excited Hydrogen was obtained by H^+ - Xe electron-transfer collisions and it is reasonable to assume that Rydberg states with a same principal quantum number are equally populated. Statistical model predicts a saturation intensity of $1.8\ 10^8\ n^{-4}\ V\ cm^{-1}$ for principal quantum number n. For a static field the amplitude required to ionise an atom in a state of principal quantum number n is $3.2\ 10^8\ n^{-4}$ V cm-1.

Saturation intensities are also useful to interpret experimental data obtained in strong field, that is for intensities larger than $I_s(T)$ for each observed process. Spatial distribution of field intensity can be characterised by $V(I_M,I)$ which is the volume where the maximum of time-dependent intensity is larger than I when the maximum intensity with respect to time and space is I_M. V depends only on the ratio I_M/I. Whenall the ions created are collected, $V(I_M,I)$ varies as $(I_M/I)^{3/2}$. When ions are collected through a thin slit, in front of the best focus of the light beam $V(I_M,I)$ varies asLn(I_M/I). In strong field, stepwise processes are dominant and multiphoton stripping is the result of successive transitions $A \rightarrow A^+, A^+ \rightarrow A^{++}$

$A^{++} \rightarrow A^{+++}$,.... Define $I_k(T)$ as the saturation intensity for transition $A^{(k-1)+} \rightarrow A^{k+}$ For a maximum intensity I_M larger than I_s, the number of ions with a charge at least k+ is proportional to $V(I_M,I_s)$. We have used this property (16) to interpret experimental results obtained by exciting various atoms (Xe, Kr, Ar, U,I) by an excimer laser at 193nm (17). In the latter experiment, ions were collected through a narrow slit in front of the light beam's best focus. For each studied atom , a log-log plot of N_k, number of ions with a charge at least k+, versus I_k, saturation intensity predicted by statistical model for the transition $A^{(k-1)+} \rightarrow A^{k+}$, fits approximately with a curve$\mathrm{Ln}(I_M/I)$, when $I_k << I_M$. This fit gives for each atom a theoretical estimate of the intensity used in the experiment. In Fig. VI, the sets of points (N_k,I_k) for Xe, Kr, Ar, U, I have been drawn on a same curve. The intensity scale is different for different atoms. Arrows mark intensity 10^{14} W cm^{-2} for each atom. The convergence of the results supports the assumption that multiphoton stripping of complex atoms can be explained by a same mechanism in light atoms as Ar or by heavy atoms as U.

In the study of multiphoton ionisation, strong field effects appear as deviation from power laws. However, it is not easily detected since, in most cases, power laws are not observed an a large range of intensity

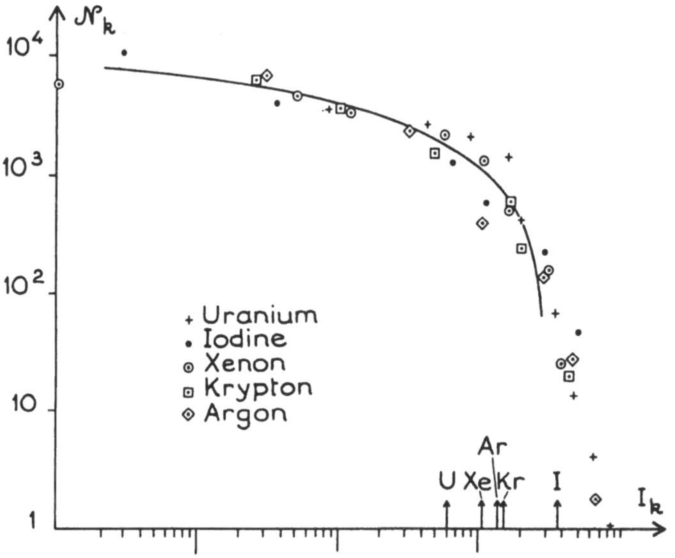

Figure VI

In the study of multiphoton ionisation, strong field effects appear as deviation from power laws. However, it is not easily detected since in most cases, power laws are not observed in a large range of intensity. Saturation can mask a departure from weak field behaviour (18). Strong field effects are essentially due to light shifts which, at lowest order are proportional to the field intensity. When ionisation is observed in a rather large range of intensity, it may happen that light shifts are small for the lowest limit of the range but large for the highest intensity. It may be the case for recent experiments carried out on Calcium (19). For excitation at 1.06μ, intensity has been varied between 7 10^{11} and 10^{13} W cm^{-2}; For excitation at 0.53μ, it has been varied between 10^{11} and 10^{12} W cm^{-2}. We reproduce on Fig.VII, the experimental curves giving the yield of Ca^+, Ca^{++}, Ca^{+++}. Saturation intensities for the formation of Ca^+ and Ca^{++} have been marked by arrows. As far as weak field yields are concerned, theoretical intensities for the formation of Ca^{++} are compatible with experimental observations while saturation intensities for the formation of Ca^+ seem to be lower than experimental ones. Due to the collection set up, we expect $V(I_M,I)$, in the latter experiment, to be close to $\mathrm{Ln}(I_M/I)$. For a given number of photons absorbed, it is possible to calculate the expected curve $N(I_M)$ which reduces to $\mathrm{Ln}(I_M/I)$ for $I_M > I_s$ (10,18). For both experiments (0.53μand 1.06μ), it is not possible to fit experimental points with such a curve. This can be taken as the indication of strong field effects: the saturation intensity which holds for low intensity part of the curve differs from the saturation intensity which governs the high intensity part of the curve. Theoretical saturation intensities can also be used to interpret the results obtained in the high intensity range. In this frame, the number of Ca^+ is predicted to be proportional to $Ln(I_1/I_2)$ once $I_M > I_2$, the number of Ca^{++} is predicted to be proportional to $Ln(I_M/I_2)$. For both wavelengths, the number of Ca^+ becomes constant in the high intensity range of the curve, while the number of Ca^{++} grows with intensity , approximately as $Ln(I_M)$. From theoretical saturation intensities, we predict that the number of Ca^+ and Ca^{++} created be equal at an intensity of $1.2\ 10^{13}$ W cm^{-2}, for excitation at 1.06μ. This is in agreement with experimental data. For excitation at 0.53μ, we predict a value of 1.5 for the ratio Ca^+/Ca^{++} while the experimental value is about 2.5 at the highest studied intensity.

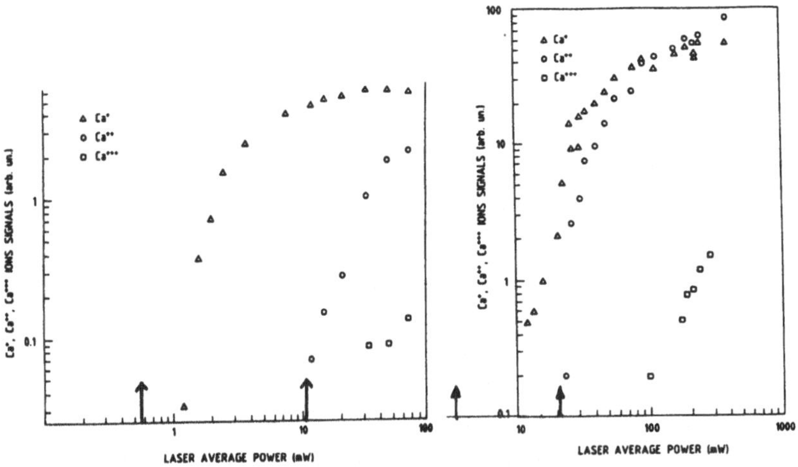

Figure VII

The good agreement between experimental data obtained for high intensities and the predictions of our model show that the behaviour of a complex atom in strong field depends essentially on the characteristics of the initial state. This conclusions set two problems to solve. At first, there is a need for non perturbative method to investigate strong field ionisation of complex atoms. Secondly, it would be worthwhile to calculate , at least in one case, the generalised cross section for single ionisation of a noble gas, in order to check whether the observed power law corresponds to weak field cross section. In this perspective, the results obtained years ago on Krypton (20) are particularly significant. Resonance features have been observed on the effective order of non-linearity, while no resonance could be detected on the ion yield.

References

1 Morellec J., Normand D., Petite G. Phys. Rev.A14 300 (1976)

2 Lompré L.A., Mainfray G., Manus C.,Thebault J. J.Physique 39 610 (1978)
Agostini P.,Georges A.T.,Wheatley S.E., Lambropoulos P.,Levenson M.D. J.PhysB 11 1733 (1978)
Crance M. J.Phys. B11 1931 (1978)
Crance M. J.Phys. B12 3655 (1979)
Crance M.,Aymar M. J.Phys. B12 3665 (1979)

Crance M. J.Phys. B13 101 (1980)

3 Petite G., Morellec J., Normand D. J.Physique 40 115 (1979)

Kruit P., Kimman J., Muller H.G., Van der Wiel M.J. Phys.Rev. A28 248 (1983)

4 Petite G.,Fabre F.,Agostini P.,Crance M.,Aymar M. Phys.Rev. A29 2677 (1984)

5 L'Huillier A.,Lompré L.A.,Mainfray G.,Manus C. J.Phys. B16 1363 (1983)

6 Luk T.S.,Pummer H.,Boyer K.,Shahidi M.,Egger H.,Rhodes C.K. Phys.Rev.Lett 51 110 (1983)

Aleksakin I.S.,Delone N.B., Zapesochnyi I.P.,Suran V.V. Sov.Phys.JETP49 447 (1979)

Feldman D.,Krautwald J.,Chin S.L.,von Hellfeld A.,Welge K.H. J.Phys.B15 1663 (1982)

L'Huillier A.,Lompré L.A.,Mainfray G.,Manus C. Phys.Rev.Lett. 48 1814 (1982)

L'Huillier A.,Lompré L.A.,Mainfray G.,Manus C. Phys.Rev. A27 2503 (1983)

Lompré L.A.,L'Huillier A.,Mainfray G.,Fan J.Y. J.Phys. B17 L817 (1984)

7 Crance M. J.Phys. B17 4323 (1984)

8 Crance M. J.Phys. B13 L421 (1980)

9 Chu S.I.,Reinhardt W.P. Phys.Rev.Lett. 39 1195 (1977)

10 Crance M.,Aymar M. J.Phys.B (1985) to be published

11 Chan F.T.,Tang C.L. Phys.Rev. 185 42 (1969)

12 Crance M. J.Phys. B17 4333 (1984)

13 Crance M. J.Phys. B17 3503 (1984)

14 Chin S.L.,Yergeau F.,Lavigne P. J.Phys. B to be published

15 Bayfield J.E.,Gardner L.D.,Koch P.M. Phys.Rev.Lett. 39 76 (1977)

16 Crance M. J.Phys. B to be published

17 Boyer K.,Egger H.,Luk T.S.,Pummer H.,Rhodes C.K. J.Opt.Soc.Am. B1 1 (1984)

18 Crance M.,Aymar M. J.Phys. B12 L667 (1979)

Lompré L.A.,Mainfray G.,Mathieu B.,Watel G.,Aymar M.,Crance M. J.Phys. B13 1799 (1980)

19 Agostini P.,Petite G. J.Phys. B17 L811 (1984)

20 Lompré L.A.,Mainfray G.,Manus C. J.Phys. B13 85 (1980)

CORRELATION IN SINGLE- AND MULTIPHOTON PROCESSES

H. Klar
Fachbereich Physik
Universität Kaiserslautern
D-6750 Kaiserslautern

1. Introduction

Atomic theory is rooted in the independent-electron model which regards each electron as moving in the combined field of the nucleus and of the average distribution of the other electrons. This picture is particularly successful as long as closed shells are considered. Correlation effects are then largely suppressed, but become important or even dominant if this constraint is relaxed. An external radiation field (laser or synchroton radiation) for instance is a very sensitive probe for correlation effects.

An early manifestation of the breakdown of the independent-electron model emerged from the observation of the first full series of doubly excited $^1P^o$ helium levels (Madden and Codling 1963, 1965). Based on the independent-electron model one would expect three Rydberg series converging to $He^+(n=2)$:

$$\left.\begin{matrix} 2snp \\ 2pns \\ 2pnd \end{matrix}\right\} {}^1P^o \quad .$$

Moreover, the two series 2snp and 2pns should show comparable intensities. The 2pnd series may be weaker because both electrons must absorb angular momentum from the radiation field, whereas in the first two series one electron remains on s-electron. The experiment instead showed one single intense series and a much weaker one. This finding was regarded as a new regularity in the joint excitation of an electron pair free from the confinement of a filled shell. An interpretation of these results (Cooper et al 1963) regarded the new doubly excited states as superpositions of 2snp and 2pns configurations

$$|(2n)sp\pm\rangle = \frac{1}{\sqrt{2}} \left\{ |2snp\rangle \pm |2pns\rangle \right\} . \qquad (1)$$

An even more conspicuous correlation effect occurs in the double photodetachment of a negative ion like H^-,

$$\gamma + H^- \rightarrow p + 2e^- \qquad (2)$$

near threshold. Uncorrelated continuum electrons described by Coulomb waves would lead for the above reaction to a total cross section being proportional to the excess energy. This linear dependence results from the phase space available for the electron pair. An experimental investigation (Donahue et al 1982) showed however a nonlinear threshold law. We shall see in this report how such correlation effects result from theory. We confine ourselves here to two-electron atoms but formal extensions of the following treatment to more particles is straightforward.

2. Hyperspherical coordinates

A key property of the wavefunctions (1) is their nodal behaviour. The plus-state has an antinode at $r_1 \approx r_2$ whereas the minus-state has a node line there. One may therefore look for collective variables which describe such a nodal picture in a simple way.

One possibility consists in introducing polar coordinates in a rectangular r_1-r_2-plot:

$$r_1 = R\cos\alpha,\ r_2 = R\sin\alpha \tag{3}$$

or

$$R = \sqrt{r_1^2 + r_2^2},\quad \alpha = tg^{-1}(r_2/r_1). \tag{4}$$

The angular positions $\hat{r}_1$ and $\hat{r}_2$ may be added,

$$\begin{bmatrix} \vec{r}_1 \\ \vec{r}_2 \end{bmatrix} = R \begin{bmatrix} \cos\alpha\ \hat{r}_1 \\ \sin\alpha\ \hat{r}_2 \end{bmatrix} . \tag{5}$$

The six coordinates (R, α, ϑ_1, φ_1, ϑ_2, φ_2) are usually called hyperspherical coordinates (Morse and Feshbach 1953, Fock 1958). Often another set of angles in eq. (5) may be advantageous (Smith 1962, Dragt 1965, Klar and Klar 1978, 1980, Pelikan and Klar 1983), the hyperradius R however is not changed in these treatments. Formal extensions to more than two electrons are straightforward (Knirk 1974, Erdêlyi et al 1953). In the following we restrict ourselves to two-electron systems and use the above Fock coordinates. The stationary Schrödinger equation reads then in atomic units

$$\left[\frac{1}{2}\left(-\frac{\partial^2}{\partial R^2}+\frac{\Lambda^2}{R^2}\right)+V(R,\ \alpha,\ \hat{r}_1,\ \hat{r}_2)-E\right]\Psi(R,\ \alpha,\ \hat{r}_1,\ \hat{r}_2)=0 \tag{6}$$

where the operator

$$\Lambda^2 = -\frac{\partial^2}{\partial\alpha^2} - \frac{1}{4} + \frac{\vec{l}_1^{\,2}}{\cos^2\alpha} + \frac{\vec{l}_2^{\,2}}{\sin^2\alpha} \tag{7}$$

is called the "grand angular momentum" (Smith 1960). The usual wave function ψ has been renormalized setting

$$\Psi = R^{5/2} \sin\alpha \cos\alpha\ \psi \tag{8}$$

to eliminate first derivatives in R and α.

The grand angular momentum consists of the sum of the squared generators for rotations in a six-dimensional space. It may also be regarded as the orbital angular momentum of a single particle (the electron pair) in six dimensions. This operator contains the ordinary orbital angular momenta $\vec{l}_1$, $\vec{l}_2$ as well as derivatives with respect to α. The eigenvalues of Λ^2 are

$$\lambda(\lambda+4) + \frac{15}{4} = (\lambda+2)^2 - \frac{1}{4} > 0 \tag{9}$$

with $\lambda = 0, 1, 2, \ldots$. These eigenvalues are highly degenerate. The eigenfunctions of Λ^2, hyperspherical harmonics, may be expressed in terms of Jacobi polynomials and ordinary spherical harmonics,

$$\phi_{nl_1l_2LM} = \cos^{l_1}\alpha \sin^{l_2}\alpha \, F(-n, n+l_1+l_2+2; l_2+\tfrac{3}{2}; \sin^2\alpha) \times \sum \langle LM | l_1m_1l_2m_2 \rangle Y_{l_1m_1}(\hat{r}_1) Y_{l_2m_2}(\hat{r}_2). \tag{10}$$

Here the radial correlation quantum number n counts the nodes in the α coordinate. The connection with λ reads

$$\lambda = 2n+l_1+l_2. \tag{11}$$

The Pauli principle allows only the subset of symmetrized hyperspherical harmonics with the symmetry

$$\phi_{nl_1l_2LM}(\tfrac{1}{2}\pi-\alpha, \hat{r}_2, \hat{r}_1) = (-)^S \phi_{nl_1l_2LM}(\alpha, \hat{r}_1, \hat{r}_2) \tag{12}$$

where S = 0, 1 is the total spin of the electron pair.
The potential energy term in eq. (6) factors out in hyperspherical coordinates,

$$V(R, \alpha, \hat{r}_1, \hat{r}_2) = \frac{1}{R} C(\alpha, \vartheta) \tag{13}$$

with

$$\vartheta = \cos^{-1}(\hat{r}_1 \cdot \hat{r}_2) \tag{14}$$

$$C(\alpha, \vartheta) = -\frac{Z}{\cos\alpha} - \frac{Z}{\sin\alpha} + \frac{1}{\sqrt{1 - \sin 2\alpha \cos\vartheta}} . \tag{15}$$

Substituting the expansion

$$\Psi(R, \alpha, \hat{r}_1, \hat{r}_2) = \sum_{nl_1l_2} F_{nl_1l_2}(R) \, \phi_{nl_1l_2LM}(\alpha, \hat{r}_1, \hat{r}_2) \tag{16}$$

into eq. (6) we obtain a set of coupled hyperradial equations,

$$\frac{1}{2}\left(-\frac{d^2}{dR^2} + \frac{(\lambda+2)^2 - \frac{1}{4}}{R^2} - 2E\right) F_{nl_1l_2}(R) + \frac{1}{R} \sum_{n'l_1'l_2'} C_{nl_1l_2n'l_1'l_2'} \, F_{n'l_1'l_2'}(R) = 0. \tag{17}$$

Here $\underline{C}$ is a real symmetric matrix formed from the function $C(\alpha, \vartheta)$ between hyperspherical harmonics.
It should be stressed that the system (17) is exact. The expansion (16) converts the Schrödinger equation into a linear, one-dimensional system without continuum since the spectrum of Λ^2 is discrete.

Basically, there are two alternative ways to solve the system (17): The first one uses an adiabatic expansion (Macek 1968). This method is now standard to attack the correlation problem, and will briefly described below. The second method uses the fact that systems of the kind (17) may be solved mathematically exactly (Fock 1958, Klar 1985) in terms of generalized power series expansions. Here however the formulation of asymptotic boundary conditions is very difficult, a matching to adiabatic channels at some fixed radius appears convenient (Feagin et al 1983).

3. Adiabatic channels

The picture of adiabatic channels is based on the fact that the Coulomb term C/R varies slowly with respect to the generalized centrifugal barrier Λ^2/R^2 along the collective coordinate R. It is therefore natural to diagonalize in a first step this part of the Hamiltonian,

$$\left[\frac{\Lambda^2}{2R^2}+\frac{C}{R}\right]\phi_\mu(\alpha,\hat{r}_1,\hat{r}_2;R) = U_\mu(R)\,\phi_\mu(\alpha,\hat{r}_1,\hat{r}_2;R). \tag{18}$$

The eigenvalues U_μ (R) are discrete, labelled by an index μ, and depend parametrically on the hyperradius R. Substituting now Macek's (1968) channel expansion

$$\Psi(R,\alpha,\hat{r}_1,\hat{r}_2) = \sum_\mu F_\mu(R)\,\phi_\mu(\alpha,\hat{r}_1,\hat{r}_2;R) \tag{19}$$

in eq. (6) we find

$$\sum_\nu -\frac{1}{2}\left(\frac{d}{dR}+\underline{P}\right)^2_{\mu\nu} F_\nu(R) + U_\mu(R)F_\mu(R) = EF_\mu(R) \tag{20}$$

where the matrix $\underline{P}$ couples the adiabatic channels,

$$P_{\mu\nu}(R) = \int d\omega\,\phi_\mu^*\,\frac{\partial}{\partial R}\,\phi_\nu = -P_{\nu\mu}(R). \tag{21}$$

Eq. (20) suggests to regard the eigenvalues U_μ (R) as potential controlling the motion of the electron pair along the coordinate R. Often the coupling matrix elements $P_{\mu\nu}$ are small in the sense

$$P_{\mu\nu}{}^2 \ll |U_\mu(R) - U_\nu(R)| \tag{22}$$

so that eq. (20) may be replaced by decoupled equations

$$-\frac{1}{2}F_\mu(R)'' + U_\mu(R)\,F_\mu(R) = EF_\mu(R). \tag{23}$$

In this approximation the whole wavefunction factorizes into a product

$$\Psi(R,\alpha,\hat{r}_1,\hat{r}_2) \simeq F_\mu(R)\,\phi_\mu(\alpha,\hat{r}_1,\hat{r}_2;R). \tag{24}$$

This adiabatic approximation is formally analogous to the Born Oppenheimer approximation in molecular physics. It must however be stressed that the problem here under consideration has no small kinematical parameter like a mass ratio which enforces the quasiseparability. This adiabatic approach rests on the slow variation of Coulomb potentials.

Let us now briefly discuss the behaviour of potentials U_μ (R) defined by eq. (18). For small values of R the centrifugal barrier dominates,

$$\lim_{R\to 0} U_\mu (R) = \frac{(\lambda+2)^2 - \frac{1}{4}}{2R^2} + O(R^{-1}). \tag{25}$$

For large values of R the normalizability of channel functions selects one-electron bound-states (Macek 1968, Lin 1974, Klar et al 1978, Pelikan et al 1983)

$$\lim_{R\to\infty} U_\mu (R) = - \frac{Z^2}{2n^2} - \frac{(Z-1)}{R} + O(R^{-2}). \tag{26}$$

Numerical work along these lines has been performed for He (Macek 1968, Lin 1974, Klar and Klar 1980), for H^- (Lin 1975, Klar and Klar 1978). This work and further applications to alcali earth atoms, to negative alcalis as well as to He^- have been reviewed by Fano (1983). More recently the system e^+H has been treated (Pelikan and Klar 1983). Photoionisation of He (Miller and Starace 1984) and two-photon absorption from H^- (Fink 1985) have also been considered.

Photoabsorption from the helium ground state leads in nonrelativistic approximation adopted here to the $^1P^o$ symmetry. The lowest channel potential carries the (1snp) Rydberg series whereas the next three channels converge to $He^+(n=2)$, see Fig. 1. Inspection of the channel functions at constant values of R indead shows for the lowest potential (μ =2) in Fig. 1 an antinode in α at $\alpha \approx 45^o$ and a node for the next channel (μ =3) at the same position. This proves the plus/minus classification by Cooper et al (1963) by direct calculation. The lower potential (μ =2) carries the 23sp+ series of autoionising states, and the next curve (μ =3) carries the 23sp- series. The weaker intensity of the minus series results from smaller dipole matrix elements computed with a wavefunction having a node line. Potential curves for H^- ($^1P^o$) converging to H (n=2) are shown in Fig. 2. For large values of R these curves are controlled by the e^--H dipole interaction. The pd-channel is entirely repulsive and carries no resonance at all. The plus-channel is more attractive than the minus-channel as in He, and has in contrast to He a barrier. The crossing near R $\approx$13.5 Bohr is important. In this region the channel coupling is large. A diabatic basis is more suitable in this situation, i.e. we assume that the curves do cross (exactly they don't do but avoid a crossing) and disregard the coupling there. The lower curve carries an infinite number of Feshbach resonances converging very rapidly to threshold such that only the lowest one is well isolated and observable in photoabsorption. The barrier above threshold carries one shape resonance. Both resonances have been observed experimentally (Bryant et al 1977),

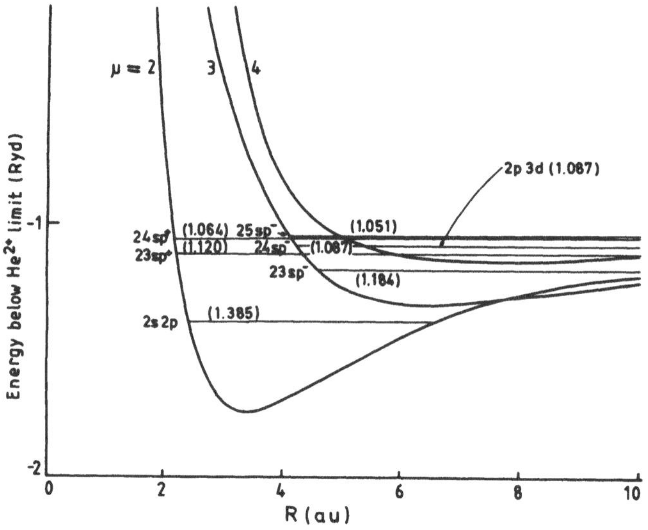

Fig. 1. Eigenvalues U_μ (R) for He $^1P^o$ channels converging to $He^+(n=2)$. Horizontal lines represent eigenvalues of uncoupled radial equations (23) (Macek 1968)

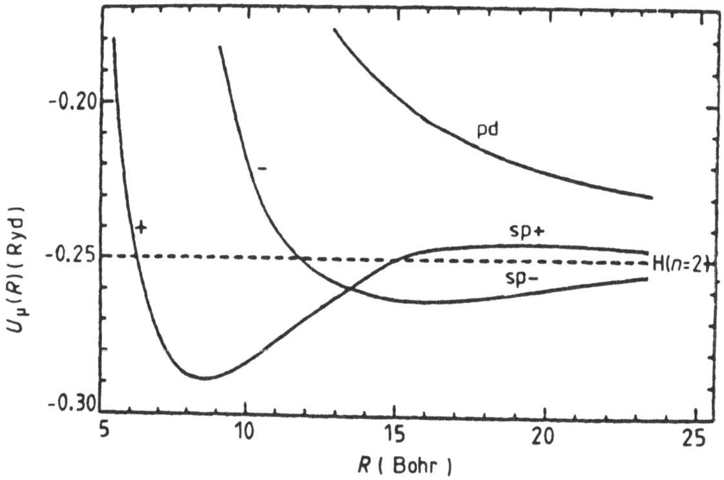

Fig.2. Potential curves $U_\mu(R)$ for H^- $^1P^o$ converging to H(n=2) (Lin 1975)

see also Fig. 3. The broad structure at about 10.98 eV is the shape resonance, the narrower structure at slightly lower energy is the Feshbach resonance.

4. Double photoionisation near threshold

An atom or negative ion may absorb one or several photons from a radiation field such that two electrons may leave the atomic system. The simplest example was mentioned in eq. (2).

In this situation two slow electrons are moving in the field of an ion or a nucleus. An adiabatic approximation breaks down here because many adiabatic channels accumulate at threshold for two continuum electrons. A theoretical treatment that describes the correlated motion of a slow electron pair in the field of an ion was first developed by Wannier (1953) in the framework of classical mechanics. From a careful inspection of the classical equations of motion he realized that double escape at low energy can be attained only if and only if both electrons remain equidistant from the ionic core and on opposite sides of it until their potential energy was fallen below their residual kinetic energy. In other words, the pathway to double escape proceeds in the subspace of coordinates

$$\{(R,\ \alpha,\ \vartheta),\ |\tfrac{\pi}{4}-\alpha|\ll 1 \text{ and } |\pi-\vartheta|\ll 1\}. \tag{27}$$

In that subspace the potential, see eq. (15) except for a factor 1/R, has the Taylor series expansion

$$C(\alpha,\vartheta) \simeq -\frac{4Z-1}{\sqrt{2}} - \frac{12Z-1}{8\sqrt{2}}\left(\frac{\pi}{4}-\alpha\right)^2 + \frac{1}{2\sqrt{2}}(\pi-\vartheta)^2 + \ldots \tag{28}$$

This expansion shows clearly that the function $C(\alpha,\vartheta)$ has a saddle point at $\alpha=\pi/4$ and $\vartheta=\pi$, the potential itself has a saddle line along R. Investigation of orbits leading to double escape near threshold requires therefore a solution within the cone $|\pi/4-\alpha|\ll 1$ and $|\pi-\vartheta|\ll 1$. The unstable motion along α depresses the probability for double escape. Trajectories outside the cone $|\pi/4-\alpha|\ll 1$ corresponding to $r_1\gg r_2$ (or $r_2\gg r_1$) lead to single escape with or without excitation. Any trajectory of the electron pair leading to double escape can be represented as a linear superposition of two basic trajectories

$$\left(\frac{\pi}{4}-\alpha\right)^2 \propto R^{2\zeta_{1\pm}} \text{ and } (\pi-\vartheta)^2 \propto R^{2\zeta_{2\pm}}. \tag{29}$$

The characteristic exponents are roots of two simple algebraic equations,

$$\zeta_{1\pm} = -\frac{1}{4} \pm \frac{1}{2}\mu,\ \zeta_{2\pm} = -\frac{1}{4} \pm \frac{i}{2}\rho \tag{30}$$

with

$$\mu = \frac{1}{2}\sqrt{\frac{100Z-9}{4Z-1}} \tag{31}$$

$$\rho = \frac{1}{2}\sqrt{\frac{9-4Z}{4Z-1}} . \tag{32}$$

As shown by Wannier the total cross section has near threshold the energy slope

$$\sigma(E) \propto E^{\frac{1}{2}\mu - \frac{1}{4}} . \tag{33}$$

Quantum mechanical formulations in WKB approximation leading to the same result (33) were presented by Peterkop (1971) and Rau (1971). The two-electron continuum wave function corresponding to Wannier's orbits has the structure

$$\Psi(R, \alpha, \vartheta) \propto R^{\nu} \exp\{iS_0(R) + iS_1(R)(\tfrac{\pi}{4} - \alpha)^2 + iS_2(R)(\pi - \vartheta)^2\}. \tag{34}$$

Photoabsorption leads to a final atomic state with definite spin/parity symmetry. Wannier considered single ionisation by electron impact and assumed that an S-state for the electron pair is dominant at threshold. Klar and Schlecht (1976) showed that ^1S and ^3S symmetries lead to different threshold laws. Wannier's law (33) is quantum mechanically based on a symmetric wave function (34) and describes therefore ^1S events. The threshold law for ^3S events reads

$$\sigma(E) \propto E^{\frac{3}{2}\mu - \frac{1}{4}} .$$

Absorption of one photon by a two-electron atom in its ground state (^1S) leads to the $^1P^o$ symmetry. Also this situation is controlled by the Wannier law (33), see also Greene and Rau (1982), and Stauffer (1983).

The Wannier law is in excellent agreement with experimental data (Donahue et al 1982). Fig. 4 shows the total cross section for the reaction (2). The nonlinear rise is evident. The result for the exponent is

$$\frac{1}{2}\mu - \frac{1}{4} = 1.15 \pm 0.04, \tag{35}$$

the theoretical value from (31) is 1.127.

The Wannier theory has not yet been fully tested by photoabsorption measurements. Such a test would require to detect the two escaping electrons in coincidence and to measure their energy distribution and their angular correlation. According to the theory the energy distribution should be flad, increasingly so for decreasing frequency of the light. The angular correlation should show a Gauß distribution,

$$I(\vartheta) = I_0 \exp\{-a(\pi - \vartheta)^2\} \tag{36}$$

where the constant a is known and proportional to $E^{-1/2}$. The main features of the Wannier theory have been summarized and compared with all available experimental data by Klar (1983, 1984).

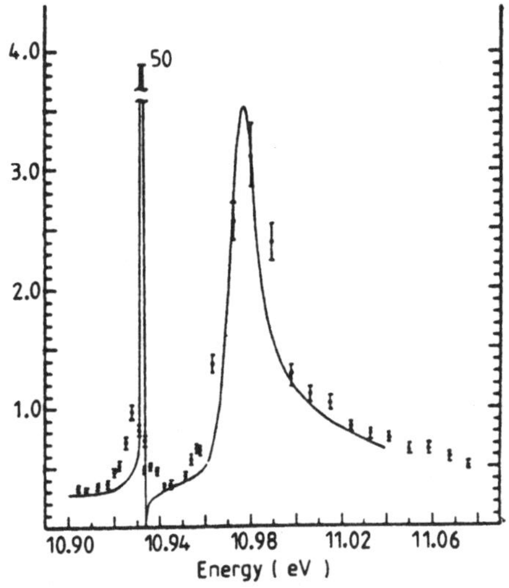

Fig. 3. Photodetachment cross section for H^- (Bryant et al 1977). The solid curve shows a calculation by Broad and Reinhardt (1976).

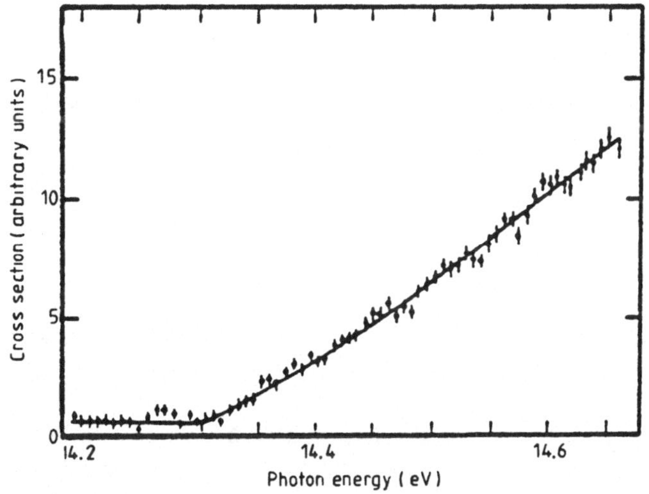

Fig. 4. Cross section for the process $h\nu + H^- \rightarrow p + 2e$ (Donahue et al 1982). The threshold is at $h\nu = 0.75 + 13.60 = 14.35$ eV.

5. Outlook

So far we have considered correlation effects which manifest themselves in lowest order perturbation theory for the radiation field. In other words: A weak laser field is a sensitive probe for these effects.

Electronic correlation in strong laser fields has been considered until now only very little. Klar et al (1984) have recently shown how an intense radiation field may be incorporated into the adiabatic channel frame of § 3. It is clear that a dipole field for instance mixes adiabatic channels of different angular momentum/ parity symmetries, i.e. whole series of autoionizing Rydberg series are mixed. The resulting potential curves which depend now on the field strength may be very different from the corresponding zero-field curve due to avoided crossings. As an example is was shown that in He^{-**} new laser-induced resonances may appear which are absent in the atomic spectrum.

An alternativ way to describe the collective motion of an electron pair in a strong radiation field is to generalize the Fock (1958) expansion incorporating the radiation field. Let us consider the time-dependent wave equation for helium in a linearly polarized dipole field

$$H \Psi = i\dot{\Psi} \tag{37}$$

with

$$H = H_0 - \varepsilon_0(z_1+z_2) \cos\omega t, \tag{38}$$

H_0 being the atomic Hamiltonian. Introducing hypershpherical coordinates as in § 2 and expanding the wavefunction now according to

$$\Psi(R, \alpha, \hat{r}_1, \hat{r}_2, t) = \sum_{nl_1l_2LMN} F_{nl_1l_2LMN} (R)\, \phi_{nl_1l_2LM} (\alpha, \hat{r}_1, \hat{r}_2) e^{-i(E+N\omega)t} \tag{39}$$

which generalizes eq. (16) a system of coupled radial equationsis obtained. This system generalizes eq. (17) and has in obvious matrix notation the form

$$\{-\frac{1}{2} \frac{d^2}{dR^2} + \frac{1}{2R^2} \underline{\Lambda}^2 - E - \underline{N}\omega + \frac{1}{R} \underline{C} - R\underline{D}\} \vec{F}(R) = 0 \tag{40}$$

where $\underline{D}$ is a constant matrix. Exact solutions of eq. (40) can be represented by generalized power series expansions at the hyperspherical origine R=0,

$$\vec{F}(R) = \sum_n \vec{\varphi}_n (s) R^n \tag{41}$$

with s=ln R (Klar and Zoller 1985). The key point is that exact solutions cannot be represented by simple Taylor expansions, i.e. by constant vectors $\vec{\varphi}_n$. The series (41) converges absolutly in each domain of the R-plane cut along the negative real axes. Because the convergence becomes however slow for larger and larger values of R these formal solutions should be matched at some finite radius to asymptotic solutions.

References:

J.T. Broad, W.P. Reinhardt, Phys. Rev. 14, 2159 (1976)
H.C. Bryant, B.D. Dieterle, J. Donahue, H. Sharifian, H. Tootoonchi, D.M. Wolfe, P.A.M. Granc, M.A. Yates-Williams, Phys. Rev. Lett. 38, 228 (1977)
J.W. Cooper, U. Fano, F. Prats, Phys. Rev. Lett. 10, 518 (1963)
J.B. Donahue, P.A.M. Granc, M.V. Hynes, R.W. Hamm, C.A. Frost, H.C. Bryant, K.B. Butterfield, D.A. Clark, W.W. Smith, Phys. Rev. Lett. 48, 1538 (1982)
A.J. Dragt, J. Math. Phys. 6, 533 (1965)
A. Erdélyi, W. Magnus, F. Oberhettinger, F.G. Tricomi, Higher Transcendental Functions, Vol. 2, McGraw-Hill, New York (1953)
U. Fano, Rep. Prog. Phys. 46, 97 (1983)
J. Feagin, J.H. Macek, A.F. Starace, Electronic and atomic collisions, XIII. ICPEAC Berlin 1983, Abstracts of contributes papers, edited by J. Eichler et al, p. 99, Int. Conf. on the Physics of Electronic and Atomic Collisions e.V., Berlin (1983)
M. Fink, to be published (1985)
V. Fock, K. Norske Vidensk. Selsksk. Forh. 31, 138 (1958)
H. Klar, Fundamental processes in energetic atomic collisions, ed. by H.O. Lutz et al, p. 483 ff., Plenum, New York (1983)
H. Klar, Electronic and atomic collisions, ed. by J. Eichler et al, p. 767 ff., Elsevier Science Publishers B.V. (1984)
H. Klar, J. Phys. A (1985 in press)
H. Klar, M. Klar, Phys. Rev. A 17, 1007 (1978)
H. Klar, M. Klar, J. Phys. B 13, 1057 (1980)
H. Klar, P. Zoller, M.V. Fedorov, Phys. Rev. A 30, 658 (1984)
H. Klar, P. Zoller, to be published (1985)
D.L. Knirk, J. Chem. Phys. 60, 66 (1974)
D.L. Knirk, J. Chem. Phys. 60, 760 (1974)
D.L. Knirk, Phys. Rev. Lett. 32, 651 (1974)
C.D. Lin, Phys. Rev. A 10, 1986 (1974)
C.D. Lin, Phys. Rev. A 12, 493 (1975)
D.C. Lin, Phys. Rev. Lett. 35, 1150 (1975)
J. H. Macek, J. Phys. B 1, 831 (1968)
R.P. Madden, K. Codling, Phys. Rev. Lett. 10, 516 (1963)
R.P. Madden, K. Codling, Astrophys. J. 141, 364 (1965)
D.L. Miller, A.F. Starace, J. Phys. B 13, L525 (1980)

E. Pelikan, H. Klar, Zeit. Phys. A 310, 153 (1983)

F.T. Smith, Phys. Rev. 118, 349 (1960)

F.T. Smith, Phys. Rev. 120, 1058 (1960)

F.T. Smith, J. Math. Phys. 3, 735 (1962)

PART III: Laser Spectroscopy

RADIATION INTERACTION OF RYDBERG ATOMS AND THE ONE-ATOM MASER

Gerhard Rempe und Herbert Walther
Sektion Physik, Universität München and
Max-Planck-Institut für Quantenoptik,
D-8046 Garching, Fed. Rep. Germany

I. Introduction - Properties of Rydberg Atoms

Rydberg atoms or Rydberg states are obtained when a valence electron of an atom is excited into an orbit with sufficiently high principal quantum number n and therefore far from the ionic core. The energy of these highly excited levels is given by the Rydberg formula; this is the reason that they are called Rydberg states of Rydberg atoms. For atoms other than hydrogen the main quantum number in the Rydberg formula has to be replaced by $n^* = n-\delta_\ell$, where δ_ℓ is the phenomenological quantum defect which considers the influence of the innerelectronic core on the valence electron. The quantum defect depends on the angular momentum l. For states of low ℓ, where the orbits of the classical Bohr-Sommerfeld theory are ellipses of high eccentricity, the penetration and polarization of the electron core by the valence electron lead to large quantum defects and strong departures from the hydrogenic behaviour. As ℓ increases, the orbits become more circular and the atom becomes more hydrogenic, δ_ℓ changing with ℓ^{-5}.

The energy changes among highly excited states are small compared with the large changes between the lower levels. Since smooth changes are characteristic of classical systems, Rydberg atoms can be expected to show classical properties. In particular, according to Bohr's correspondence principle, the frequency of electromagnetic radiation emitted for transitions between neighbouring states approaches the frequency at which the electron circulates around the ionic core. This suggests that many properties of these atoms can be understood on simple classical terms. Nevertheless, some very surprising properties of Rydberg atoms have recently been found, which has led to a steady increase in the number of experiments being performed on these atoms. The main reasons for the growing interest are:

a) The outer electron is a very good probe for the interatomic potential; quantum defects due to the penetration and polarization of the electron core are therefore being investigated as well as fine-structure and hyperfine-structure splittings.

b) Radiation interaction is large and different from that of ground state atoms owing to the large matrix elements for transitions to neighbouring levels; radiation-induced effects overcome spontaneous emission. Therefore Rydberg atoms in high-n states become sensitive to black-body radiation, and maser emission with only a small absolute number of radiators can also be observed. Observation and study of these effects allows testing of fundamental theories on light-matter interaction, which is not possible with ordinary atoms.

c) Collisional interaction becomes very important owing to the size of the atoms; its influence shows strong dependence on the main quantum numbers. It is thus found, for example, for the collisional angular momentum mixing in the low-n region that the cross-section increase is proportional to the geometric size of the atoms, i.e. n^4. As the size of the Rydberg orbit increases further, the electron distribution becomes very diffuse and the cross-section decreases.

d) The binding energy of the outer electron of a Rydberg atom is very small. Therefore external electric and magnetic fields show a very large influence even at small field strengths. The observation of similar effects for ground state atoms would require fields which are not attainable in the laboratory.

<u>Table I:</u> Scaling laws for properties of Rydberg atoms

Energy:	$E_n = R/(n-\delta_\ell)^2 = R/n^{*2}$ δ_ℓ quantum defect, n^* effective quantum number R Rydberg constant
Radius:	$\langle r\rangle \sim n^{*2}$
Lifetimes:	$\tau \sim n^{*3}$ (low angular momentum states) $\tau \sim n^{*5}$ (high angular momentum states)
Fine-structure interval:	$\Delta E \sim 1/n^{*3}$

In Table I the scaling laws for the properties of Rydberg atoms are compiled: The radius of the charge distribution of the valence electron scales as n^{*2} and for $n^* = 50$ the linear dimension of the atom is already comparable with the wavelength of light in the visible region and competes with the size of larger biomolecules.

The electric polarizability for the quadratic Stark effect increases as n^{*7} and the diamagnetic interaction as n^{*4}. This allows one to perform experiments at field strengths high enough to make the interaction energy in the external electric or magnetic field comparable with or larger than the Coulomb energy of the atom. For practical reasons the corresponding field strengths for ground state atoms cannot be reached in the laboratory. The study of highly excited atoms in external electric and magnetic fields is therefore interesting in itself. (For reviews see References [1-4])

The sensitivity of Rydberg atoms to external electric fields also means that the atoms already ionize in rather weak fields. This opens the possibility of a very effective detection, as will be discussed later.

The large Rydberg atom orbitals are characterized by natural lifetimes much longer than the ones of less excited atoms. In the case of hydrogen Rydberg states, the dependence of the lifetime on n can be obtained by fully quantum mechanical radiation rate calculations involving hydrogenic coulombic wavefunctions. For Rydberg states of other species the lifetimes (and the other radiative parameters) scale not exactly as a power of n but rather as a power of n^*. The n^* scaling law can be determined using Bates and Damgaard type of calculations[5]. The lifetimes scale either with n^{*3} (when ℓ is small) or with n^{*5} (when $\ell \cong n$).

The rate of spontaneous emission of radiation for a transition from a state n to n' is given by the Einstein A coefficient:

$$A_{n\to n'} = 16\pi^3\upsilon^3 \langle r_{nn'}\rangle^2/3\varepsilon_o hc^3,$$

where υ is the transition frequency and $\langle r_{nn'}\rangle$ the matrix element of the electric dipole operator between the initial n and the final state n'. For the case $n' \ll n$ one has a small matrix element for dipole transitions $\langle r_{nn'}\rangle$ owing to the small overlap of the radial wave functions for n and n', $A_{n\to n'} \sim n^{-3}$. If n' is close to n, the energy difference $E_n - E_{n'} \sim n^{-3}$ and $\langle r_{nn'}\rangle^2 \sim n^4$, and so $A_{n\to n'}$ becomes proportional to n^{-5}.

Since the matrix element $\langle r_{nn'}\rangle$ ($n \cong n'$) scales with n^2, this leads to rather high transition probabilities for induced transitions. Rydberg atoms therefore strongly absorb microwave or far-infrared radiation. As a consequence, blackbody radiation may cause strong mixing of the states. This is especially the case for states with high angular momenta since the spontaneous lifetimes for these are much larger than for low ℓ states and the induced transitions can therefore be saturated more easily.

We now wish to discuss the scaling laws related to blackbody-induced effects. The induced transition rate due to blackbody radiation is proportional to $<r_{nn'}>^2 S_\upsilon$, where S_υ is the energy flux of the blackbody radiation per unit band width and unit surface area. At low frequencies (Rayleigh-Jeans limit) S_υ changes as υ^2. Considering the distance between the Rydberg states to scale as n^3 (here again we perform the discussion with n instead of n*), it is therefore found that S_υ is proportional to n^6. Since $<r_{nn'}>^2 \sim n^4$, it follows that the induced transition rate behaves as n^{-2}. Important in experiments is the ratio between the induced transition rate and the spontaneous rate, which changes as n^{-3} for low ℓ and as n^{-5} for high ℓ. This means that for a given atom and a given temperature there exists an n, above which the blackbody-induced rate overcomes the spontaneous rate.

The sensitivity of Rydberg atoms to blackbody radiation can also be explained in the following terms. The blackbody radiation energy density can be expressed in terms of the average number of photons per mode n. For the Rayleigh-Jeans limit this gives $n = kT/h\upsilon$. At 300 K it follows that $kT/h \cong 6 \times 10^{12}$ Hz, this means that for frequencies larger than kT/h, where $n \ll 1$, no significant blackbody influence can be observed. However, for a Rydberg state with a transition frequency to a neighbouring state at 10^{11} Hz, for which $n = 40$, the blackbody-induced transition rates can be orders of magnitude larger than the spontaneous rates.

In addition to population changes induced by the blackbody radiation, energy shifts of the atomic levels also occur. Their magnitudes depend on the match of the atomic frequencies with the blackbody frequencies and the strength of the coupling of the Rydberg atom to the blackbody radiation[14,15,16].

In absorption spectroscopy with classical light sources only those Rydberg states could be investigated which can be optically excited directly from the ground state. For spectra of atoms with one valence electron this means that only the ^2P series can be studied. The alternative method of populating Rydberg states in an electric discharge and observing the spectrally resolved fluorescence is not practical: At the required particle densities collisional deactivation is much more probable than radiative decay, since it is a result of the large collisional cross-sections of Rydberg states and of the long lifetimes. However, in atomic beam experiments where one can reach collision-free conditions it is possible to use electron bombardment or charge exchange collisions to populate Rydberg states. One drawback of this excitation process, however, is that it is not state-selective.

Most of the limitations discussed above were overcome after the invention of frequency-tunable lasers, little more than a decade ago. This lead to a renaissance of the spectroscopy of highly excited atomic states. The use of lasers to populate high-lying atomic levels in one, two or three excitation steps considerably increased the number of atomic states accessible to experiments. In particular, states with the same parity as the ground state could be reached. For atoms with large ionization potentials it can be advantageous to combine the collisional excitation of metastable states and subsequent laser excitation to Rydberg states[11,17,18]

For high-resolution spectroscopy of Rydberg atoms a low atomic density is also required in order to avoid collisional broadening or a collisional shift. This excludes absorption measurements from the onset since there larger densities are required. The alternative method of observing the fluorescence, however, is not suitable either for $n \geqq 15$ since the n^3 dependence of the radiative lifetime implies a corresponding decrease of the fluorescence intensity. Most experiments, therefore, exploit collisional, photo or field ionization to detect Rydberg atoms.

Field ionization was first observed in hydrogen, where about 10^6 V/cm has to be applied to ionize the $n = 4$ states. In the range of $n = 30$ an electric field of about 300 V/cm is enough to reach the onset of field ionization. The superposition of the atomic Coulomb potential and the linear slope potential of the externaly applied field results in a potential structure having a saddle point. The simplest approach to field ionization is to say that states above the saddle point fully ionize and states below are stable. For a state with principal quantum number n there exists a critical field defined by the onset of field ionization. Using the simple potential picture, one obtains[1]:

$$E_{crit} \sim 1/n^4$$

As long as spectroscopic information about the unperturbed Rydberg atom is wanted, the excitation of the Rydberg state and field ionization have to be separated in time. If an electric field ramp is applied, after a time lag with respect to pulsed laser excitation, Rydberg atoms in different n-states will ionize at different electric fields, i.e. at different times, and can thus be discriminated.

II. Influence of Blackbody Radiation on Rydberg Atoms

The influence of blackbody radiation on Rydberg atoms was first demonstrated in lifetime measurements. For instance, Gallagher et al.[6] observed that the measured lifetimes of the 16p and 17p states of Na are three times shorter than expected; the shorter lifetime was supposed to be due to blackbody interaction. Haroche et al.[7] found a population transfer to nearby levels which could not be explained by spontaneous decay. More direct evidence of interaction with blackbody radiation was observed later[8-10].

The influence of blackbody radiation is demonstrated in Fig. 1. For this measurement the 5s23f state of the Sr atom was excited by a pulsed dye laser[11]. The Rydberg atoms were detected by field ionization. For this purpose a field ramp was used so that the different Rydberg states were successively ionized starting with the levels closest to the ionization limit. The field ramp was started for the first measurement 1μs after the laser excitation. The measurements performed with larger time delays 2,6 and 12 μs clearly show the increasing population change due to the strong interaction with blackbody radiation (see also Ref.[8]).

A consequence of the long radiative decay time of Rydberg levels and the very large value of the electric dipole matrix elements is that the saturating power for transitions between closely lying Rydberg levels is very small. The corresponding saturating power fluxes are proportional to n^{-10} for low and to n^{-14} for high angular momentum states. A very vivid way of describing the behaviour is to express the saturation power flux in terms of number of photons per surface of the size λ^2 and per lifetime, (the size λ^2 corresponds to the resonant cross-section). For $n \cong 30$ one obtains 10^2 and 1 for low and high angular momentum states, respectively. This means that for high angular momentum states a single photon is required (in the chosen units) to saturate the transition to a neighbouring Rydberg level[12].

There are many applications of wide ranging importance for detectors in the submillimeter region, e.g. infrared and radio-astronomy, diagnostics of plasmas for nuclear fusion, stratospheric monitoring and materials research. The investigation of new principles for detectors is therefore as important as the further development and improvement of known detector principles. The ultimate sensitivity obtainable for detection of any radiation is of course reached when single photons can be monitored with a high probability and when the noise of the signal is only determined by the quantum noise of the radiation. The quantum noise limit of the radiation can be reached in the visible or near infrared spectral range since there the available photomultipliers allow the photocurrent of a single photon to be amplified to a value exceeding the noise of the dark current and the amplifier.

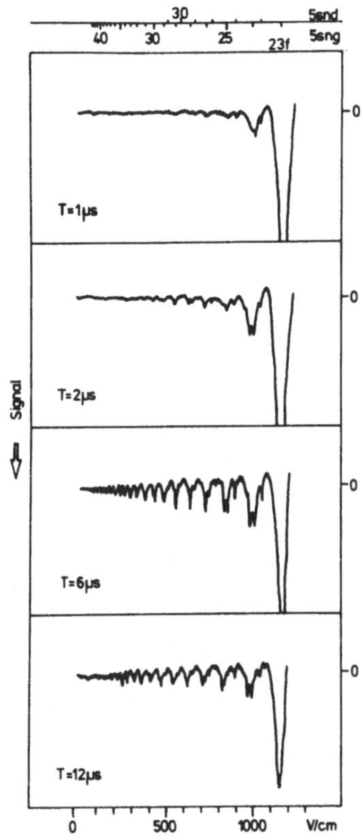

Fig. 1. Blackbody induced transitions between Rydberg states

Ducas et al.[13] demonstrated the very sensitive detection of low-power far infrared laser radiation at 600 GHz by inducing transitions between Rydberg states of sodium atoms. To check the ultimate sensitivity of the Rydberg detector to microwave or far infrared radiation, two improvements of the previous experiments have to be effected. First, the population of the Rydberg atoms must be performed by continuous wave lasers in order to increase the duty cycle and, in addition, the walls of the surrounding chamber have to be cooled to a low temperature, so that the influence of the thermal background radiation is minimized. In the following an experiment of this type will be described[9].

Sodium atoms of an atomic beam were excited to high-lying ^2P-states in two steps via the $3^2P_{3/2}$ intermediate state. The $3^2S_{1/2}$, $F = 2 \rightarrow 3^2P_{3/2}$, $F = 3$ hyperfine transition (589 nm) is saturated with circularly polarized light by means of a single mode continuous wave dye laser stabilized to this transition. A fraction of the atoms in the $3^2P_{3/2}$, F = 3 state is then excited to the n^2D state by means of a multimode dye laser whose cavity length is wobbled in order to obtain a more homogeneous intensity distribution over the laser line width of 100 GHz at a wavelength of about 4150 Å.

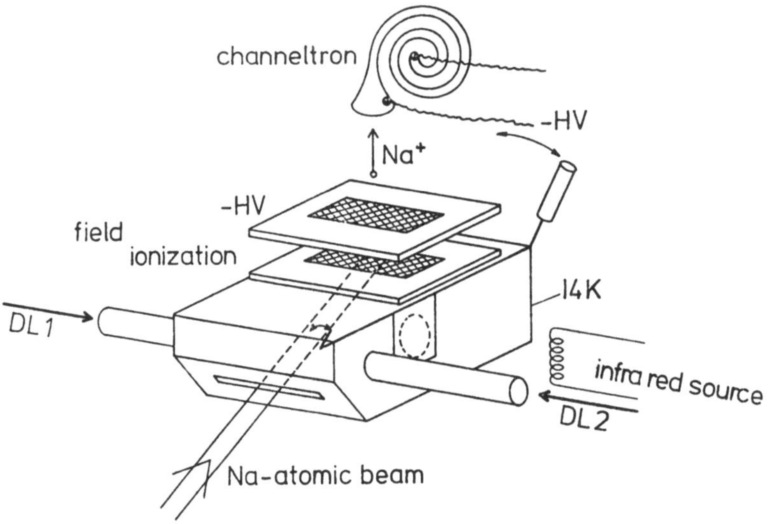

Fig. 2. Experimental setup for the investigation of the interaction of blackbody radiation with Rydberg atoms. The arrows marked by DL1 and DL2 indicate the dye laser beams for the first and second excitation step. For details see Reference 9.

The interaction region is surrounded by a box cooled down to 14 K in order to keep the background of blackbody radiation as low as possible (Fig. 2). The slit where the atomic beam enters the cooled box is covered with a wire mesh to reduce microwave and far infrared radiation emitted by the atomic beam oven. After a path of 20 mm the excited atoms leave the box through a second wire mesh which also acts as one plate of the capacitor used for field ionization. If an atom is ionized, the ion is accelerated and leaves the capacitor through the mesh in the negative plate and is detected by a channeltron multiplier.

A small flap is mounted at one side of the cooled box. By opening this flap, the Rydberg atoms can be exposed to the radiation of a heated wire. The microwave transitions induced in the Rydberg atoms by the thermal radiation of the wire are then monitored via field ionization. For the experiments described here the atoms were excited to the 22^2D state. The strength of the electric field was adjusted, so that atoms in the 22^2D state are not ionized. However, any transition induced to a higher state is detected through the ion signal.

The lifetime of the 22^2D state is 10 µs. Only about 5 % of the initially excited Rydberg atoms therefore reach the field ionization region. The $\Delta n = 1$ transition to the 23^2P state is most likely to be induced. The 23^2P state has a lifetime of about 100 µs. All atoms excited by the microwave radiation therefore reach the field ionization region. The ratio between the rate of the detected ions and the rate of absorbed photons gives the quantum efficiency of the device of 3×10^{-3}.

As a result, the noise equivalent power (NEP) of the detector is $10^{-17}W/(Hz)^{1/2}$, using an output bandwidth of 1 Hz. The NEP of this preliminary setup favourably compares with the NEP of other detectors, which is at least one order of magnitude larger. The NEP was calculated by assuming a background noise which is equal to the signal at 310 K. The noise of the signal is not statistical, a major contribution comes from slow intensity and frequency fluctuations of the second laser indicating that a better NEP can be achieved by stabilizing the second laser.

Another considerable source of background signal is the black-body radiation at 14 K, which is always present inside the cooled box. This background signal is about five times larger than the signal resulting from the radiation of the heated wire at a temperature of 300 K. If the box is cooled with liquid helium, this background can be reduced by a factor of six. With a stabilized single mode laser for the second excitation step it will thus be possible to obtain a NEP of 10^{-19} $W/(Hz)^{1/2}$. It has been demonstrated experimentally that the same value for the NEP can be obtained if coherent amplification of the microwave signal is performed in a Rydberg maser[38].

III. Radiation Interaction of Rydberg Atoms in Cavities - Test System for Simple Quantum Electrodynamical Effects

The invention of the maser has generated a great deal of interest in theoretical models describing the interaction of two-level atoms with a single mode of an electromagnetic field[19-21]. Although the first models treated a purely academic problem, modified versions were stimulated. These then led to an understanding of a major part of the experimentally observed phenomena, including the even larger variety of effects found after the laser was invented. In the experiments it was always necessary to have large numbers of atoms and photons. This was due to the facts that it was impossible to detect small amounts of atoms, and that the small size of the transition matrix elements resulted in atom-field coupling times much longer than other characteristic times of the system, such as atomic relaxation or the interaction time with the field.

This situation completely changed when the advent of frequency-tunable lasers allowed study of the highly excited Rydberg states of atoms, for three reasons. First, these states are very strongly coupled to the radiation field. Second, the transitions to neighboring levels are in the region of millimeter waves, which allows one to build cavities with low-order modes being sufficiently large to ensure rather long interaction times. Finally, the Rydberg atoms have long spontaneous lifetimes, so that only the coupling with the selected cavity mode is important. The theories of the one-atom, one-mode problem predict a few interesting and basic effects that can now be studied experimentally. They result from modifications of the boundary conditions for the vacuum field around the atom due to the presence of the cavity walls. The following are examples: the spontaneous emission of a single atom in the cavity is expected to be drastically modified as compared with its behaviour in free space; spontaneous emission is enhanced in a resonant cavity and suppressed if the cavity is off-resonant. Furthermore, if the cavity finesse is very high, an oscillatory energy exchange between a single atom and the cavity mode is expected to occur. If a field is initially present in the resonator, being produced either by a thermal or a coherent source, various types of dynamics are predicted, such as e.g. disappearance and revival of the Rabi nutation. In the following several phenomena observable with Rydberg atoms are discussed in more detail.

Single Atom in Resonant Cavity - Modification of Spontaneous Emission Rates

The energy levels of the combined two-level atom and field system can be described in the dressed atom picture[22,44]. The lowest energy of the system is represented by $|g,0\rangle$ describing the atom in its ground state $|g\rangle$ with no photon in the cavity. The higher energy levels are separated by the energy of a photon. The states $|\pm n\rangle$ are a superposition of the states $|e,n\rangle$ (e stands for excited atomic state and n for the photon number) and $|g,n+1\rangle$ of the system without interaction between the cavity field and the atom. At resonance:

$$|\pm n\rangle = [|e,n\rangle \pm |g,n+1\rangle]/\sqrt{2}.$$

The energy separation between the levels $|+n\rangle$ and $|-n\rangle$ is $2\hbar\,\Omega\sqrt{n+1}$; there is a small change proportional to $\sqrt{n}$ when the field strength is increased.

In a realistic description of the interaction the dissipative processes also have to be considered. Since Rydberg atoms have lifetimes longer than the atom-field interaction time, their relaxation can generally be neglected. However, the relaxation of the cavity field is important: the harmonic oscillator representing the field is coupled to a thermal reservoir at temperature T representing, for example, the cavity walls. The thermal equilibrium of the field mode is obtained in the characteristic time Q/ω, where Q is the quality factor of the cavity.

The behaviour of an atom entering an empty cavity (i.e., at T = 0 K) in the excited state $|e\rangle$ depends on the relative size of Ω and ω/Q. If $\Omega > \omega/Q$ (small damping of the cavity), the probability of finding the atom in the state $|e\rangle$ undergoes a damped oscillation. This regime can be considered as a self-induced Rabi nutation in the field of the single photon emitted and reabsorbed by the atom. If $\Omega < \omega/Q$, the probability decreases exponentially at a rate $\Gamma_{cavity} = 4\Omega^2 Q/\omega$. There is a cavity-enhanced decay rate which is related to the spontaneous rate in free space Γ_{spont} in the following way:

$$\Gamma_{cavity} = (3/4\pi^2) \cdot Q\lambda^3 \, \Gamma_{spont} / V,$$

where V is the volume of the cavity and λ the wavelength of the radiation. This relation was predicted long ago by Purcell[23]. Physically, the cavity enhances the strength of the vacuum fluctuations at the resonance frequency; as a consequence the transition rate is increased. ($\Gamma_{cavity}/\Gamma_{spont}$ is obtained when the number of oscillator modes per unit frequency interval in a resonant cavity is divided by the corresponding value in free space.)

The opposite effect, the decrease of the decay rate, is obtained when the cavity is detuned. If the transition frequency of the atom lies below the fundamental frequency of the cavity, spontaneous emission is significantly inhibited. In an ideal case no mode is available for the photon and therefore spontaneous emission cannot occur [24].

To change the decay rate of an atom, in principle no resonator has to be present; any conducting surface near the radiator affects the mode density and, therefore, the radiation rate. Parallel-conducting planes can somewhat alter the emission rate but can only reduce the rate by a factor of 2 because of the existence of TEM modes which are independent of the separation.

To demonstrate experimentally the modification of the spontaneous decay rate, it is not necessary to go to single-atom densities in both cases. The experiments where the spontaneous emission is inhibited can also be performed with higher densities. However, in the opposite case, when the increase of the spontaneous rate is observed, a large number of excited atoms increases the field strength in the cavity and the induced transitions disturb the experiment.

The first experiments on the inhibited spontaneous emission were performed by Drexhage[25]. The fluorescence of a thin dye film near a mirror was investigated. Drexhage observed an alteration in the fluorescence lifetime arising from the interference of the molecular radiation with its surface image. An experiment with Rydberg atoms was recently performed by Vaidyanathan, Spencer and Kleppner [26]. They observed a wavelength-dependent cutoff in the absorption of blackbody radiation by Rydberg atoms arising from a discontinuity in the density of modes between parallel conducting plates. Absorption at a wavelength of 2/3 cm by atoms between planes 1/3 cm apart was measured at a temperature of 180 K. The discontinuity in the absorption rate occurred when the absorption wavelength was varied across the cutoff of the parallel-plate modes. The experiment was performed with Na atoms and the transition employed was 29d → 30p. For the tuning of the atomic resonance across the cutoff frequency a small electric field was applied to the parallel plates.

Inhibited spontaneous emission was observed clearly for the first time by Gabrielse and Dehmelt[45]. In these nice experiments on a single electron stored in a Penning trap it was observed that the cyclotron excitation shows a lifetime which is up to 10 times larger than that calculated for a cyclotron orbit in free space. The electrodes of the trap form a cavity which decouples the cyclotron motion from the vacuum fluctuations leading to a longer lifetime.

The first observation of enhanced atomic spontaneous emission in a resonant cavity was published by Goy, Raimond, Gross and Haroche[27]. Their experiment was performed with Rydberg atoms of Na excited in the 23S state in a niobium superconducting cavity resonant at 340 GHz. By taking advantage of the very strong electric dipole of these atoms and of the high Q value of the superconducting resonator cavity-tuning-dependent shortening of the lifetime was observed. This cooling, necessary for superconducting operation, also had the advantage of totally suppressing the black-body field effects ($n = 0$) required to test purely spontaneous emission effects in the cavity.

It was shown that the partial spontaneous emission probability on the 23S → 22P transition in Na is increased from its free space value Γ_{spont} = 150 s^{-1} up to $\Gamma_{cavity} = 8 \times 10^4\ s^{-1}$. This enhanced rate is still 35 times smaller than the damping rate $\omega/Q = 2.8 \cdot 10^6 s^{-1}$ of the field in the cavity.

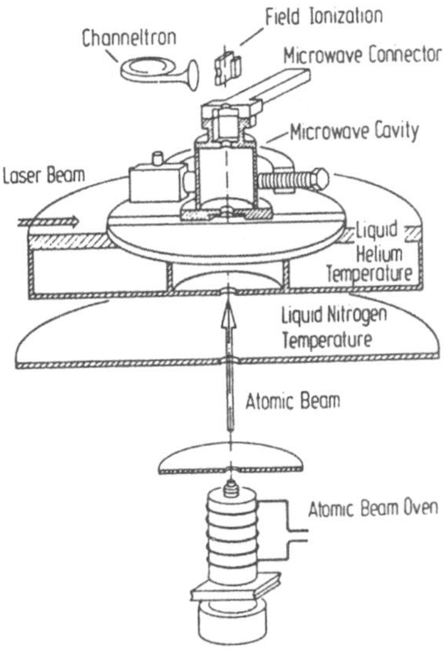

Fig. 3. Experimental setup of the single-atom maser. The length of the cylindrical superconducting niobium microwave cavity is about 24 mm. The upper part is cooled to liquid-helium temperature. Fig. 3 is taken from Ref. [28].

This means that the photon emitted in the mode is absorbed in the mirrors much faster than the atoms decay. The atoms in the Rydberg states were detected by applying an electric field increasing in time for ionization. The two adjacent states were therefore subsequently ionized. The average number of atoms in the cavity was as low as 1.3.

With a tenfold increase in Q, the values of Γ_{cavity} and ω/Q would be of the same size, so that the emitted photon would be stored in the cavity long enough for the atom to reabsorb it. This would approach the regime of quantum mechanical oscillations between a two-level atom and a single electromagnetic field mode mentioned at the beginning of this section. The self-induced single-photon Rabi nutation is much more difficult to observe than the collective Rabi oscillation (which will be described later) because it occurs at a rate $\sqrt{N}$ times smaller (N is the number of atoms in the cavity) and thus requires the atom to be kept in the cavity for much longer times. The first experiment to observe single-atom and single photon interaction was performed recently by Meschede, Walther and Müller[28]. This Rydberg maser experiment again employs an atomic beam to ensure collision free conditions for the highly excited Rydberg states. A diagram of the vacuum chamber with the atomic beam arrangement and microwave cavity is shown in

Fig. 3. These parts are mounted inside a helium bath cryostat. Rubidium atoms were used for the experiment. The atomic beam oven is carefully shielded from the cavity by copper plates cooled by water, liquid nitrogen and liquid helium. The beam passes through small apertures into the liquid-helium-cooled part of the apparatus. There the atoms are pumped by the laser radiation to the upper maser level and enter the cavity. Behind the cavity the atoms are monitored by field ionization and subsequent detection of the ejected electrons in a channeltron electron multiplier.

The Rydberg states were excited with the frequency-doubled light of a commercial continuous wave ring dye laser (Coherent CR 699-21). The second harmonic was generated by means of a temperature-stabilized ADA crystal. The upper maser level was the $63p_{3/2}$ of ^{85}Rb. The finestructure splitting between $63p_{3/2}$ and $63p_{1/2}$ of ^{85}Rb amounts to 396 MHz. It is, therefore, no problem to excite a single fine-structure level of ^{85}Rb with the narrow-band ultraviolet radiation ($\Delta\upsilon \cong 2$ MHz).

The atomic beam passes through the cylindrical cavity along its axis, where only the TE_{1np} and TM_{1np} modes possess a non-vanishing transversal electric field. For our experiment the TE_{121} mode was used. The variation of the field strength of this mode over the cross-section of the atomic beam (0.5 mm) is less than 2 %. The frequency of the maser transition (21506.51(5) MHz) was determined in separate double-resonance experiments. This was necessary since the tuning range of the cavity is very small, requiring that the cavity resonance be nearly equal to the atomic resonance.

The TE_{121} mode has a plane field distribution and is doubly degenerate in an ideal cylindrical cavity. The degeneracy is removed by slightly deforming the circular cross-section into an oval shape, which then determines the direction of polarization of the field mode. The deformation is achieved by squeezing the cylinder both with a screw and a piezo-electric (PZT) crystal ($\Delta\ell \cong 4$ µm/1500V at 2 K) for fine tuning. This causes the one degenerate resonance to be shifted towards higher frequencies, whereas that for orthogonal polarization shifts to lower values. The field polarization important for our experiment is coupled to an external waveguide so that the cavity performance can be tested. The upper frequency branch used in our experiment can be mechanically tuned by about 15 MHz; the piezo drive can sweep the frequency by 0.5 MHz/1500 V.

The temperature of the cavity could be varied from 4.3 to 2.0 K, corresponding to Q factors of 1.7×10^7 and 8×10^8, respectively. To shield the 300 K thermal radiation of the test equipment, a cooled flap with a temperature of less than 10 K was inserted into the waveguide at the upper end of the cavity. The thermal background field inside the niobium cavity is therefore essentially determined by the temperature of the walls.

The average number of photons of the black-body radiation is given by $\bar{n} = [\exp(h\nu/kT)-1]^{-1}$, being about $\bar{n} = 3.8$ at 4.3 K and 1.5 at 2 K. Field ionization is the standard technique for detecting Rydberg atoms. It is possible to distinguish between Rydberg states belonging to different main quantum numbers. If the maximum field strength is chosen properly (≅ 20V/cm), atoms in the 63p state are ionized and detected by counting the electrons with a channeltron electron multiplier. At the same field strength, atoms in the neighboring 61d level are ionized with a smaller probability (≅ 15 % of 63p state).

Transitions from the initially prepared $63p_{3/2}$ state to the $61d_{3/2}$ level are thus detected by reduction of the electron count rate. The detector is placed 63 mm downstream from the point where the atoms are excited. Considering the lifetime of the $63p_{3/2}$ level, one finds that ≅ 70 % of the excited atoms reach the detector.

In the case of the measurements at a cavity temperature of 2 K (Fig. 4), a reduction of the $63p_{3/2}$ signal could be clearly seen for fluxes as small as 800 atoms/s. An increase of the flux causes power broadening and finally asymmetry and a small shift. The shift has to be attributed to the ac Stark effect, caused predominantly by virtual transitions to the $61d_{5/2}$ level, which is only 50 MHz away from the maser transition. The fact that the field ionization signal at resonance is independent of the particle flux (between 800 and 22 x 10^3 atoms/s) indicates that the transition is saturated. This fact and the observed power broadening show that there is multiple exchange of photons between the Rydberg atoms and the cavity field.

With an average transit time of the Rydberg atoms through the cavity of 80 μs, one calculates for a flux of 800 atoms/s a probability of 0.06 of finding a Rydberg atom in the cavity. According to Poissonian statistics this implies that more than 99 % of the events are single-atom. This clearly demonstrates that single atoms are able to maintain continuous oscillation of the cavity. Since the transition is saturated, half of the atoms initially excited in the $63p_{3/2}$ state leave the cavity in the lower $61d_{3/2}$ maser level. The decay to other levels can be neglected for the average transit time of 80 μs. The energy radiated by these atoms is stored in the cavity field for the characteristic cavity decay time, increasing the average field strength.

The average number of photons left in the cavity by the Rydberg atoms is given by

$$n = \tau_d \cdot N/2,$$

where τ_d is the characteristic decay time of the cavity and N the number of Rydberg atoms in the upper maser level entering the cavity per unit time. For the highest particle flux used in our experiment, $N = 22 \times 10^3$ atoms/s, one calculates $n \simeq 55$ photons at 2 K ($\tau_d \simeq 5$ ms).

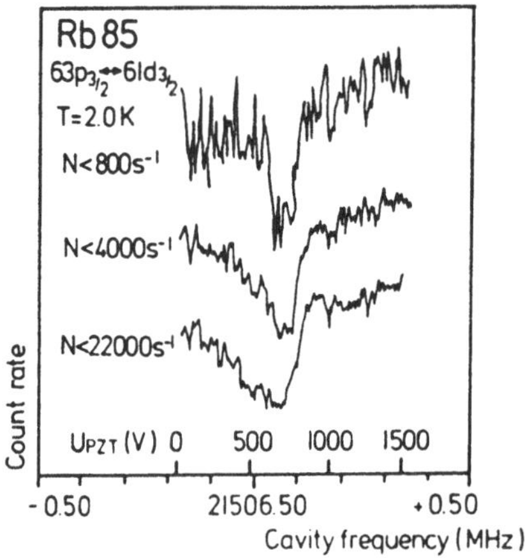

Fig. 4: Maser resonance at a cavity temperature of 2 K
The figure is taken from Ref. [28].

At 2 K the average number of black-body photons is $n \simeq 1.5$. In the case of $N \simeq 800$ atoms/s one obtains $n \simeq 2$, which means that the energy of the radiation generated by the Rydberg atoms in the cavity is about the same size as that of the blackbody radiation.

When the squares of the half-widths $\Delta\upsilon$ of the signal curves are plotted versus the Rydberg atom flux, a straight line is obtained as expected (Fig. 5). This line intersects the $(\Delta\upsilon)^2$ axis at a finite value, from which the number of blackbody photons originally in the cavity can be evaluated. The result (3 ± 1) is in reasonable agreement with the value given above. It follows that as the atomic flux decreases the thermal radiation becomes the dominant part of the field.

The coupling constant between the atom and radiation is big enough to allow multiple exchange of photons between the cavity mode and a single Rydberg atom to occur under these conditions. It follows that under the conditions shown in Fig.4

the atom performs on the average about 5 to 20 Rabi periods when passing through the cavity.

Because of the velocity distribution of the atoms it is not possible to observe the Rabi nutation directly. At present a Fizeau-type velocity selector being inserted between the atomic beam oven and the cavity. This will enable us to observe the Rabi nutation of the atoms direct, because then we will have a fixed interaction time of the atom with the cavity field. Changing the selected velocity leads to a different interaction time and leaves the atom in another phase of the Rabi cycle when it arrives at the detector. In this way more detailed studies of the atom-field interaction than in the present experiment will be possible. In particular, there is a good chance that experimental observation of the predicted disappearance and revival of the Rabi nutation will also become possible. These effects will be discussed in more detail in the next section.

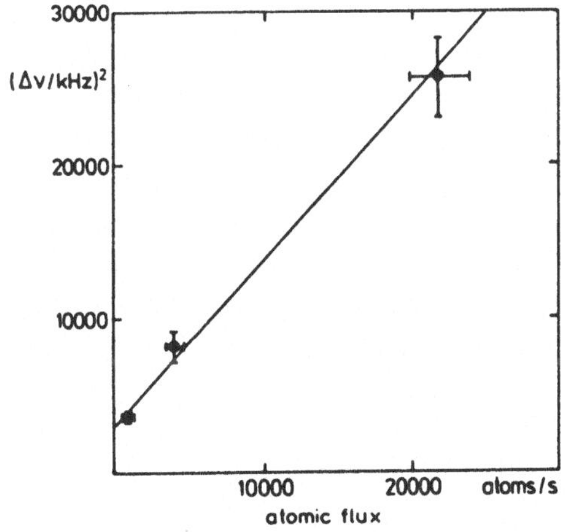

Fig. 5. Squares of the half-widths of the maser resonances vs atomic flux. From the intersection of the straight line with the $(\Delta\upsilon)^2$ axis, the number of blackbody photons in the cavity can be evaluated. This was done with use of the Rabi frequency Ω = 43 kHz, which was evaluated from the maser data at 4.3 K. Fig. 5 is taken from Ref. [28].

Single Atom in Resonant Cavity - Disapearance and Revival of Optical Nutation

At temperatures T > 0 K the cavity also contains thermal photons. The effects described above therefore become more complicated since the atom evolves through an oscillatory or irreversibly damped transient regime towards a final state distribution corresponding to the thermal equilibrium. The transient behaviour is again dependent on whether there is weak ($\Omega >> \omega/Q$) or strong damping ($\Omega << \omega/Q$) of the cavity. In the first case the transient regime can be described by a sum of elementary Rabi oscillations in a field in which the number of photons is a random quantity following the Bose-Einstein statistics. The distribution of Rabi frequencies results in an apparently random oscillation which for large n values very quickly collapses and then revives again (Faist et al.[29]; Meystre et al.[30]; Eberly et al.[31]; Knight and Radmore, et al.[32]. This behaviour is typical of a chaotic quantum field; a semi-classical description of a random Gaussian field does not give this result. This interesting phenomenon was always thought of as being incapable of experimental observation. However, the possibilities now opened up by Rydberg atoms bring us close to its realization. Fig. 6 shows the Rabi-nutation resulting in a cavity at 2 K. Plotted is the probability P for the population of the upper (P = 1) and the lower maser level (P = 0). P for a chaotic thermal field is given by:

$$P(t) = \frac{1}{2}(1-\exp(h\upsilon/kT))\sum_n \exp\left(-\frac{h\upsilon n}{kT}\right)(1+\cos(2\Omega\sqrt{n+1}\,t))$$

The single photon Rabi-frequency used for the calculation of Fig. 6 corresponds to the conditions of the one-atom maser experiment described above.

Quantum collapses and revivals also occur when the cavity contains a coherent field or a mixture of chaotic and coherent fields, whereby the time intervals between collapses and revivals depend strongly on the mixture between the two. Therefore the statistics of very few photons in the cavity can be investigated.

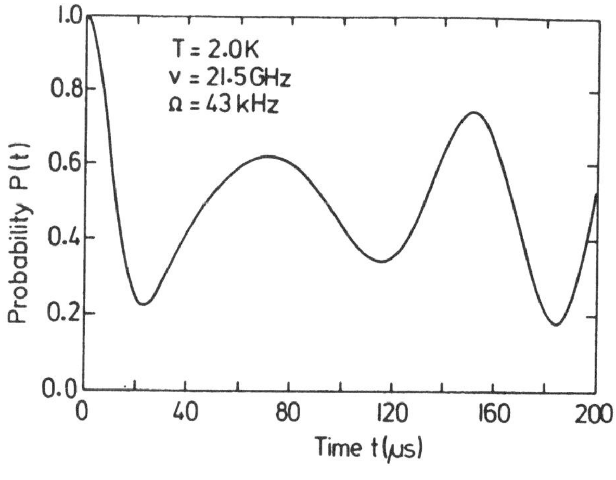

Fig. 6. Rabi-nutation in a chaotic field formed by an average number of 1.5 photons

N Atoms in Resonant Cavity - Collective Behaviour

The generalization of the single-atom effects described in the previous sections to N two-level atoms can be based on the ladder of equidistant non-degenerate states, the so-called symmetrical Dicke states[43]. Such states, where J + M atoms are excited in level $|e\rangle$ and J - M in level $|g\rangle$, are written formally

$$|JM\rangle = S\,|\underbrace{e,e,\ldots e}_{J+M};\underbrace{g,g,\ldots g}_{J-M}\rangle$$

where S is the symmetrization operator. (M = J and M = -J correspond to the totally excited and de-excited states, respectively.) The analysis of the atomic system by Dicke states is related to the atomic indiscernibility with respect to the single mode of the cavity. The N + 1 states $|JM\rangle$ describe situations in which strong correlations exist between the dipoles of different atoms resulting in a collective behaviour of the atoms in the cavity. Again the strong atom-to-field coupling of the Rydberg atoms is a big advantage, so that the experimental verification of the phenomena is much simpler than for "ordinary" atoms. The effects observed are cooperative features which cannot be interpreted in terms of an independent atom model: collective oscillations and superradiance, when the system is initially in the upper level, and collective absorption, when the system starts in the lower level.

For discussion of the phenomena two cases have again to be considered depending upon whether the Rabi frequency $\sqrt{N}\,\Omega$ is larger or smaller than the reciprocal of the cavity damping time ω/Q. In the case without black-body photons (T = 0 K), and $\omega/Q = 0$, and with all the atoms in the excited state, the spontaneous emission causes the atomic system to cascade through the ladder of eigenstates. The field strength in the cavity is increased and the photons are reabsorbed. The subsequent oscillations can be interpreted as a Rabi nutation in the field radiated by the atoms and stored in the cavity (Bonifacio and Preparata[33]; Scharf[34]). The oscillations show a rather complicated beating pattern for small N values[35].

For larger N values, the number of states to keep track of becomes prohibitive. Fortunately, the system can then be described in a classical way by using the concept of Bloch vector (see, for example, Allen and Eberly[20]; for the relation between the quantum mechanical and Bloch vector approaches see, for example, Bonifacio et al.[36]).

In the case of strong cavity damping the energy decays with a rate

$T_R^{-1} = 4\ \Omega^2 NQ/\omega$[33]. The value corresponds to $N\cdot\Gamma_{cavity}$ where Γ_{cavity} is the single-atom cavity-enhanced decay rate as discussed in the previous section.

The experimental observation of the above-mentioned effects was performed by Haroche and co-workers with an atomic beam of alkalis excited by pulsed lasers in the Rydberg states. Either the upper or the lower level of a millimeter-wave transition was in resonance with a mode of a cavity surrounding the atoms. The relatively long wavelength of the transitions allows all atoms to be excited in a region of constant field amplitude.

The Rydberg atoms are monitored by field ionization after the atoms have passed the cavity. In this way the number of atoms in the upper or the lower level of the microwave transitions was measured. In order to reconstruct the atomic evolution during the time the atoms spend in the cavity a small electrode producing an inhomogeneous electric field at a preset time t was inserted into the cavity. The Stark shift produced by this field suddenly brings the atoms out of the cavity resonance. Therefore the atom-cavity coupling is interrupted and the detector measures the state the atom had at time t. By varying t, the dynamics of the atom-cavity interaction can be reconstructed.

Actual experiments were performed with cavities at T = 300 K whereas the theories deal with systems at T = 0 K. In fact, it can be shown that, as long as N is larger than the number of black-body photons in the cavity, the thermal-field contributions rapidly become negligible (Raimond et al.[37]). Blackbody effects are relevant only at the onset of the emission, when the emitted field is still much smaller than the thermal one. With respect to fluctuations there is no difference since thermal and vacuum fields have the same statistical nature.

The experiments in a moderate Q cavity, typically $Q \approx 10^4$ (Moi et al.[38], Raimond et al.[37]), give the predicted cavity-assisted overdamped superradiance. This superradiant Rydberg "maser" is characterized by an extremely low inversion density threshold ($N \approx 10^4$ atoms). The inverted medium emits a short burst of radiation and decays within a few hundred ns to the lower state of the transition (In the experiments mostly $nS_{1/2} \rightarrow (n-1)\ P_{1/2,3/2}$ or $nS_{1/2} \rightarrow (n-2)\ P_{1/2,3/2}$, $n \approx 30$). This maser emission was also detected by using Schottky heterodyne receivers (Moi et al.[38,39]). The latter detection technique is of course considerably less sensitive than the one based on atomic field ionization, which actually allows one to count the atoms which have radiated inside the cavity during a given time interval, and hence the emitted photons.

Raimond et al.[37,40] succeeded in measuring the probability distribution P(N,t) that N atoms have been deexcited at time t; the measured transition was 29 $S_{1/2} \rightarrow 28\ P_{1/2}$ of Na at 162.4 GHz. At short times, P(N,t) appears to obey a Bose-Einstein-type statistics, which is typical of a linearly amplified blackbody field. At later times the amplification process becomes nonlinear and the distribution evolves into a broad bell-shaped Poisson-like function typical of a coherent process. These measurements represent the first direct and quantitative test of the theory in a system in which superradiance is not complicated by propagation and diffraction effects, as it normally is in the optical domain.

As discussed above, in the high-Q regime ($Q \simeq 10^6$) one expects to observe an oscillatory exchange of energy between the atoms and the cavity field which can be described as a self-induced Rabi-nutation of the atomic system. The experimental observation was performed by Kaluzny et al.[41]. The transition investigated was 36 $S_{1/2} \rightarrow 35\ P_{1/2}$ of the Na atom. In order to remove the twofold degeneracy in the upper and lower levels and to study a true two-level atom transition, a small constant magnetic field was applied along the cavity axis and the cavity is tuned to resonance with 36 $S_{1/2}$, $m_J = + 1/2 \rightarrow 35\ P_{1/2}$, $m_J = 1/2$ transition at about 82 GHz.

The emission of the N atoms in the cavity occurs faster than it would in free space, essentially owing to the cavity enhancement effect. When N was sufficiently high (N > 20 000), oscillations in the atomic population evolution become clearly observable. This collective self-nutation regime has also been discussed in the context of superradiance theories. It is then generally referred to as the "ringing" regime of superfluorescent emission. In the case of free-space superradiance this phenomenon has not yet been clearly observed since the simple Rabi nutation is then masked by multimode diffraction and propagation effects.

N Atoms in Resonant Cavity - Collective Absorption of Blackbody Photons

In the previous section the case where the N atoms were initially in the excited Dicke state $|J, +J\rangle$ was discussed. In the following, the N atoms are now assumed to enter the cavity in the lowest state $|J, -J\rangle$; furthermore, it is assumed that the cavity field is in thermal equilibrium at a temperature $T \neq 0$ K.

The thermal photons represent a Bose-Einstein distribution with an average photon number $n \neq 0$. As the time evolves, the atoms gain energy at the expense of the mode which is then supplemented by the thermal reservoir. The time constant for reaching thermal equilibrium depends on the values of N, n and ω/Q. Since the atomic energy diagram consists of non-degenerate equidistant levels with the same spacing as the field levels, the atoms will obviously reach an equilibrium described by a Boltzmann law quite similar to the Bose-Einstein distribution of the photon number in the field mode. (The only difference between the two distributions is that the number of levels for the atomic system is finite; this changes the normalization of the Boltzmann distribution.) As a consequence, the number of absorbed photons is limited to ΔN, no matter what value N has, and is equal to the average blackbody photon number per mode (as soon as $N > n$): $\Delta N = n = [\exp(h\omega/kT) - 1]^{-1}$, which is close to $kT/h\omega$ in the Rayleigh-Jeans limit.

The energy absorbed by N atoms in the cavity is not identical with the sum of energies that would be absorbed by N independent atoms. In this process the atomic sample evolves in a collective mode and behaves as a single quantum system exhibiting basic effects of Bose-Einstein statistics and Brownian motion (Raimond et al.[42]). A detailed study of the pulse-to-pulse random variations of ΔN around ΔN should allow one to probe the fluctuations of the cavity mode and to reconstruct their Bose-Einstein distribution. There is a connection between the absorption and emission of N atoms in a cavity: The atomic indiscernibility, which is responsible for superradiance when the system is initially excited, leads to a kind of "subabsorption" (Raimond et al.[42]) when it starts from its lower state.

The experimental demonstration of the effects just described was first performed by Raimond et al.[43]. The 30 $P_{1/2} \rightarrow$ 30 $S_{1/2}$ transition of Na was used in this experiment. The atoms were excited by a pulsed dye laser in a semifocal Fabry-Perot millimeter-wave cavity (~ 134 GHz) and detected by field ionization after the Na beam left the cavity. The cavity was coupled to a small, electrically heated tungsten wire whose temperature could be varied between 300 and 1600 K. For a resonant cavity an increase of the number of excited atoms is observed for small N values, but for $N \gtrsim 5000$ ΔN reaches the limit expected from theory. The experiment demonstrated that, although the limiting value is independent of cavity characteristics, it changes with the radiation temperature. This could be demonstrated by increasing the temperature of the tungsten wire in the cavity. The experiment clearly demonstrated that the average number of excited atoms is exactly equal to the mean number n of blackbody photons in the mode. The Rydberg atoms thus constitute an absolute radiation thermometer for the thermal field, the temperature being directly related to a particle number.

References

1. S. Feneuille, P. Jacquinot, Adv. Atom. Mol. Phys. 17, 99 (1981).
2. D. Kleppner, in "Laser-Plasma Interactions", Les Houches XXXVI, edited by R. Balian pp. 733-784, North-Holland, Amsterdam 1982.
3. D. Kleppner, M.G. Littman, M.L. Zimmerman, in "Rydberg States of Atoms and Molecules" edited by R.F. Stebbings and F.B. Dunning, pp. 73-116, Cambridge University Press, Cambridge 1983.
4. D. Delande, J.C. Gay, Comments At. Mol. Phys. 13, 275 (1983).
5. D.R. Bates, A. Damgaard, Philos. Trans. Roy. Soc. London, 242, 101 (1949).
6. T. F. Gallagher, and W.E. Cooke, Phys. Rev. Lett. 42, 835 (1979).
7. S. Haroche, C. Fabre, P. Goy, M. Gross, and J.M. Raymond, in "Laser Spectroscopy IV", edited by H. Walther, and K.W. Rothe, Springer Series in Optical Sciences, Vol. 21, Springer Verlag, Berlin, Heidelberg, New York 1979.
8. E. J. Beiting, G.F. Hildebrandt, F.G. Kellert, G.W. Foltz, K.A. Smith, F.B. Dunning, and R.F. Stebbings, J. Chem. Phys. 70, 3551 (1971).
9. H. Figger, G. Leuchs, R. Straubinger, H. Walther, Opt. Comm. 33, 37 (1980).
10. P.R. Koch, H. Hieronymus, A.F.J. Van Raan, W. Raith, Physics Letters 75 A, 273 (1980).
11. G. Rempe, Diplomarbeit, Ludwig-Maximilians Universität München, 1982.
12. S. Haroche, in "Atomic Physics 7", edited by D. Kleppner, and F.M. Pipkin, Plenum Press, New York, London 1981, p. 141.
13. T.W. Ducas, W.P. Spencer, A.G. Vaidyanathan, W.H. Hamilton, and D. Kleppner, Appl. Phys. Lett. 35, 382 (1979).
14. W.E. Cooke, T.F. Gallagher, Phys. Rev. A 21, 588 (1980).
15. T.F. Gallagher, W. Sandner, K.A. Safinya, W.E. Cooke, Phys. Rev. A 23, 2065 (1981).
16. J.W. Farley, W.H. Wing, Phys. Rev. A 23, 2397 (1981).
17. R.F. Stebbings, C.J. Latimer, W.P. West, F.B. Dunning, T.B. Look, Phys. Rev. A 12, 1453 (1975).
18. L. Barbier, R.J. Champeau, Journal de Physique 41, 947 (1980).
19. E.T. Jaynes, F.W. Cummings, Proc. IEEE 51, 89 (1963).
20. L. Allen, J.H. Eberly, "Optical Resonance and Two Level Atoms", Wiley, New York (1975).
21. P.L. Knight, P.W. Milonni, Phys. Rev. 66C, 21 (1980).
22. S. Haroche, Ann. Phys. (Paris) 6, 189 (1971).
23. E.M. Purcell, Phys. Rev. 69, 681 (1946).
24. D. Kleppner, Phys. Rev. Lett. 47, 233 (1981).
25. K.H. Drexhage, in "Progress in Optics" edited by E. Wolf, Vol. 12, pp 165 North-Holland, Amsterdam (1974).
26. A. Vaidyanathan, W. Spencer, D. Kleppner, Phys. Rev. Lett. 47, 1592 (1981).
27. P. Goy, J.M. Raimond, M. Gross, S. Haroche, Phys. Rev. Lett. 50, 1903 (1983).
28. D. Meschede, H. Walther, G. Müller, Phys. Rev. Lett., 54, 551 (1985)
29. A. Faist, E. Geneux, P. Meystre, A. Quattropani, Helv. Phys. Acta 45, 946 (1972).
30. P. Meystre, E. Geneux, A. Quattropani, A. Faist, Nuovo Cim. 25B, 521 (1975).
31. J.H. Eberly, N.B. Narozhny, J.J. Sanchez-Mondragon, Phys. Rev. Lett. 44, 1323 (1980).
32. P. Knight, P.M. Radmore, Phys. Rev. A 26, 676 (1982).
33. R. Bonifacio, G. Preparata, Phys. Rev. A 2, 336 (1970).
34. G. Scharf, Helv. Phys. Acta 43, 806 (1970).
35. S. Haroche, in "New Trends in Atomic Physics", Proceedings of the Summer School Session XXXVIII, edited by G. Grynberg, R. Stern, North-Holland, Amsterdam 1982.
36. R. Bonifacio, D.M. Kim, M.O. Scully, Phys. Rev. 187, 441 (1969).
37. J.M. Raimond, P. Goy, M. Gross, C. Fabre, S. Haroche, Phys. Rev. Lett. 49, 1924 (1982).
38. L. Moi, P. Goy, M. Gross, J.M. Raimond, C. Fabre, S. Haroche, Phys. Rev. A 27, 2043 and 2065 (1983).

39. L. Moi, C. Fabre, P. Goy, M. Gross, S. Haroche, P. Eucrenaz, G. Beaudin, B. Lazareff, Opt. Comm. 33, 47 (1980).
40. J.M. Raimond, P. Goy, M. Gross, C. Fabre, S. Haroche, "Laser Spectroscopy VI", edited by H.P. Weber and W. Lüthy, Springer Series in Optical Sciences, Vol. 40, pp 237 - 241, Springer, Berlin.
41. Y. Kaluzny, P. Goy, M. Gross, J.M. Raimond, S. Haroche, Phys. Rev. Lett. 51, 1175 (1983).
42. J.M. Raimond, P. Goy, M. Gross, C. Fabre, S. Haroche, Phys. Rev. Lett. 49, 117 (1982).
43. R.H. Dicke, Phys. Rev. 93, 99 (1954).
44. S. Haroche, C. Fabre, J.M. Raimond, P. Goy, M. Gross, L. Moi, Journ. de Physique 43, C2 - 265 (1982).
45. G. Gabrielse, H. Dehmelt, Proc. Natl. Acad. Sciences, to be published

PLANETARY ATOMS

W.E. Cooke
Physics Department
University of Southern California
Los Angeles, CA 90089-0484

L.A. Bloomfield, R.R. Freeman
AT&T Bell Laboratories
Murray Hill, NJ 07974

J. Bokor
AT&T Bell Laboratories
Holmdel, NJ 07733

ABSTRACT

We have experimentally observed the systematic onset of strong mixing of confirugations in core-excited autoionization states of Ba as a function of excitation level of the 'core' electron relative to the Rydberg electron. This behavior was originally suggested by Percival in his discussion of 'Planetary Atoms'.

INTRODUCTION

In 1977, I.C. Percival coined the term "planetary atoms" to describe those atoms which had at least two highly excited elctrons.[1] Percival used a semiclassical quantization theory to give some basic characteristics to these unsolved, three body quantum mechanical systems. Others were attempting to solve this problem from a purely quantum mechanical approach. Herrick and Sinanoglu[2] used group theory methods to classify energy levels for the He $3\ell3\ell'$ manifold. Their approach was strongly suggestive that these atoms vibrated and rotated with the electrons localized on opposite sides of the nucleus - more like a "molecular atom". Fano and others[3] analyzed the problem using hyperspherical coordinates ad found that, at least for the case of near threshold double ionication, both electrons were highly correlated in angle as they moved along a "Wannier ridge", a local maximum in the potential energy surface.

There are some characteristics of planetary atoms that all the theoreticians agree upon. (1) These states will typically have high energies, usually in the xuv, since two electrons are both excited. (2) In some of the planetary states the electrons will be correlated to produce long lifetimes, generating nearly bound states embedded in the ionization continuum. (3) Most of the planetary states will have little overlap with the ground state. (4) Single photon excitation will not efficiently excite these states (because of the low overlap with the ground state). (5) The structure of these states will depend primarily on the charge of the parent ion, and not its structure (again because of the lack of overlap with the ground state).

However, because of these characteristics, very little experimental observation of planetary atoms has been possible. Percival suggested that electron impact excitation mught be the only efficient way to populate planetary atoms, although the immense phase space available would make it difficult to excite specific planetary states.[1] Recently, Buckman et al[4] have demonstrated the advantages and difficulties of this excitation method. They have used electron impace to excite doubly excited states of He^-, and have observed some interesting departures from normal Rydberg interval scaling. But is is very difficult to identify confidently the resonances they observe, and their energy resolution is limited to tens of meV, typical of electron monochromators.

Our approach has been to use multiphoton excitation, usually employing three different wavelength photons. In this way, we have maintained standard optical energy resolutions of fractions of cm^{-1} and we have retained the specificity of dipole selection rules. In fact, by using a multiphoton version of Isolated Core Excitation (ICE)[5] we have an even greater ability to identify specific configurations than is usually obtained in optical spectra.

EXPERIMENTAL METHOD

The idea behind multiphoton ICE is simply illustrated in figure 1. First, we use two lasers to excite barium atoms in an atomic beam from their ground state to a highly excited, or Rydberg, state. We use barium atoms because of their convenient visible transitions; the planetary barium atom should be insensitive to the xeonon-like Ba^{++} core. These Rydberg states consist of primarily one single configuration which has one valence electron in the ground, 6s, state and one electron in a highly excited $n\ell$ state (where ℓ=0 or 2 by

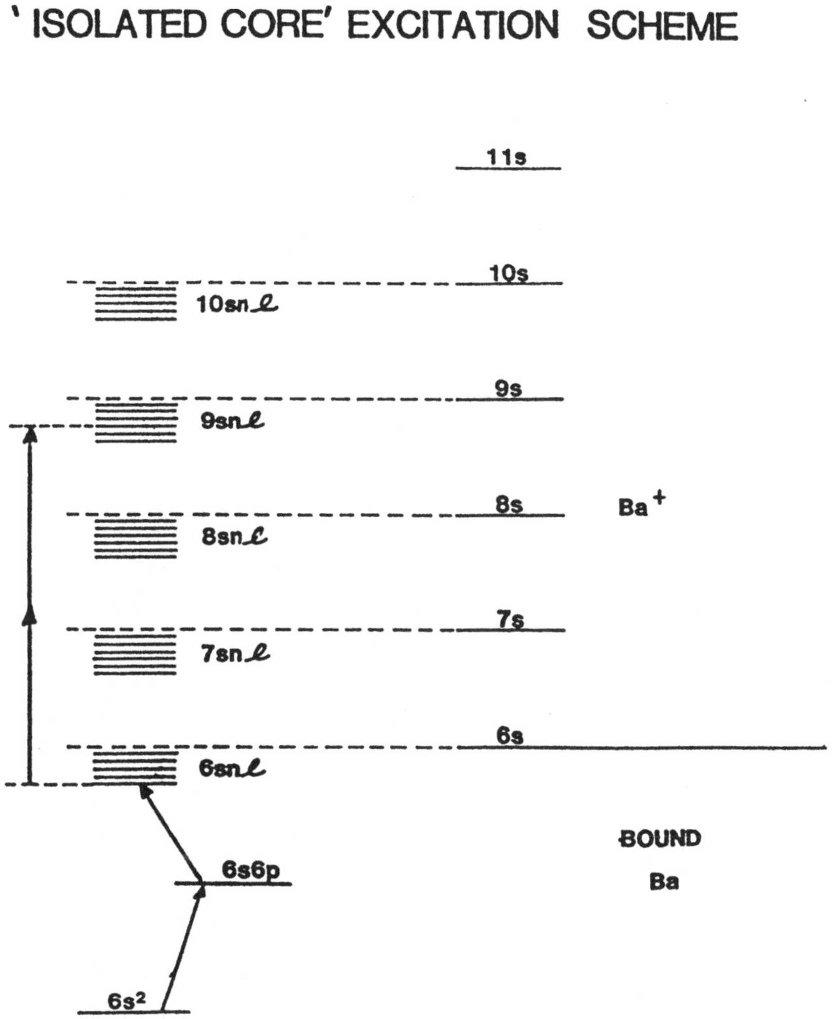

Figure 1. Schematic of Excitation Process. Two lasers are used to excite selectively a particular 6snℓ Rydberg state. Subsequently, the transition 6s → ms is driven in a two-photon process resulting in the population of states msnℓ.

dipole selection rules). These bound states have been well characterized spectroscopically[6] and are easily resolvable with our pulsed lasers. This initial excitation isolates the core, insofar as further optical fields only affect the remaining 6s core electron. The Rydberg electron is moving too slowly to respond to a rapidly oscillating field. (Rydberg photoionization cross-sections decrease rapidly as $1/n^3$.) Furthermore, since the Rydberg electron is localized far from the 6s electron, the 6s electron responds to frequencies very near the those that excite the 6s core electron to an ms excited state. This method relies upon the basic concepts of the independent electron model, and therefore gives very simple lineshapes only when that model is applicable. If an atom is not well described by the independent electron model, then this excitation scheme produces a more complex spectra. Thus the complexity of the spectra is a direct measure of the breakdown of the independent electron model.

The core excitation can be accomplished stepwise,[7] with two different lasers, although it has been more convenient to use two photons from a single uv laser in most cases. Spectra are obtained by tuning the core laser frequency and monitoring the production of doubly excited states, while the first lasers remain fixed to a specific Rydberg state. This produces spectra for the $6sn\ell \rightarrow msn'\ell$ transitions, where n and n' need not be the same, although ℓ does not change. The n change can be seen as a type of shake-up of the Rydberg electron resulting from core excitation. The changed core produces a different potential at small r for the Rydberg electron so that its initial Rydberg state is not orthogonal to the excited core Rydberg state. The initial wavefunction is then projected onto several final, excited core Rydberg wavefunctions:

$$\langle n_i | n_f \rangle = \frac{\sin\pi(n_f^* - n_i^*)}{\pi(n_i^* n_f^*)^{3/2}(W_f - W_i)} \qquad (1)$$

where W_i and W_f are the initial and final binding energies. However, since we use narrow linewidth lasers, and energy must be conserved, only one state is produced in this shake up. The relative strengths of different shake-up transitions can be measured by tuning the core laser over the entire spectrum. In the usual shake-up process, the core excitation is so rapid that the uncertainty principle allows all of these states to be populated at once.

The detection of a $6sn\ell \rightarrow msn'\ell$ transition is not trivial for any value of m since the atom will ionize if it absorbs one or both core transition photons. Consequently, we have used two different techniques to detect the two photon core transitions. For the lower core states (m = 7 or 8) we have analyzed the energy of the electrons produced by ionization.[7] Some of the doubly excited states will autoionized to produce Ba+(6s) ions, leaving the remainder of the energy with the free electrons. By only looking for electrons with enough energy to represent two photon absorption, we eliminated the single photon absorption background. For the electron energy analysis, we have used a simple drift tube with a retarding voltage to discriminate against the slower electrons. Our tube is mounted inside a solenoid (≈5 Gauss) that guides the electrons to a microchannel plate detector.

For the higher core excitations, this technique is not efficient since the atom most often autoionizes to produce one of the many excited states of the Ba^+ ion. The availability of excited ion states increases as m^3, with most of the states resulting in very slow electrons. Our second technique is to detect the excited ions by further ionizing them with the absorption of yet an additional core photon. We have easily separated the Ba^{++} ions from the Ba^+ ions using a quadruple mass filter. This technique also provides an immediate laser wavelength calibration, since there are always some $Ba^+(6s)$ ions in our chamber which are three-photon ionized when the core laser is tuned exactly to the Ba^+ 6s → ms transition.

RESULTS AND DISCUSSIONS

The planetary atoms we have observed using these methods fall into two groups: states where $n \gg m$; and statest where $n \approx m$. In the first case, one expects that the independent electron model will still be a good approximation; in the second case it is not clear what to expect. Figure 2 shows a typical spectra for the 6s15s→10sn's transitions, which are examples of the first case. The very narrow peak at 2857Å corresponds to the laser calibrating ionic transition referred to above. The two broader features are transitions to doubly excited staes with values of n' to 16 and 17. The shift of these lines form the ionic transition allows us to determine how much the binding energy of the Rydberg electrons changed by exciting the core. This binding energy, W, can then be easily related to the doubly excited state's effective quantum number, n^*, or quantum detect, δ

$$W = 1/n^{*2} = 1/(n-\delta)^2 \text{ (a.u.)} \qquad (2)$$

It should be noted that equation (2), like equation (1) only gives information about n^*, not about n. For our spectra, we have checked that both equations (1) and (2) give the same result for n^*, although

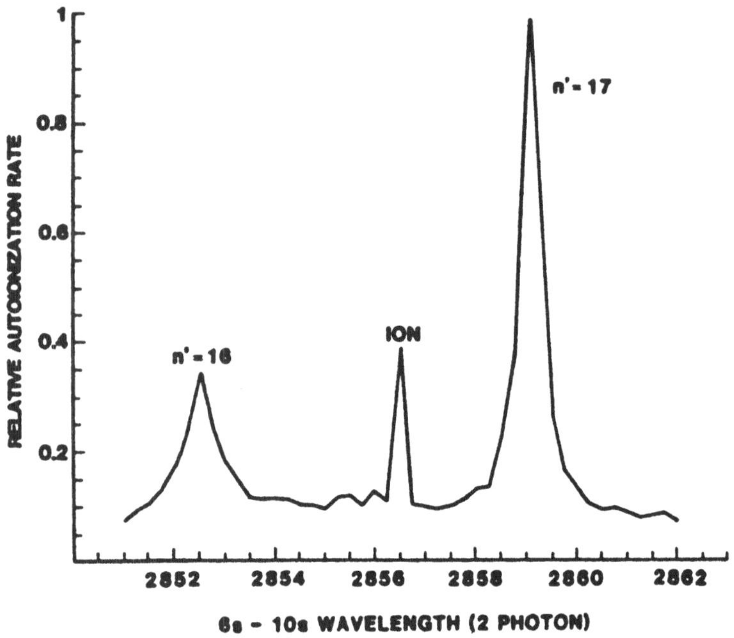

Figure 2. Excitation Spectrum of the 6s 15s → 10sn's Transition. This is a "normal" shake-up spectrum, resulting in the population of n'=17 and n'=18.

equation (2) is typically more accurate since the positions are insensitive to laser power variations.

We do not observe large changes in the binding energy of the Rydberg electron because its binding energy is simply related to its most probable location - its classical turning point - and our core excitation cannot perturb the Rydberg electron much. The small change is, in fact, a direct result of the interaction between the core and Rydberg electrons. In addition to binding energies, the spectrum of figure 2 also shows the autoionization rates of the two states as the widths of their excitation lines. Since the lineshapes are symmetric, their linewidths can be easily measured.

When the Rydberg electron is inside the ms electron, its wavefunction oscillates faster so that there is a difference between the number of nodes it has (n) and the number of nodes a hydrogenic wavefunction of equivalent energy would have (n*). The quantum defect is a direct

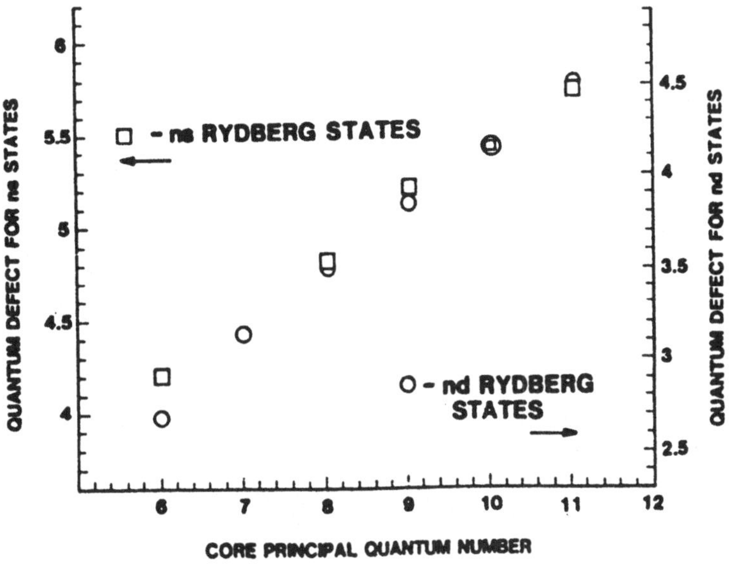

Figure 3. Quantum Defects of the ms and md states as a function of core principal quantum number.

measure of this. Figure 3 shows how the measured quantum defect of $msn\ell$ states increases with increasing m.[8] Since this effect just depends on the extent of the ms wavefunction, it is independent of ℓ (as long as m is large enough so that the $n\ell$ wavefunction penetrates it). As noted above, only n^* and not n can be measured from the spectra, thus the integer part of δ is undetermined. For figure 3, we have chosen the integer part of δ to produce a monotonically increasing function.

We have done an approximate calculation of the shielding effect by using wavefunctions with no oscillations, but that incorporate the proper $r^{-3/4}$ amplitude variation. This crude estimate predicts a linear increase, as observed. The two cases of Rydberg electrons, ns or nd, do have different quantum defects for m=6, but this is primarily due to the xeon-like core of Ba++; note that the relative difference between quantum defects will become negligible as m becomes increasingly larger.

To analyze the autoionization widths of the $msn\ell$ states, it is convenient to consider scaled widths, i.e. autoionization widths multiplied by $(n^*)^3$. Since a Rydberg electron has a classical period proportional to $(n^*)^3$, this corresponds to the probability of autoionization per Rydberg electron orbit which should remain constant for an entire Rydberg series with a specific core configuration. A quantum mechanical argument also predicts this $(n^*)^3$ scaling, as a result of the normalization of the Rydberg wavefunction near the core. But there are few simple arguments for rules concerning the scaling of these autoionization widths as the core quantum number, m, is changed. One might expect that the scaled autoionization widths would increase dramatically for two reasons: (1) the core volume increases very rapidly with increasing m (as m^6), so that classically the core is a bigger target for the Rydberg electron; and (2) the number of final continuum states available increases as m^3. However, one could also argue that because the core size is increasing the magnitude of an $1/r_{12}$ interaction must decrease in inverse proportion to the distance scale increase.

We have attempted to estimate the autoionization rates for these states by solving two electron Hartree-Fock equations (with a model potential for the Ba++ core). Autoionization results from matrix elements such as $\langle msns|1/r_{12}|m's\varepsilon s\rangle$. It is instructive to create "autoionization potentials" by integrating only the core electrons coordinate. This function shows explicitly where the important region for autoionization is. Generally, function looks like a square window which extends from the origin out to the lesser of the two core wavefunction edges. The amplitude of this function has the largest amplitude for m'=m-1. The larger amplitude and width of this function strongly suggests that m'=m-1 will be the dominant decay channel as mentioned above. These effects also predict a dramatic increase in autoionization as m increases.

Figure 4 summarizes our data on the scaling of autoionization widths times $(n^*)^3$ as a function of the core electron quantum number for msns and msnd doubly excited states.[8] The msnd states show an increase of over an order of magnitude in the probability of autoionization per Rydberg orbit at m is increased from 7 to 11. However, the msns states show virtually no change in their scaled autoionization rates!

There is a classical argument for the insensitivity of the msns states to the core size. If one has a single classical solution to the two electron problem, one can generate an entire set of related solutions by scaling the radial coordinates of both the Rydberg and the core electron by some factor γ^2. The geometry of this solution would be

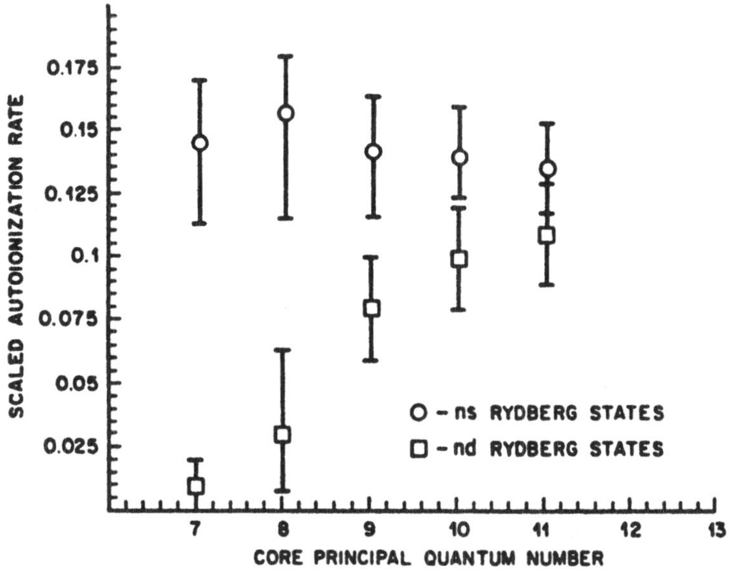

Figure 4. Core Scaling of Autoionization of ms and md states. The md states show increasing autoionization width with increasing core, while the ms states are independent of the core principal quantum number.

idential to that of the first solution, only its distance scale and time scale would be different. By Keppler's laws, the time scale would be increased by a factor γ^3. Consequently, the probability of autoionization per Rydberg electron orbit, which is a purely geometric factor with no time dependence, must be invariant under this distance scaling. This classical scaling also requires a scaling of the individual electrons' angular momenta, since both distance <u>and</u> time were scaled. Specifically, each electron's value of ℓ must be scaled by γ. For s states with $\ell=0$, this scaling has no effect, and we can conclude that the classical autoionization rate per Rydberg electron orbit should be independent of m for msns series with m<<n. Although the data support this prediction, we do not know of an equivalent quantum mechanical argument suggesting this behavior.

An alternate explanation of figure 4 suggests that the msns behavior is normal while the msnℓ behavior is unexplained. The wavefunctions produced using the hyperpherical coordinate method[9] would generally suggest little dependence of the autoionization rate on m. If this is true, then it is possible that the mdnℓ autoionization rates started at an anomalously low value and will saturate at the same value as the msns states. More data at higher m values is required to check this claim.

The second class of planetary atoms, those with m≈n have been much more difficult to observe and characterize. Since significant deviations from the independent electron model are expected, but the form of these deviations is not known, it is not clear where to search for such states. Moreover, it is expected that the interactions between the electrons will tend to mix a specific ms^2 configuration with a band of states that increases rapidly with m.[10] We have observed indirect evidence of this insofar as our signal became progressively weaker as we decreased n to values close to m, although the complications of our detection system do not allow a definitive measurement of oscillator strength to these states.

However, for states where n is within 3 of m, and m is large (10 or 11), we have observed striking, new structure. In these cases, $m^* \approx n^*$ due to the increased quantum defect of the Rydberg electron. Figure 5 shows a typical 6s12s→10s13s, excitation profile. Instead of a single line, the complex structure represents severe configuration mixing of the 10s13s, $^1S_0$ state with one or more Rydberg series. When an autoionizing state mixes with a Rydberg series, the result is a modification of the excitation profile such that additional resonances occur at the locations of the Rydberg states.[11] Near the center of the profile these resonances appear as dips while in the wings of the profile they appear as peaks. The relative strength of the configuration can be determined from the intensity of the structure compared to the unmodified background.

It is possible to determine what series are causing the structure in this spectra <u>because</u> we are using barium and not helium. The Ba+ ion is isoelectronic with C_s, and although the Ba++ core electrons are not active, they nevertheless produce quontum defects for the Ba+(nℓ) states. Thus, the quantum defects of the f, g and h states are 0.81, 0.017, and 0.005 respectively. From the spacings of the structure, we can determine that the interfering sequence converges to the 6g or 6h limit, with the 6h being most likely. Thus, the structure in figure 5 suggests that the 10s13s, $^1S_0$ states is mixed with the 6hnh Rydberg sequence as strongly as it is mixed with all other continuum configurations combined! The 6hnh Rydberg series is particularly

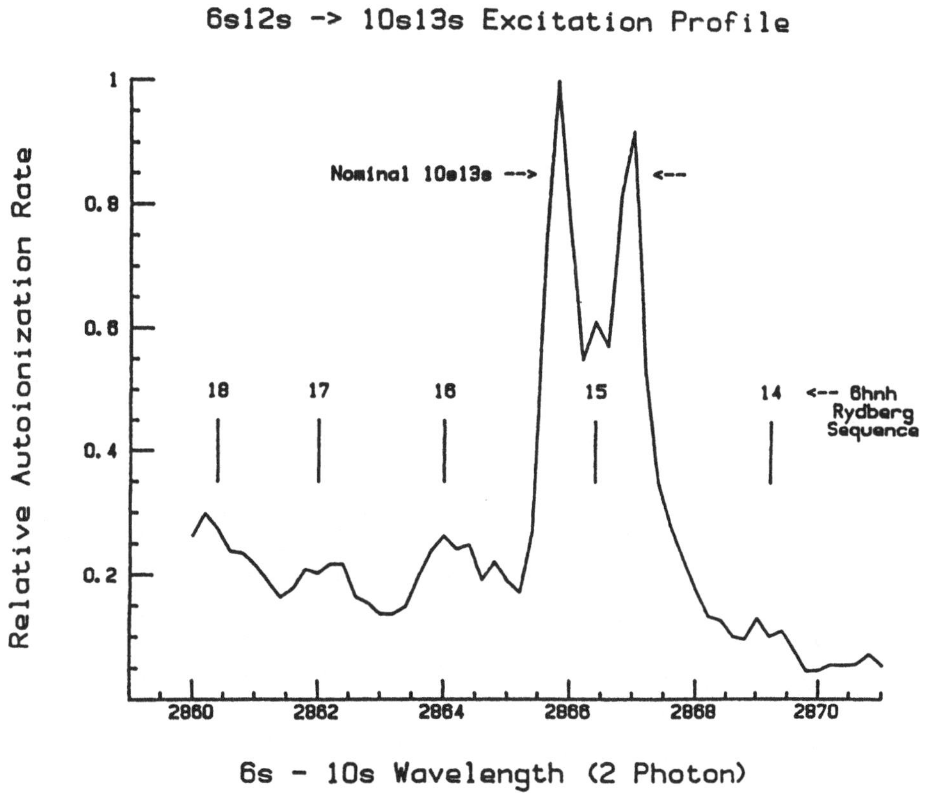

Figure 5. Excitation spectrum of 6s 12s → 10s 13s Transition. This spectrum shows substantial mixing with the 6hnh Rydberg series. This mixing of the highest available angular momenta for each individual electron is indicative of the onset of correlated behavior.

significant because it represents the circular ($\ell = n-1$) state of the ion, i.e. the interaction has mixed in the highest values of ℓ possible, even though this requires a r^5 multipole interaction. This contrasts with the theoretical work on the "Wannier ridge" type states which sould mix only with states of ℓ as high as $.5\sqrt{m}$.[12]

CONCLUSION

In conclusion, we have developed an efficient, selective method for populating "planetary" atoms. By exciting msns and msnd states with n>>m, we have observed simple behavior of the parameters associated with autoionizing Rydberg states, although much remains to be understood about the origins and the calculation of these parameters. We have also excited states with n=m and find that, if m is large, new structure suggests a surprisingly strong mixing of states with high values of individual ℓ. Current theoretical models do not appear to describe the structure and mixings we are beginning to observe.

ACKNOWLEDGEMENTS

The portion of this work conducted at USC was supported in part by the National Science Foundation under grant PHY82-01688. One of us (WEC) wishes to acknowledge support from the Alfred P. Sloan Foundation.

REFERENCES

1. I.C. Percival, Proc. R. Soc. Lond. A 353, 189 (1977).
2. D.R. Herrick and O. Sinanoglu, Phys. Rev. A 11, 97 (1975).
3. U. Fano, J. Phys. B. 7, L401 (1974); see also a review: U. Fano, Rep. Prog. Phys. 46, 97 (1983).
4. S.J. Buckman, P. Hammond, F.H. Read and G.C. King, J. Phys. B 16, 4039 (1983).
5. S.A. Bhatti and W.E. Cooke, Phys. Rev. A 28, 756 (1983).
6. J.R. Rubbmark, S.A. Borgstrom and K. Bockasten, J. Phys. B 10, 421 (1977).
7. R.M. Jopson, R.R. Freeman, W.E. Cooke and J. Bokor, Phys. Rev. Lett. 51, 1640 (1983).
8. L.A. Bloomfield, R.R. Freeman, W.E. Cooke and J. Bokor, Phys. Rev. Lett. 53, 2234 (1984).
9. J. Macek, Private Communication (November, 1984).
10. H. Taylor, Private Communication (March, 1984).
11. W.E. Cooke and S.A. Bhatti, Phys. Rev. A. 26, 391 (1982).
12. A.R.P. Rau, J. Phys. B 17, L75 (1984).

LASER SPECTROSCOPY OF HIGHLY EXCITED HYDROGEN ATOMS IN ELECTRIC AND MAGNETIC FIELDS

H. Rottke, A. Holle, and Karl H. Welge

Fakultät f. Physik, Universität Bielefeld, D-4800 Bielefeld 1, FRG

Abstract The first part of the paper provides a review of systematic studies carried out in our laboratory on the Stark effect of highly excited H-atoms around the ionization threshold. Excitation is performed from individual Stark sublevels in the H(n = 2) state with π and σ polarized laser radiation, resulting in eight essentially different ionization spectra. The experimental results fully agree with theory.

In the second part we report first experiments on the diamagnetism of the H-atom around threshold, exciting the atoms, as in the electric field case, in two steps through the H(n = 2) state. Individual Paschen-Back levels in n = 2 are selectively excited in the first step. Spectra have been obtained with final, ionizing states m_l = 0, -1, -2 at fields B ≤ 6T at excitation energies from the l-mixing regime, that is Rydberg states n ≥ 23, through the ionization limit into the continuum. Quasi-Landau resonances have been observed for the first time with the H-atom.

Introduction

The electronic structure and motion of atoms and molecules in highly excited states around the ionization threshold in the presence of external electric and magnetic fields is essentially governed by two intimately related special and characteristic circumstances. Firstly, the external forces are comparable with, or larger than the binding forces of the excited electron, so that they can no more be treated as perturbations of the internal interactions. This strong mixing of internal and external forces occurs generally already at fields of laboratory strength with atomic systems around threshold. Secondly, because of the strong force mixing the symmetry of the system in high discrete bound and unbound continuum states, that is at large distances from the ionic core, is profoundly determined and altered by the external fields. The consequence of both features is that such systems can no more be described and understood by common theoretical approximation and perturbation schemes. Also, the problem may no more be exactly separable in any coordinate system, as is the case with magnetic fields where the diamagnetic interaction dominates the linear Zeeman effect and mixes strongly with the binding potential. Aside from the basic significance of the strong mixing interactions, they are of wider interest and relevance for other areas in physics, for instance Rydberg states, ionization and recombination processes, plasmas, and astronomical objects with superstrong fields.

The physics of weakly bound, highly excited atoms in external fields has gained increasing interest in recent years for a number of different reasons. It obtained renewed attention by the discovery of field induced quasi-Landau resonances in the photoabsorption cross section of atoms in magnetic fields by Garton and Tomkins (1). Complementary oscillatory resonance structures were later observed in electric fields first by Freeman et al. (2). Further stimulation was very much provided by the progress in experimental techniques, particularly laserspectroscopy which made state selective excitation and investigation of atoms and molecules possible (3).

The first observations of the field induced resonances in the ionization-cross sections have been followed by extensive experimental and theoretical work in this field (3). However, with very few most recent exceptions, all experimental studies have been performed with non-hydrogenic atoms, while on the other hand a vast amount of theoretical work is based on the hydrogen atom, serving as prototype and basis with its purely Coulombic potential. The application of laserspectroscopy methods to the hydrogen atom, otherwise widely used in this field (3), has been prevented by experimental obstacles in the state selective excitation of the atom. In electric fields, Koch and collaborators (4) have performed experiments on the H atom Stark effect of quasi-stable states. They employed a fast atomic beam laserspectroscopy. Bergeman et al. (5) have carried out more recently work in very high fields ($\sim$ 3 MeV/cm) in low Rydberg states ($n \simeq 4$) with high energy ($\sim$ 800 MeV) beams. Nayfeh et al. (6) have carried out experiments with a multiphoton excitation technique, at electric field strengths in the kilovolt-range.

No previous experiments are known with the H-atom in strong magnetic fields. First experiments by us have been reported recently (7).

In this paper we report experiments carried out in our laboratory with the H atom in electric and magnetic fields. We briefly summarize in the first part results of systematic investigations on the Stark effect. For details we refer to our previous publications (9, 8). The electric field work has reached a sort of concluding point in so far as the experimental results are found to fully agree with theory. This must be expected because the non relativistic hydrogenic Stark problem is separable in parabolic coordinates and thus in principle quantitatively solvable to any degree of accuracy (11, 10). The situation is basically different for the diamagnetism of the H atom since the problem is not separable. First results of our experiments in magnetic fields have been recently reported, including the first observation of quasi-Landau resonances (7).

Experimental

For details of the experimental procedure we refer to previous publications (9, 8). The basis of the experiments, in both electric and magnetic fields, is the two-step excitation with pulsed (∿ 8 nsec) tunable vacuumultraviolet (vuv) and ultraviolet (uv) laser radiation,

$$H(n = 1) + vuv \rightarrow H(n = 2) + uv \rightarrow H^{*}, \qquad (1)$$

in crossed laser-atomic beam arrangements. The beams crossed each other at the center of a uniform electric field between two parallel field electrodes. Ionization of the H^* atoms was monitored by detection of the electrons formed. Keeping the vuv in resonance with the (n = 1) → (n = 2) transition, ionization spectra were taken by scanning the uv laser wavelength.

In the two-step excitation method the Stark splitting in the n = 2 state (12) had to be carefully taken into account. At the field strengths applied the four Stark sublevels in n = 2 are of practically pure parabolic character. In common parabolic quantum number notation (11, 10) they are identified by $|1, 0, 0\rangle$, $|0, 1, 0\rangle$, $|0, 0, |\pm 1|\rangle$, $|0, 0, |\mp 1|\rangle$. The resolution in the (n = 1) → (n = 2) step, given by the laser bandwidth and the Doppler linewidth, was such that each level was excited individually, the first two with π polarized and the second two with σ polarized vuv.

In the magnetic field experiments field strengths B ≤ 6T have been applied. The atomic beam passed through the field region parallel to the field axis. It was intersected at the center of the field perpendicularly by the two vuv and uv laser beams. An electric field, parallel to the magnetic field, was turned on after the laser excitation pulse (∿ 1 μs delay), to ionize the H^* atoms and to extract the electrons onto a surface barrier detector, placed behind the positively charged field electrode grid.

Electric Field

From each of the four Stark levels in n = 2, ionization spectra have been taken and investigated with π and σ polarized uv, resulting in eight essentially different types of spectra. Fig. 1 shows two examples with $|1, 0, 0\rangle$ as initial level at F = 5714 V/cm with, respectively, π and σ excitation. They span the energy range from the classical saddle point defined by $E_{sp} = -2\sqrt{F}$ a.u., through

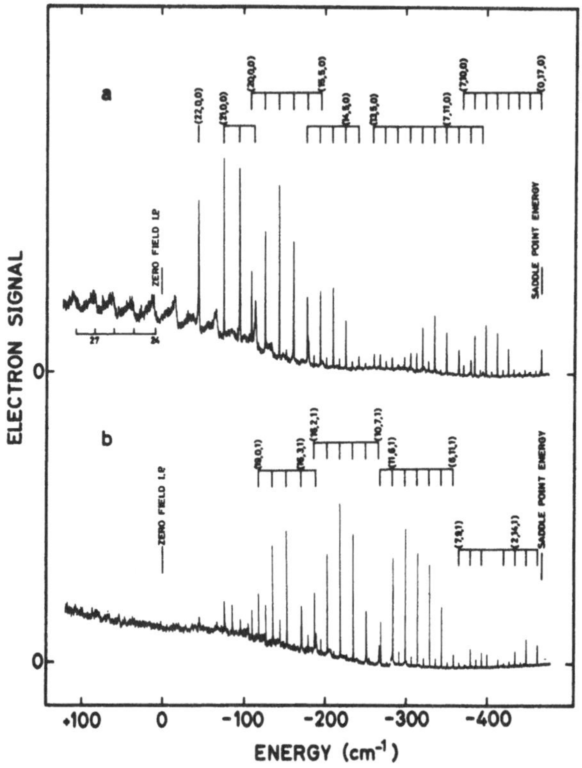

Figure 1. Photoionization spectra of the H atom in an electric field F = 5714 V/cm. Excitation from the H(n = 2) Stark sublevel $|1, 0, 0\rangle$ with π and σ polarized radiation, spectrum (a) and (b), respectively.

the zero-field limit (E = 0) into the continuum (E > 0). They exhibit all essential features of the H-atom Stark effect in the field ionization region: In the range $-|E_{sp}| < E < 0$ we observe the sharp line structures of quasi-stable states superimposed on a continuum background. The lines are labelled by the parabolic quantum number of the final states, $|n_1^f, n_2^f, |m^f|\rangle$, grouped according to $n = n_1 + n_2 + |m| + 1$, i.e. the state manifolds originating from a given principle quantum number n. The identification of the states has been achieved by calculating the absolute state energies by theoretical procedures previously developed (13). The field-ionization lifetimes of the quasi-stable states range from $> 2 \times 10^{-6}$ sec, which is the detection limit in these experiments, to $> 10^{-12}$ sec, where the states form the continuum (14). Lifetime measurements (8,9) agree with the theoretical dependence of the stability of the quasi-stable states on the quantum numbers (n, n_1, n_2, $|m|$).

The spectrum in Fig. 1a taken from $|1, 0, 0\rangle$ with π polarized uv, shows the field induced oscillatory resonances in the continuum ionization cross section. They were not observable in any of the other seven spectra, that is the modulation degree there was utmost of the order of a few percent. Fig. 2 shows the dependence (open circles) of the energy spacing between two adjacent maxima (or minima)

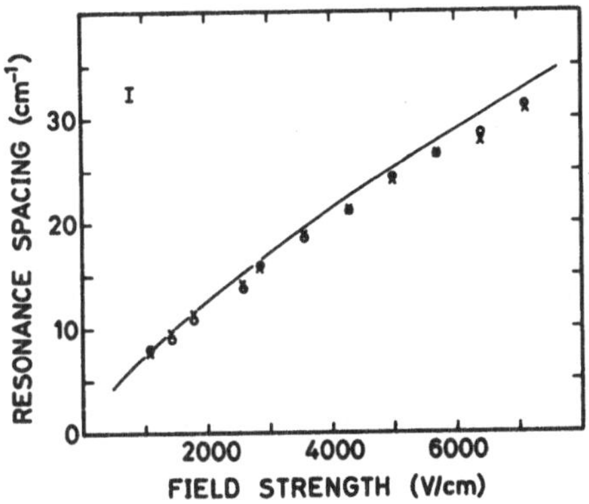

Figure 2. Energy spacing of field induced oscillations as function of the electric field strength; two adjacent maxima closest to the zero-field ionization threshold. Open circles: experimental results; solid curve: $F^{3/4}$ dependence; crosses: theoretical calculations (see text). Vertical bar: precision limit.

closest to E = 0, measured as a function of the field strength. The solid line represents the $F^{3/4}$ dependence expected from the simplest WKB approximation procedure (15, 9). The crosses indicate results from a more exact calculation (see below).

The field induced resonances have been theoretically treated by various methods (15,16,17,18,19). Following WKB procedures (16, 17, 18) we have calculated total excitation-ionization cross sections at $E \geq 0$ for all eight spectra from the four initial n = 2 levels. Fig. 3 shows two examples for the $|1, 0, 0\rangle$ and

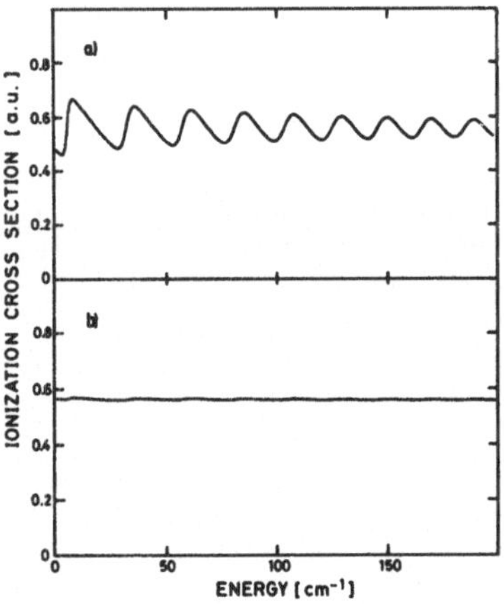

Figure 3. Total ionization cross sections from initial Stark sublevel:
(a) $|1, 0, 0\rangle$ in n = 2;
(b) $|0, 1, 0\rangle$ in n = 2
with π polarization. Theoretical calculation (ref. 9)

$|0, 1, 0\rangle$ initial states with excitation by π polarized ionizing radiation. The theoretically obtained results agree quantitatively well in all essential features of the resonances, that is their degree of modulation, their shape and spacing, with the experiments.

As has been shown previously (15, 3) the spacing can be readily obtained from the WKB quantization condition for the electronic motion in the bound potential in the parabolic coordinate, ξ:

$$\int_{\xi''}^{\xi'} (E/2 - m_1^2/4\xi^2 + Z_1/\xi - F\xi/4)^{1/2}\, d\xi = (n_1^f + 1/2)\pi \qquad (2)$$

ξ'' and ξ' are the turning points of this motion. For the H atom the separation constant Z_1 can be set $Z_1 = 1$ to a good approximation (9). For given n_1^f, m_2^f, and F one thus derives the maximum (or minimum) resonance energy positions. Results obtained for the spectrum in Fig. 1a agree well with the experiment. The field dependence of the spacing has been derived by means of equ. (2), however, not setting E = 0 as previously (15) but calculating directly the actual energies of two adjacent maxima closest to E = 0. The results shown in Fig. 2 by the crosses agree very well with the measurements.

Magnetic Field

Because of the two-photon excitation through the n = 2 state the magnetic structure in this state has to be taken into account. At the fields applied ($B \leq 6T$) the Paschen-Back effect dominates as indicated in Fig. 4. The three m_1 components, $m_1 = 0, \pm 1$, are individually excited by employing π or σ polarized vuv and a

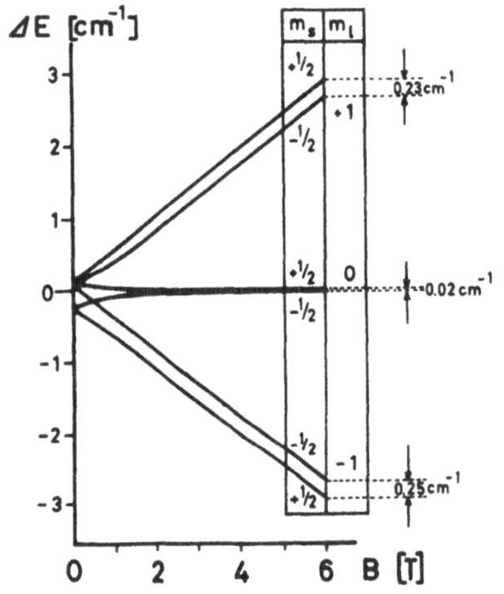

Figure 4. Paschen-Back splitting in the $H(n = 1) \rightarrow H(n = 2)$ transition as function of the field strength.

total resolution (laserbandwidth and effective Doppler width) of $\sim 0.2\ \mathrm{cm}^{-1}$ in the first step. The m_s sub-components were not resolved. They did not effect the final state excitation, as indicated by preliminary experiments with somewhat higher resolution in the first step. From the n = 2 levels $m_l = 0, \pm1$ the following final-state transitions are allowed with π and σ polarized radiation: $(m_l = 0, \pm1 \xrightarrow{\pi} m_l^f = 0, \pm1)$; $(m_l = 0 \xrightarrow{\sigma} m_l^f = \pm1)$; $(m_l = -1 \xrightarrow{\sigma} m_l^f = 0, -2)$; $(m_l = +1 \xrightarrow{\sigma} m_l^f = 0, +2)$. According to Clark and Taylor (20) in σ-transitions (i.e. $\Delta m = \pm1$) the ones to $m_l^f = 0$ levels have smaller oscillator strengths than the ones to $m^f = \pm2$ (the last two of the cases above) by about an order of magnitude. If we neglect the weaker transitions the following final states with $m_l = 0, \pm1, \pm2$ are excited predominantly. Since parity is conserved all final states are of even parity (p^f = even).

Fig. 5 shows three examples of overview spectra with relatively low resolution

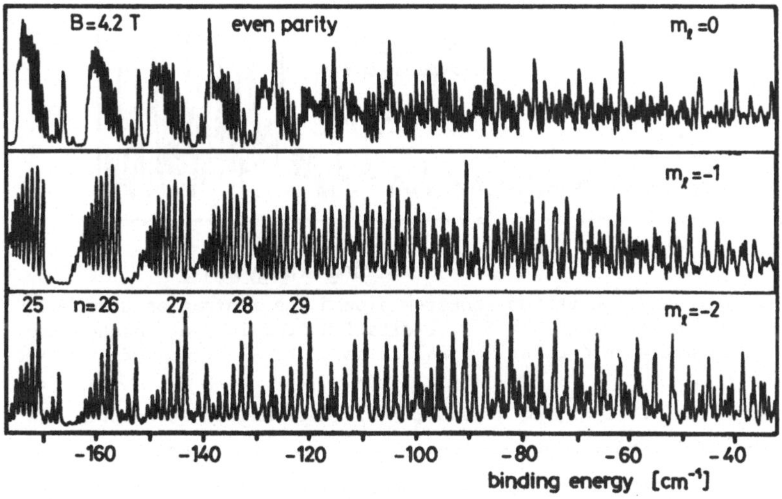

Figure 5. Hydrogen atom Rydberg spectra in a magnetic field with dominant diamagnetic interaction in the l-mixing and lower n-mixing regime. Excitation from levels in the Paschen-Back manifold of n = 2: $m_l = 0 \xrightarrow{\pi} m_l^f = 0$; $m_l = -1 \xrightarrow{\pi} m_l^f = -1$; $m_l = -1 \xrightarrow{\sigma} m_l^f = -2$ (0). Field strength B = 4.2 T.

taken at B = 4.2 T. They are excited through the transitions (from top to bottom): $m_l = 0 \xrightarrow{\pi} m_l^f = 0$; $m_l = -1 \xrightarrow{\pi} m_l^f = -1$; $m_l = -1 \xrightarrow{\sigma} m_l^f = -2$. In the last case a small admixture is present with final states $m_l^f = 0$. Transitions to final states $m_l^f = +1, +2$ yield correspondingly the same types of spectra. At low energies n-manifolds with l-mixing are still individually distinguished, going over with increasing energy into the n-mixing regime. Sections of these spectra have been taken and investigated (8) at higher resolution in regions of individual n = manifolds. Analysis of, for instance, the manifolds n = 23, 24 and 25 at B = 6 T, showed the splitting in lines identified by the k quantum number (21, 20).

According to selection rules (21) final states with k of even values are observed, i.e. for given n and m_l a series of states k = 0, 2, ... n - $|m_l|$ - 1 is obtained. The observed spectral intensity distributions depend essentially on m_l^f, as seen in Fig. 5. The spectral intensity distributions are found to qualitatively agree with the ones theoretically calculated by Clark and Taylor (20). According to the diamagnetic interaction the energy position of final levels shifts quadratically with the field strength (3 , 21). We have investigated the splitting at fields up to B = 6 T in the energy region from -160 cm^{-1} to -110 cm^{-1}, covering Rydberg states n = 26 to 32. As to be expected lines with given k quantum number are observed to shift proportinal to B^2, as in previous experiments with non-hydrogenic atoms (21). Fig. 6 shows a spectrum around threshold at B = 6 T with final states m^f = -2, that is the same final state character as in the

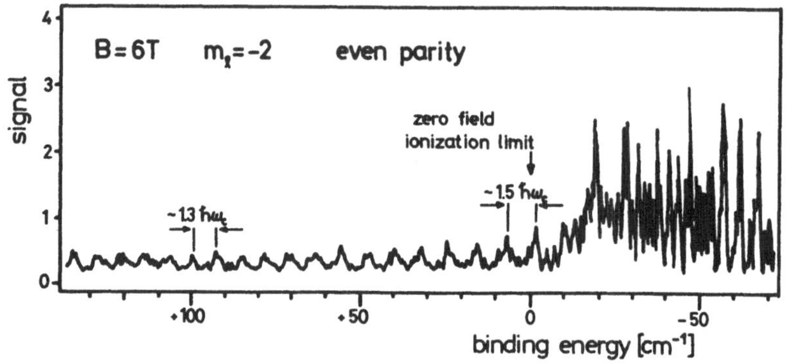

Figure 6. Hydrogen atom ionization spectrum around the ionization threshold (E = 0) in a magnetic field of strength B = 6 T. Excitation from initial Paschen-Back level m_l = -1 in n = 2 state with σ polarized radiation to final states of m_l^f = -2.

third spectrum in Fig. 5. Starting around E = 0, quasi-Landau structures are observed. At threshold the energy spacing is ∿ 1.5 $\hbar\omega_c$, (ω_c = e/m B, the cyclotron frequency), gradually decreasing at higher energy, as to be expected from theory (15, 3) and previous observations with other atoms (1, 3).

Conclusion

The experiments in electric fields agree fully with theory, so that the non-relativistic Stark problem of the highly excited H atom has come to a kind of concluding point.

The work demonstrates the feasibility of experiments with the hydrogen atom around threshold in magnetic fields. First results have been obtained in the l-mixing region and in the quasi-Landau regime. They agree with existing theory. Objectives of current work with the H atom are: a) the n-mixing region beow the ionization limit and the transition to the quasi-Landau regime with higher resolution,

b) the shape and modulation of the quasi-Landau resonances, and c) experiments in combined electric-magnetic fields.

References

1. W.R.S. Garton and F.S. Tomkins, Astrophys. J. 158, 839 (1969)
2. R.R. Freeman, N.P. Economou, G.C. Bjorklund, and K.T. Lu, Phys. Rev. Lett. 41, 1463 (1978); R.R. Freeman and N.P. Economou, Phys. Rev. A20, 2356 (1979)
3. Recent reviews of the field for instance: (a) C.W. Clark, K.T. Lu, and A.F. Starace, "Progress in Atomic Spectroscopy" Part C, ed. H.J. Beyer and H. Kleinpoppen; Plenum, N.Y. (1984).
   (b) J. Physique, Colloque C-2 (1982) on "Atomic and Molecular Physics close to the Ionization Threshold in High Fields";
   (c) S. Feneuille and P. Jacquinot, Adv. At. Mol. Phys. 17, 99 (1981).
4. P.M. Koch, Phys. Rev. Lett. 41, 99 (1978); P.M. Koch and D.R. Mariani, J. Phys. B13, L 645 (1980); Phys. Rev. Lett. 46, 1275 (1981); H.J. Silverstone and P.M. Koch, J. Phys. B12, L 537 (1979).
5. T. Bergeman, C. Harvey, K.B. Butterfield, H.C. Bryant, D.A. Clark, P.A.M. Gram, D. MacArthur, M. Davis, J. Dayton, and W.W. Smith, preprint (1984)
6. W.L. Glab, K. Ng, D. Yao, and M.H. Nayfeh, preprint, Phys. Rev. (1985)
7. H. Rottke, A. Holle, and Karl H. Welge, in "Atomic Excitation and Recombination in External Fields"; ed. M.H. Nayfeh and C.W. Clark, Harwood Acad. Publ. (1985)
8. K.H. Welge and H. Rottke, in "Laser Techniques in the Extreme Ultraviolet", p.p. 213-219 (1984); AIP Conf. Proceedings No. 119
9. H. Rottke and K.H. Welge, submitted for publ. in Phys. Rev. A (1985)
10. H.A. Bethe and E.E. Salpeter, "Quantum Mechanics of One and Two Electron Atoms", Springer (1957)
11. L.D. Landau and E.M. Lifshitz, "Quantum Mechanics", Pergamon Press (1965)
12. G. Lüders, Am. Phys. 8, 301 (1951)
13. H.J. Silverstone and P.M. Koch, J. Phys. B12, L 537 (1979); H.J. Silverstone, Phys. Rev. A18, 1853 (1978), H.J. Silverstone, B.G. Adams, J. Cizek, and P. Otto, Phys. Rev. Lett. 43, 1498 (1979); R.J. Damburg and V.V. Kolosov, J. Phys. B14, 829 (1981).
14. R.J. Damburg and V.V. Kolosov, J. Phys. B12, 2637 (1979); ibid. B14, 829 (1981)
15. A.R.P. Rau, J. Phys. B12, L 193 (1979); A.R.P. Rau and K.T. Lu, Phys. Rev. A21, 1057 (1980)
16. V.D. Kondratovich and V.N. Ostrovsky, Sov. Phys. JETP 52, 198 (1980)
17. D.A. Harmin, Phys. Rev. A24, 2491 (1981); ibid. A26, 2656 (1982)
18. E. Luc-Koenig and A. Bachelier, J. Phys. B13, 1743 (1980); ibid. B13, 1769 (1980)
19. W.P. Reinhardt, J. Phys. B16, L 635 (1983)
20. C.W. Clark and K.T. Taylor, J. Phys. B15, 1175 (1982)
21. D. Kleppner, M.A. Littman, and M.L. Zimmerman; in "Rydberg states of atoms and molecules"; ed. R.F. Stebbings and F.B. Dunning; Cambridge Univ. Press, p.p. 73-116 (1983)

DYNAMIC PROCESSES IN MOLECULAR RYDBERG STATES

Hanspeter Helm
Molecular Physics Department
SRI International, Menlo Park CA 94025

Laser excitation in fast beams of triplet hydrogen allows the preparation of short-lived molecular states which are coupled to the continuum of their fragmentation or autoionization products, under conditions where all quantum numbers and the total energy of the system are specified. The decay of the system can then be monitored by detecting the dissociation or autoionization products, and their translational energy and quantum state can be determined. The experiments are performed under conditions nearly free of Doppler broadening employing single molecule detection. This allows experimental determination of precise molecular term energies, and of natural lifetimes of the excited states. As a consequence we may specify exact boundary conditions of the dynamic path that connects both the molecular and the product frame.

INTRODUCTION

The development of the laser has had a profound effect on the applicability of fast molecular beams for the study of dynamic processes in molecular systems. Photofragment spectroscopy of molecular ions is a specific example of such an application[1] which has provided detailed insight into the bound and continuum region of molecular systems and has allowed the experimental definition of nonadiabatic coupling to continuum states in great detail. This technique combines the virtues of three fields: - molecular spectroscopy of bound states, - photodissociation through states that lie energetically in the continuum, and - translational spectroscopy of the fragmentation products.

This synthesis of experimental approaches combines the classical approach of molecular spectroscopy where bound states of a molecule are emphasized, with scattering experiments, in which the continuum wavefunction is the prime observable, by defining the dynamic path that connects these two ends of the spectrum. Very recently we have applied this technique to the study of dynamic processes of Rydberg states in neutral molecules of hydrogen. Our first results in this direction indicate the great potential of this experimental approach also in the study of slow atom-atom and electron-molecule interactions.

EXPERIMENTAL

The virtues of spectroscopy in fast beams of charged and neutral molecules are associated with the possibility of the precise definition of the molecular species under study and the reduction of inhomogeneous broadending.[1] The former is related to the fact that a fast molecular ion beam can be mass selected with a suitable velocity or momentum analyzer, and, if a neutral beam is required, neutralized in a suitable charge transfer gas. This seemingly complex way of preparing the sample molecule offers the possibility of selectively preparing molecules in states other than the ground electronic state, in non-statistical vibrational distributions, or radicals purified from their chemical precursors or reaction products. Metastable molecular hydrogen in the triplet state is such an example (see Figure 1).

Electron impact ionization of H_2 in a low pressure ion source leads to the formation of H_2^+, populated in all vibrational levels. This broad distribution reflects the unequal equilibrium internuclear separations of ground state H_2 and H_2^+, since Franck-Condon arguments determine the ionization event. The ions are then accelerated into a beam at keV energies and separated from ions other than mass two in a Wien-filter. As a consequence a well-collimated beam of H_2^+ with known translational energy can be prepared in a high vacuum environment. For spectroscopic studies of neutral molecular hydrogen this ion beam can be neutralized in a charge transfer cell. If alkalis are used[2] as charge-transfer agent a dominant product in the charge transfer process is the near resonant formation of H_2 in the metastable $c^3\Pi_u^-$ state. Again Franck-Condon factors influence the reaction and since H_2^+ is prepared in a wide range of vibrational levels by electron impact, so is the resultant neutral state. Also, since little momentum is transferred by the exchanged electron, a neutral beam with qualities

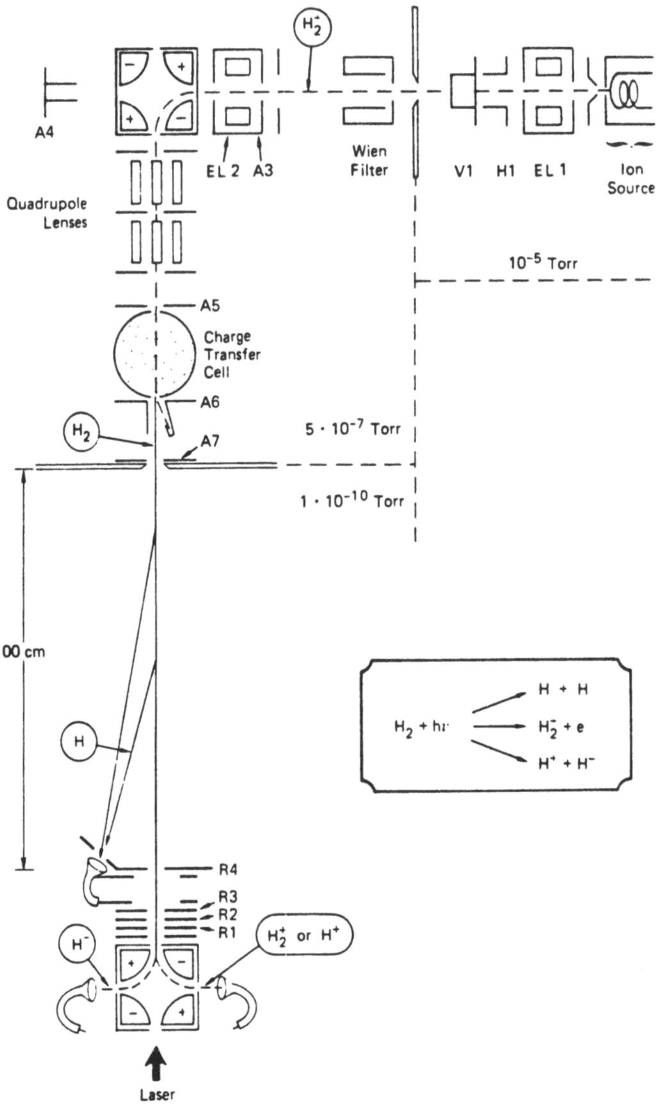

FIGURE 1 SRI Fast Molecular Neutral and Ion Beam Spectrometer

comparable to the primary ion beam results particularly in the case of near resonance. This beam can now be stripped from residual ions and apertured to suppress dissociative charge-transfer products which appear with a near isotropic angular distribution. In this manner a fast beam of up to typically 10^9 molecules per second travelling at speeds of the order of .1% of the speed of light can be prepared. Note that the pressure in this beam amounts to as little as 10^{-13} Torr. Nevertheless laser spectroscopy is readily performed with such beams as will be shown below.

An intrinsic second advantage of spectroscopy in a fast beam lies in the dramatic reduction in the inhomogeneous broadening of optical transitions. When an ion cloud is accelerated to high energy a compression of relative velocities occurs when viewing the beam coaxially. If the angular spread of the ion beam is kept low by proper ion-optical focussing the residual spread of relative velocities in the beam is reduced to that of a gas sample at a few degrees Kelvin. For the metastable hydrogen beam formed in the apparatus shown in Figure 1 we have observed a residual Doppler spread, $\delta\lambda/\lambda < 2.10^{-7}$. However, when viewing the beam coaxially, a significant coaxial Doppler shift arises which is readily calculated.

An additional beneficial property of the beam can be employed in the study of dynamic processes of molecules which are induced by photoexcitation. This advantage derives from the fact that the center-of-mass of the products of a process such as dissociation, autoionization, ion-pair formation or detachment retains, but for a negligible photon recoil, the velocity of the parent. Since the parent is fast, at keV energies, efficient single-particle detection techniques can be employed to monitor the reaction products and measure their translational energy, thus enabling to monitor single-molecule processes. The application of fast beams to photofragment spectroscopy and photoinduced autoionization of neutral molecules is very recent, still at a developmental stage, but with high promises. In what follows we describe this new technique for the example of metastable hydrogen molecules.

PHOTODISSOCIATION

A novel approach to the study of dissociative processes in fast neutral beams has become possible with an ingeneous time- and position-sensitive detector which has been developed at FOM.[2] In these studies the fast beam of excited neutral molecules is photodissociated with a tunable dye-laser in a crossed beam configuration and the resulting photofragments are detected with a time- and position-sensitive detector, which allows the measurement of the momentum distribution of fragment pairs arising from a single dissociation event.[3] A schematic diagram of this experimental arrangement is shown in Figure 2. The fast neutral beam is crossed with the intracavity beam of a cw dye laser and then stopped in a V-shaped beam flag which shadows the inactive portion of the detector. The flight path from the photon interaction region to the detector is of the order of 150 cm.

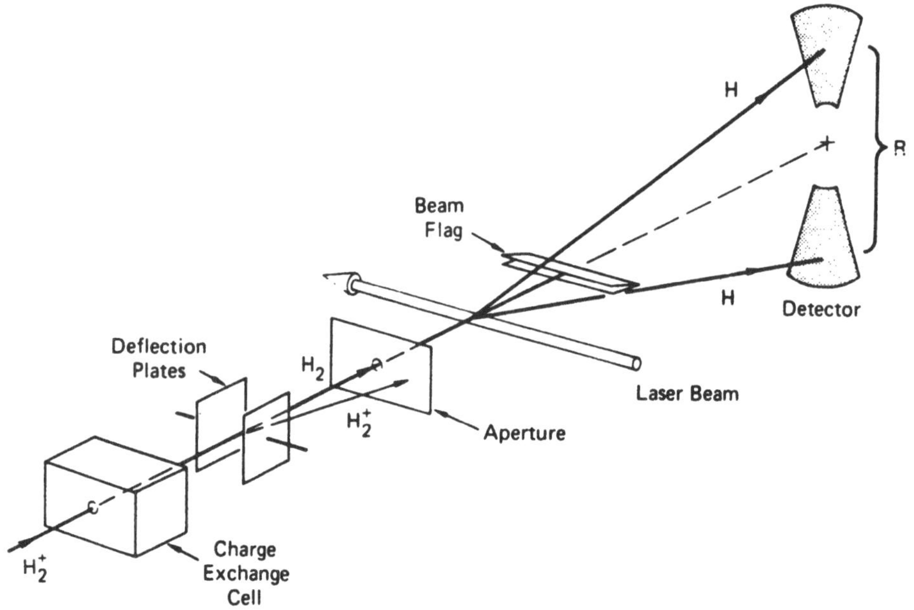

FIGURE 2 Schematic Diagram of the FOM Neutral-Beam Photofragment Spectrometer with Time- and Position-Sensitive Detector.

The multichannel plate detector consists of two opposing sectors of opening angle 20° which allow separate detection of the two photofragments produced in a single photodissociation event. The detector and its associated electronics permit the measurement of the spatial separation, R, of the two fragments at the channel-plate surface with a precision of typically 70 μm by measurement of the center of charge of each electron cloud emitted by the channel plates with a multianode system and the FOM charge division[4] method. The flight-time difference between these two fragments can be measured with a precision of 500 ps with the fast current pulse induced in the supply lines to the output face of the channel plates when a particle is detected. The spatial separation, R, that can be measured with the current device lies between 1 and 4 cm. At 7.5-keV parent-beam energy, this separation corresponds to c.m. energy releases in the range from 80 meV to 1.25 eV when dissociation occurs perpendicular to the parent-beam direction.

By tuning the laser and monitoring the coincidence count rate at the detector one may obtain absorption spectra of transitions in the neutral beam which lead to photodissociation. At an intracavity power of 20 W (multimode) counting rates of typically 10^4 fragment pairs/s are obtained on strong transitions in the 3d←2p systems if a primary beam current of 10^{-10}A of H_2^+ is charge-exchanged in a 1 cm long cell of 10^{-3} Torr of Rb. Space-and time-resolved spectra of photofragments can then be recorded with the dye-laser set to an absorption transition. If the dissociation is instantaneous, on a time scale short compared to the flight time between the excitation region and the detector, the time and spatial coordinates of the fragment pair carry all the information required to calculate the center-of-mass angle and energy under which the fragment pair emerges from the dissociation event.

In what follows we are primarily concerned with the energy release in the c.m. frame. It can be obtained from fragment pairs which appear coincident in time at the two detector halves. Such fragments are formed when dissociation occurs very nearly perpendicular to the molecular beam axis. Under these conditions the c.m. separation energy, W, is related to the measured spatial separation of photofragments, R, by the equation

$$W = E_0(R/L)^2 \quad ,$$

where E_0 is the parent-beam energy and L is the distance from the photon interaction region to the detector.

Figure 3 shows three spatial spectra of correlated fragment pairs, H + H, which illustrate three different mechanisms of photodissociation that are possible from the $H_2 c^3\Pi_u$ state.[3]

(1) Bound-bound-free photodissociation.--The excitation of bound triplet gerade states from $c^3\Pi_u$ gives rise to radiation[5,6] into the continuum of the $b^3\Sigma_u^+$ state. Excitation of a single rovibrational level in the $i^3\Pi_g$ state will give rise to a continuum distribution of photofragment energies which reflects the overlap of the bound-state vibrational wave function with the continuum wavefunction of the $b^3\Sigma_u^+$ state.

A number of such bound-bound-free transitions were observed belonging to the $i^3\Pi_g \leftarrow c^3\Pi_u$, $g^3\Sigma_g^+ \leftarrow c^3\Pi_u$, $j^3\Delta_g \leftarrow c^3\Pi_u$ and $h^3\Sigma_g^+ \leftarrow c^3\Pi_u$ systems.[7] Figure 3(a) shows an example of the continuous fragment-energy distribution which is obtained when we pump the R1 line of the i←c transition in the (3,3) band. The measured energy distribution represents the lower-energy portion of the total distribution produced, fragments with separation energies > 1.25 eV falling outside the current detection geometry and time window. The small structure which appears in the continuum energy distribution in Figure 3 (a) arises from an underlying bound-free photodissociation which is discussed below.

(2) Bound-quasibound photodissociation.--The $i^3\Pi_g$ and $h^3\Sigma_g^+$ states correlate to the energetically higher lying 3p and 3s united atom limits. This correlation leads to initially repulsive electronic states, the repulsion overcome by the Rydberg character imposed by the H_2^+ core at molecular distances. As a result these states develop intrinsic barriers in their potential energy curves which can support quasibound vibrational levels lying above the asymptotic dissociation limit H(1s) + H(2ℓ). We observed several transitions to quasibound levels which we have assigned to the

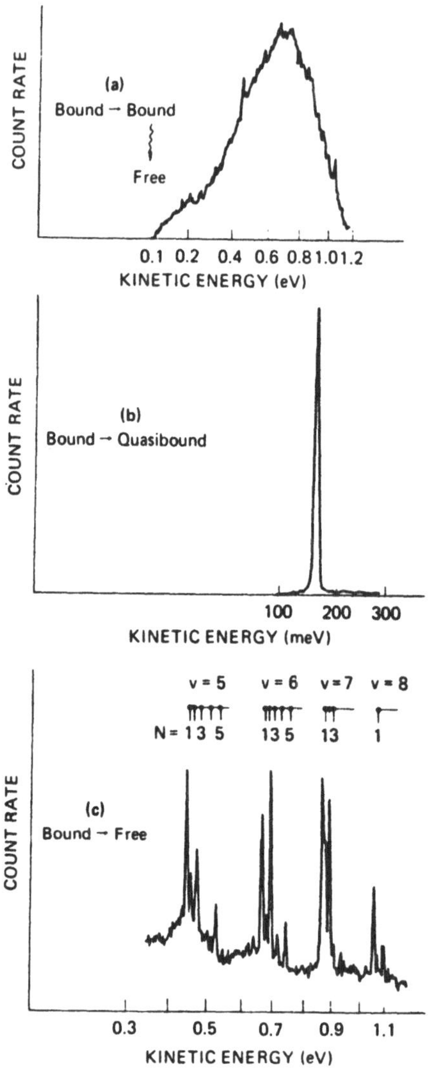

FIGURE 3 Photofragment Kinetic-Energy Spectra Observed in Photodissociation of H_2 $c^3\Pi_u^-$.

(4,4) and (5,5) bands of the i←c system. Figure 3(b) shows as an example the kinetic-energy spectrum of photofragments produced by pumping the R1 line of the (4,4) band (W ~ 160 meV).

(3) Bound-free photodissociation.--Underlying the discrete photodissociation spectrum a continuous background of laser-induced

dissociation was observed which arises from bound-free photodissociation (see Figure 4). At a fixed laser wavelength, bound-free transitions will produce photofragments at discrete energies $W = h\nu - D_{v,N}$ where $D_{v,N}$ is the dissociation energy of the rovibrational level with quantum numbers v,N from which the optical absorption occurs. Bound-free photodissociation at fixed wavelength leads to a fragment energy distribution that reflects directly the lower-state rotational and vibrational spacings. Because of the very high energy resolution of the time and position-sensitive detector we were able to resolve this distribution experimentally. Figure 3(c) shows such an energy distribution obtained at a fixed frequency near 16480 cm^{-1}, where no noticeable peak occurs in the absorption spectrum. As indicated in this figure, bound-free transitions are observed from the vibrational levels v" = 5, 6, 7, and 8 of the $c^3\Pi_u$ state, with individual rotational levels being resolved in the kinetic-energy spectrum. The intensity distribution over the individual rovibrational bound-free transitions reflects the population in the lower-state levels multiplied by the v- and N-dependent photodissociation cross section.

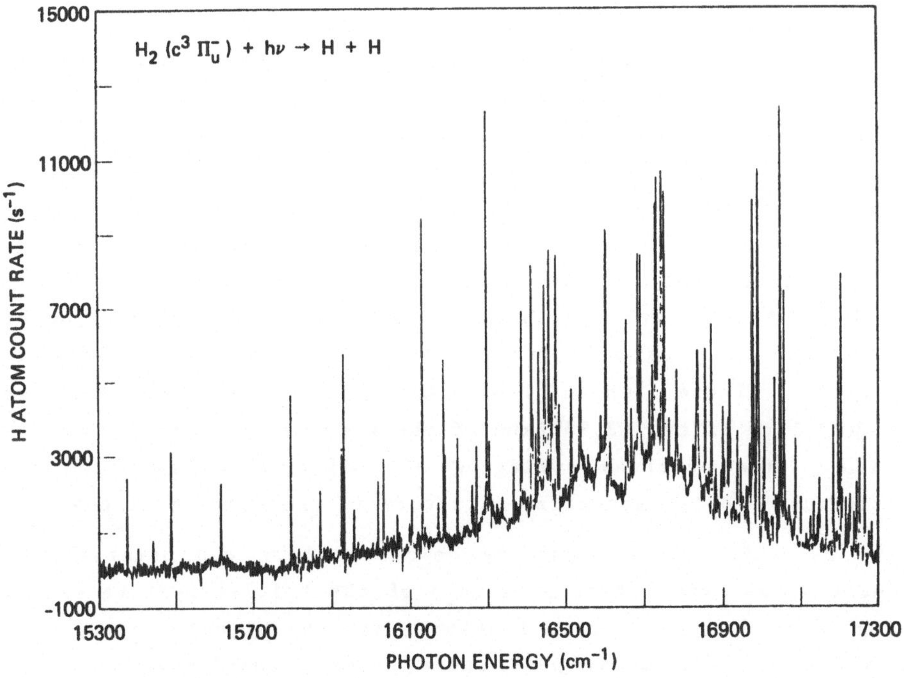

FIGURE 4 Low-Resolution Photodissociation Spectrum of H_2 $c^3\Pi_u^-$.

Note that the upper state involved in the direct dissociation from v" = 8, and 9 of the c-state lies at energies above the H_2^+v=0 ionization limit. However, autoionization was not observed to be a strong competitor to the dissociative path due to the unfavorable overlap between the continuum wavefunction of the dissociative channel with the vibrational wavefunction of the lowest vibrational level in H_2^+.

The data in Figure 3 demonstrate that a time and position sensitive detector extends the capabilities of photofragment spectroscopy to neutral molecules, permitting very high resolution, at the level of meV, in the translational energy of neutral fragments. The detection arrangement necessarily requires a crossed laser-neutral beam configuration. For such a configuration the ultimate optical apparatus resolution is determined by the angular divergence of the fast neutral beam and the precision at which the laser and neutral beam can be held perpendicular. For a neutral beam divergence of 1 mrad the residual Doppler broadening amounts to 3 GHz at 7.5 keV beam energy and 6000 Å for perfect perpendicular alignment.

The intrinsic high optical resolution of a fast beam can more easily be exploited in a coaxial configuration,[8] though at the expense of simultaneous determination of the translational energy of neutral fragments. Figure 4 gives the absorption spectrum from the c-state which leads to photodissociation in the visible wavelength range, obtained in the coaxial configuration shown in Figure 1. Here only a small portion of photofragments are detected, namely those which happen to fall into the solid angle seen by the off-axis channeltron. The low efficiency of the small solid angle is compensated for by the fact that the photon interaction region is now 100 cm long. If a single frequency laser is used an optical resolution of below 100 MHz can be achieved in this configuration. This is illustrated in Figure 5 which gives a high resolution spectrum of the of the Q(1) line of the (2,2) band of the g ← c system at 6134.28 Å. The observed splitting of this line arises from the fine-structure and the hyperfine-structure in the c and g states.

The photodissociation studies described here can naturally be extended to probe the higher n members of the triplet Rydberg series by using shorter wavelength excitation. For the higher n members at least one additional decay path opens, namely autoionization. In principle also ion-pair formation into $H^+ + H^-$ is possible via spin-orbit coupling of the triplet Rydberg states to the $^1\Sigma_g^+$ terms.

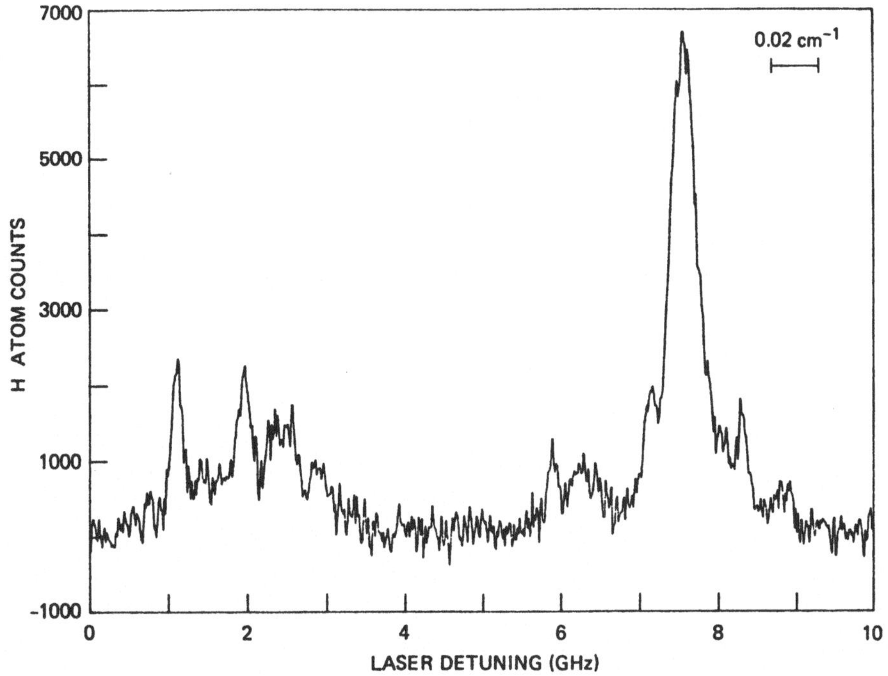

FIGURE 5 High-Resolution Photodissociation Spectrum Near 6134.28 Å.

AUTOIONIZATION

Using the experimental setup shown in Figure 1 we have obtained the excitation spectrum of high lying Rydberg states by detecting the appearance of H_2^+ formed in autoionization and field ionization[9] (see Figure 6). In these experiments a pulsed doubled dye laser pumped by a doubled Nd:YAG laser was used. The molecular hydrogen ions formed along the beam path are separated from the neutral beam using the electrostatic quadrupole deflector at the end of the beam line. This device acts as a low resolution energy analyzer of the ion kinetic energy. Hence, an unambiguous identification of the reformed H_2^+ can be made and the molecular ion can be detected on a channeltron, free from any interference of fast neutrals, or possible atomic ion fragments. Since the pressure in the photon interaction region is of the order of 1.10^{-10} Torr only insignificant collisional ionization of the H_2 beam molecules occurs and the experiment is essentially free of background.

Three ionization processes appear in Figure 6. The dominant discrete transitions arise from excitation of the nd Rydberg series which undergoes vibrational autoionization. A continuous background underlies the spectrum, gradually increasing from 3500 Å due to direct photoionization. Finally, near 3400 Å several intense peaks appear which are due to field ionization of high n members (n = 28 to 36) which lie slightly below the autoionization threshold. The field ionization observed here occurs at the entrance of the second quadrupole field where the beam experiences a transverse electric field of ~ 1.1 kV/cm.

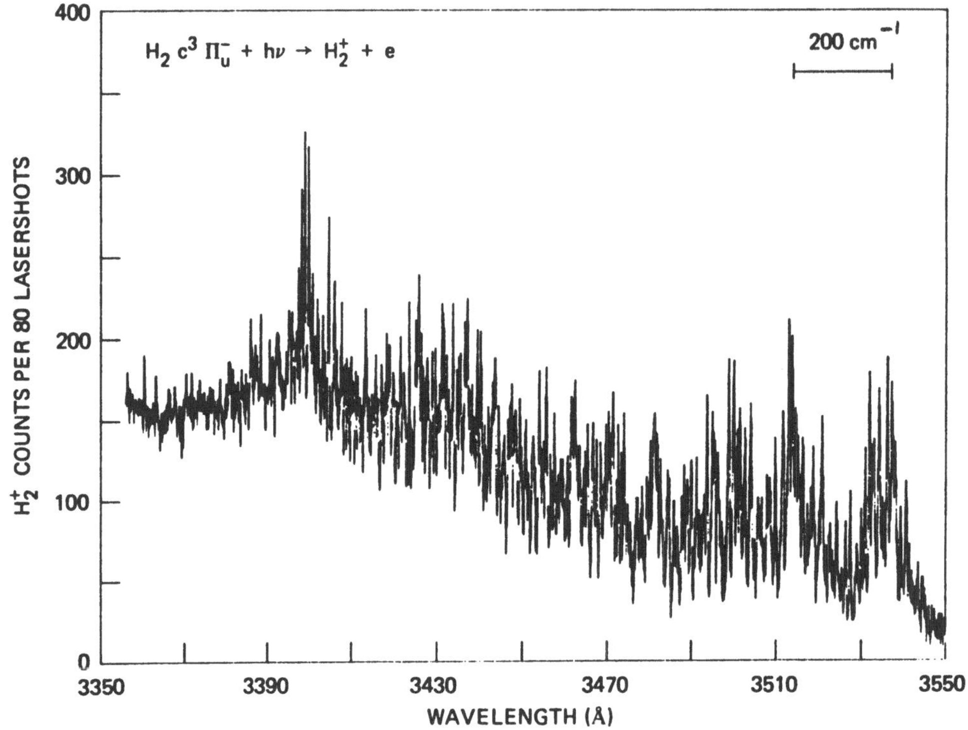

FIGURE 6 Low-Resolution Autoionization Spectrum of Rydberg States Excited From H_2 $c^3\Pi_u^-$.

Figure 7 shows a portion of the spectrum near 3400 Å obtained with a pressure tuned narrow-band (0.1 cm^{-1}) pulsed dye laser. The higher resolution spectrum shows that each low resolution peaks splits into several peaks, each group representing primarily contributions of a single value of n. Each individual peak shows additional

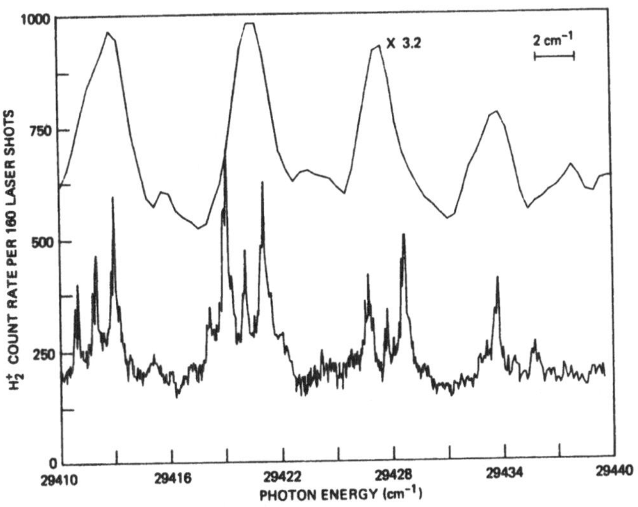

FIGURE 7 Medium-Resolution Autoionization Spectrum Near 3400 Å.

structure at the level of 0.2 cm^{-1} (6 GHz) indicating the fine-structure splitting in the c-state as seen in more detail in Figure 5. A detailed analysis of the spectra is currently being performed and will be reported elsewhere.

CONCLUSION

In summary, a fast beam of molecules in a high vacuum environment allows the selective excitation of molecules at sub-Doppler resolution. If this excitation occurs to a state which is coupled to a continuum, a dynamic rearrangement of molecular coordinates may occur. As a consequence the system might dissociate, or autoionize. The characteristic time in which this rearrangement occurs is reflected by the energetic width of the excited state and therefore in the absorption line profile. Since the continuum products can be efficiently detected and their translational energy and internal energy can be determined, a near complete description of a dynamic process which transforms matter can often be achieved.

ACKNOWLEDGEMENTS

It is a pleasure to acknowledge the substantial contributions to the experiments discussed here by Drs. N. Bjerre, D. P. de Bruijn, and R. Kachru. The author also wishes to thank the National Science Foundation for financial support through the grant NSF PHY-8411517.

REFERENCES

1. H. Helm, "Ion Photofragment Spectroscopy," in Electronic and Atomic Collisions (Ed., J. Eichler, I. V. Hertel, and N. Stolterfoht) Elsevier Science Publishers, p. 275 (1984).
2. D. P. de Bruijn, J. Neuteboom, V. Sidis, and J. Los, Chem. Phys. 85, 215 (1984).
3. H. Helm, D. P. de Bruijn, and J. Los Phys. Rev. Lett. 53, 1642 (1984).
4. D. P. de Bruijn and J. Los, Rev. Sci. Instrum. 53, 1020 (1982).
5. E. E. Eyler and F. M. Pipkin, Phys. Rev. Lett. 47, 1270 (1981).
6. E. E. Eyler and F. M. Pipkin, J. Chem. Phys. 77, 5315 (1982), and Phys. Rev. A 27, 2462 (1983).
7. Hydrogen Molecule Wavelength Tables of Gerhard Heinrich Dieke, edited by H. M. Crosswhite (Wiley-Interscience, New York, 1972).
8. N. Bjerre and H. Helm in preparation.
9. R. Kachru and H. Helm in preparation.

FOUR WAVE FREQUENCY MIXING IN GASES

C.R. Vidal

Max-Planck-Institut für extraterrestrische Physik
8046 Garching, F.R.G.

Abstract

Four wave frequency mixing in gases is now understood in a quantitative manner and has been used for the design of coherent infrared as well as vacuum uv sources. A brief review of the fundamental processes is given. The vacuum uv sources have already extended laser spectroscopy into the vacuum ultraviolet spectral region below 200 nm. Several examples of state selective spectroscopy of small molecules in the vacuum ultraviolet spectral region are presented.

1. Introduction

In recent years four wave parametric processes have been successfully used to extend the tuning range of lasers into the vacuum ultraviolet as well as into the infrared. In the visible and the infrared spectral region this has generally been done by sum or difference frequency mixing in suitable nonlinear crystals. In the vacuum ultraviolet spectral region, however, most solids become opaque and can no longer be used. In this case gaseous nonlinear media offer an alternative with several advantages:

1. Extended spectral regions of high transparency for the ultraviolet radiation can be provided.
2. The anomalous dispersion can be used for phase matching of gaseous two-component systems (1,2) and for providing large column densities as required for efficient nonlinear media.
3. Large nonlinear susceptibilities can be achieved by a suitable resonant enhancement (3-5).
4. The material properties such as the linear and nonlinear susceptibilities can be accurately calculated from first principles allowing a reliable test of the theoretical model.

In view of the fact that a comprehensive review of four wave frequency mixing in gases has recently been given by the author (6), this contribution will only be a brief summary pointing out the most important aspects of two-photon resonant and of nonresonant frequency mixing in gaseous nonlinear media (6-8). In recent years this technique has become increasingly interesting as a means of extending high resolution laser spectroscopy into the vacuum ultraviolet spectral region (9) where many atoms and a number of the most important small molecules such as H_2, CO, NO, and others have some of their most prominent absorption features.

In presenting the fundamental physical processes of four wave frequency mixing in gases it is useful to distinguish three regimes which differ with respect to the electric field amplitudes to be considered:

1. small signal limit,
2. onset of saturation,
3. high intensity saturation.

With growing electric field amplitudes an increasing number of nonlinear processes have to be taken into account for a quantitative description of four wave frequency mixing in gases.

2. Small Signal Limit

In the small signal limit the intensity of the resulting wave is proportional to the product of the intensities of the incident waves. For the third harmonic generation this yields the well-known cubic dependence for the harmonic intensity. In the small signal limit it basically suffices to know the phase matching curve which primarily depends on the complex linear susceptibilities, and to know the conversion efficiency which depends on the absolute value of the corresponding nonlinear susceptibility.

The phase matching curve is also affected by the properties of the incident beams and of the nonlinear medium. For focused beams the intensity distribution has to be considered which is typically characterized by the confocal parameter of a Gaussian beam. For practical applications the confocal parameter should be comparable or larger than the length of the nonlinear medium (10-12). An excessive focusing does not improve the power conversion efficiency any more. It only gives rise to additional perturbing nonlinear processes which are associated with the increased intensity. A further modification of the phase matching curve originates from the line profiles involved (13,14) and the mode structure of the incident beams(15).

Finally, one has to take into account that the nonlinear medium which in many cases is provided by a concentric heat pipe oven (16,17), does not have a rectangular density profile as assumed in most theoretical models. Instead, density gradients at the end of the metal vapor zone, which give rise to an asymmetry of the phase matching curve, have to be considered (12,18,19).

In the small signal limit the optimum conversion efficiency is achieved, if the optical depth for one of the incident waves or the outgoing wave approaches a value of the order of unity (6,18). The large column density associated with this experimental condition, generally requires a very careful phase matching and hence a highly homogeneous nonlinear medium. Excessively large column densities, on the other hand, do not gain any further improvement because the power generated per unit length is eventually balanced by the power absorbed. In this case the conversion efficiency turns out to be independent of the column density (6).

3. Onset of Saturation

In the small signal limit of the preceding section the intensity of the harmonic wave follows a simple power law over a wide range of input intensities until eventually the harmonic intensity starts to level off long before energy conservation leads to a depletion of the fundamental waves. This onset of saturation can be explained by additional nonlinear polarizations which give rise to field dependent changes of the refractive index. They are responsible for a destruction of the phase matching condition. For gaseous media this was shown in nonresonant systems (12) as well as in two-photon resonant systems (20). For the nonresonant case it is the Kerr effect which causes an intensity dependent change of the refractive index. For the two-photon resonant situation it is the two-photon absorption which induces an intensity dependent redistribution of the population densities (20). In both cases the effective overall refractive index of the nonlinear medium is modified.

Because of the intensity dependent changes of the refractive index, the profile of the incident beam has to be considered in space and time in order to obtain the conversion efficiency of a particular nonlinear medium. For nonresonant cases the field dependent changes of the refractive index occur instantaneously, whereas for resonant situations the population densities inside the nonlinear medium have to be calculated which have been accumulated at some point in space and time. In this manner not only an intensity dependent, but also a time dependent change of the refractive index has to be considered. In the small signal limit the intensity distribution is taken care of by a constant factor in the conversion efficiency.

4. High Intensity Saturation

Due to the large number of different nonlinear processes the high intensity saturation regime is the most difficult one to analyze. In the high intensity saturation regime one is still primarily dealing with the same nonlinear polarizations which have already caused the onset of the saturation. They are, however, strongly enhanced. In addition, a variety of higher order nonlinearities become important which in part may be viewed at as field dependent corrections of the preceding third order nonlinearities. On the other hand, some nonlinearities give rise to a new class of higher order parametric processes.

All of these processes show up most clearly in two-photon resonant situations where they are resonantly enhanced. They can be detected, for example, by measuring the conversion efficiency around a particular two-photon resonance as a function of the input intensity. These conversion profiles show several very characteristic features (21). One observes a pronounced broadening due to power broadening and a narrow dip which causes a minimum of the conversion efficiency right on the two-photon resonance. In this case the highest conversion efficiency occurs right next to a two-photon resonance. The latter effect was first predicted theoretically (22) and is due to a bleaching of the two-photon transition.

As a further phenomenon self-(de)focusing of the incident beams can be observed which originates from the same field dependent changes of the refractive index which already destroyed the phase matching condition near the onset of the saturation. A theoretical treatment of this effect is exceedingly difficult because also the radial derivatives of the Laplacian operator inside the basic equations of nonlinear optics have to be considered which otherwise can be neglected.

As a consequence of the strong two-photon absorption a severe redistribution of the population densities occurs inside the nonlinear medium and may give rise to population inversions. Any resulting stimulated emission is therefore collinear with the incident beams and leads to additional parametric processes. They can be as intense as the original sum frequency wave (21).

In the high intensity saturation regime additional effects will generally occur such as higher harmonics of the incident waves, ionization processes and level shifts due to the AC Stark effect. All of these phenomena give rise to a highly intricate behaviour of the conversion efficiency. For practical applications, however, they are of little interest if a monochromatic vacuum uv source of high conversion efficiency has to be designed which can be applied, for example, to high resolution laser spectroscopy in the vacuum ultraviolet spectral region. In this case the most favorable condition is achieved with a phase matched system of optimum column density at an input intensity for which the onset of saturation occurs.

5. Spectroscopic Applications

In several laboratories tunable and narrow band coherent vacuum uv sources have already been built. They consist typically of two dye lasers which are pumped, for example, by a nitrogen laser or by an excimer laser or by a frequency tripled Nd laser. One dye laser is tuned to a particular two-photon resonance of the nonlinear medium, whereas the other dye laser provides the tunability of the resulting sum frequency wave in the vacuum ultraviolet spectral region. Using air-spaced etalons for achieving a small linewidth, the laser systems are most conveniently pressure tuned. In this manner the various dispersive elements are synchronized over the entire tuning range. The nonlinear medium frequently consists of a modified concentric heat pipe oven (17) containing, for example, a magnesium krypton or a strontium xenon mixture.

Vacuum uv systems have so far been designed to cover a spectral region from about 100 to 200 nm depending on the dyes used. A line width of 0.1 to 0.3 cm^{-1} in the vacuum uv has been achieved. The number of photons can be about 10^{11} to 10^{13} per shot at a repetition rate of up to 20 Hertz. The spectral brightness of these systems exceeds the one of any synchrotron and of laboratory light sources by several orders of magnitude (23). Also the spectral resolution $\lambda / \Delta \lambda$ exceeds the one of the best conventional instruments by almost an order of magnitude. Further improvements are expected once dye lasers are used which approach the transform limited linewidth.

Third harmonic generation and sum frequency mixing offer the additional advantage of allowing a wavelength calibration of the vacuum uv radiation by calibrating the wavelength of the fundamental beams in the visible part of the spectrum where secondary length standards are significantly more accurate. So far the iodine spectrum as measured by Gerstenkorn and Luc (24) was used in our laboratory as a calibration spectrum. The wavelengths of the fundamental waves were determined from a least squares fit of the iodine spectrum with a typical standard error of 0.003 cm^{-1} corresponding to an accuracy of 0.01 cm^{-1} or 2 parts in 10^7, in the vacuum uv.

In several laboratories high resolution laser spectroscopy in the vacuum uv region (9) has been initiated. The following methods have already been demonstrated where particular emphasis has been given to the spectroscopy of small molecules:

1. Absorption and excitation spectroscopy (9).
2. Frequency selective excitation spectroscopy on the CO intercombination bands (25) and on several electronic transitions of the NO molecule (26).
3. Laser induced fluorescence spectroscopy on many levels of the first excited states of the NO molecule (26).
4. Time selective fluorescence spectroscopy on the NO molecule (26).
5. Life time measurements on selected levels of the excited triplet states of the CO molecule.
6. Two step excitation spectroscopy on the CO molecule using a vacuum uv system and an additonal laser in the visible part of the spectrum. With this technique the lithium fluoride cutoff can be circumvented in a cell confined by magnesium fluoride windows. The experiments have shown collision induced transitions in the A-state of the CO molecule, predissociation, isotope shifts, perturbations, and fine structure splittings of excited states in CO (27).
7. Ionization and photofragment spectroscopy (9).

References

1. S.E. Harris and R.B. Miles: Appl. Phys. Lett. 19, 385 (1971)
2. J.F. Young, G.C. Bjorklund, A.H. Kung, R.B. Miles and S.E. Harris: Phys. Rev. Lett. 27, 1551 (1971)
3. D.M. Bloom, J.T. Yardley, J.F. Young and S.E. Harris: Appl. Phys. Lett. 24, 427 (1974)
4. R.T. Hodgson, P.P. Sorokin and J.J. Wynne: Phys. Rev. Lett. 32, 343 (1974)
5. K.M. Leung, J.F. Ward and B.J. Orr: Phys. Rev. A 9, 2440 (1974)
6. C.R. Vidal: "Four wave frequency mixing in gases", to be published in "Tunable Lasers" edited by L.F. Mollenauer and J.C. White in Topics in Applied Physics (Springer, Heidelberg, 1985)
7. C.R. Vidal: Appl. Opt. 19, 3897 (1980)
8. W. Jamroz and B.P. Stoicheff: "Generation of tunable coherent vacuum ultraviolet radiation," in Progress in Optics edited by E. Wolf, (North Holland, Amsterdam, 1983), Vol. 20, pp. 326-380.
9. C.R. Vidal, "Advances in Vacuum Ultraviolet Spectroscopy", LASER 84, San Francisco 1984
10. J.F. Ward and G.H.C. New: Phys Rev. 185, 57 (1969)
11. G.C. Bjorklund: IEEE J. Quant. Electr. QE-11, 287 (1975)
12. H. Puell, K. Spanner, W. Falkenstein, W. Kaiser and C.R. Vidal: Phys. Rev. A 14, 2240 (1976)

13. A. Stappaerts, G.W. Bekker, J.F. Young and S.E. Harris: IEEE J. Quant. Electr. QE-12, 330 (1976)
14. C. Leubner, H. Scheingraber and C.R. Vidal: Opt. Commun. 36, 205 (1981)
15. Y.M. Yiu, T.J. McIlrath and R. Mahon: Phys. Rev. A 20, 2470 (1979)
16. C.R. Vidal and F.B. Haller: Rev. Scient. Instr. 40, 3370 (1969)
17. H. Scheingraber and C.R. Vidal: Rev. Scient. Instr. 52, 1010 (1981)
18. H. Scheingraber, H. Puell and C.R. Vidal: Phys. Rev. A 18, 2585 (1978)
19. H. Junginger, H. Puell, H. Scheingraber and C.R. Vidal: IEEE J. Quant. Electr. QE-16, 1132 (1980)
20. H. Puell, H. Scheingraber and C.R. Vidal: Phys. Rev. A 22, 1165 (1980)
21. H. Scheingraber and C.R. Vidal: IEEE J. Quant. Electr. QU-19, 1747 (1983)
22. H. Scheingraber and C.R. Vidal: Opt. Commun. 38, 75 (1981)
23. K. Radler and J. Berkowitz: J. Opt. Soc. Am. 68, 1181 (1978)
24. S. Gerstenkorn and P. Luc: "Atlas du spectre d'absorption de la molecule d'iode", editions du CNRS, 15, Quai Anatole France, Paris, 1978
25. P. Klopotek and C.R. Vidal: Can J. Phys. 62, 1426 (1984)
26. H. Scheingraber and C.R. Vidal: J. Opt. Soc Am. B to be published
27. C.R. Vidal, P. Klopotek and H. Scheingraber: AIP Conf. Proc. 119, 233 (1984)

PART IV: Laser Cooling and Trapping

COHERENCE OF STATES IN TRAPPED IONS

Th. Sauter, W. Neuhauser, and P.E. Toschek
I. Institut für Experimentalphysik, Universität Hamburg
D-2000 Hamburg 36, F.R. Germany

When dealing with the subject of this paper, it is obviously appropriate to start with a twofold motivation by commenting to the questions "Why trapping ions?" and "Why to consider coherence of states?" Without going into details of the preparation of localized particles and their use for spectroscopy |1|, let us select a few points of view.

1. Ion trapping

Trapping of charged particles has recently brought about a couple of remarkable observations, as the precision measurement of g-2 of a single trapped electron |2|, the demonstration of optical cooling of a cloud of ions |3,4|, the localization and detection, by resonance fluorescence, of single ions |5|, and ultra-precise double-resonance measurements |6,7|. These observations have manifestly proved the benefits of trapping (and cooling) particles - in particular *individual* particles - for studying their interaction with light:

- The particles can be localized collision-free, in ultra-high vacuum, for long periods of time.
- The interaction with the light field can last almost arbitrarily long, such that there is no equivalent of "transit-time broadening" of spectral lines well-known from the interaction of particle beams with light.
- For particles at rest there is neither Doppler broadening of first *nor* of higher order, a convenience for which is no match with other schemes of interaction.

There are two kinds of traps suitable for tight localization as well as long-time trapping. Both make use of a quadrupole field of axial symmetry (Fig. 1):

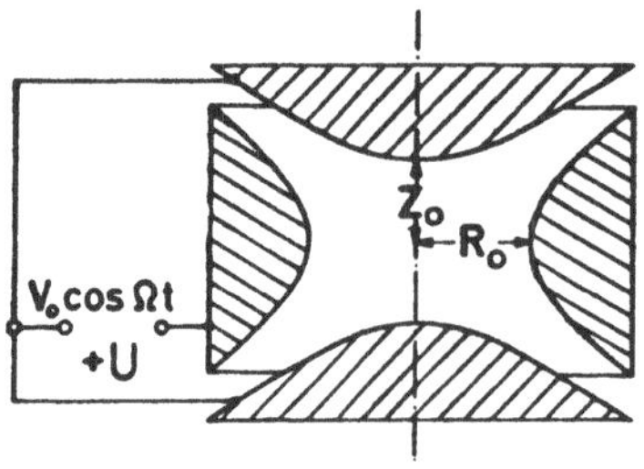

Fig.1: Cross section of ion trap. Electromagnetic ("Penning") trap: $V_0 = 0$, $V_0 \neq 0$, magnetic field $B = B_z$ = const. Electrodynamic ("Paul") trap: $V_0 \neq 0$, $B = 0$.

(*i*) In the first version |8|, a dc electric field repels the charged particles from the cap electrodes, i.e. there is focusing in z direction. In fact, the electric potential is

$$(1)\quad \Phi_T = A(r^2 - 2z^2) \quad ,$$

and the axial motion is harmonic,

$$(2)\quad \ddot{z} + \omega_z^2 z = 0$$

$$\omega_z^2 = 4eA/M \quad ,$$

with e,M being the charge and mass of a trapped particle, respectively. To overcome the defocusing in the x-y plane, a dc magnetic field curves the particle trajectories. In this "Penning" trap the Lorentz force balances the combined centrifugal and radial electric forces,

$$(3)\quad eB(v/c) = Mv^2/r + (M/2)\omega_z^2 r \quad .$$

With $v/r = \omega$, this is

$$(4)\quad 2\omega(\omega_c - \omega) = \omega_z^2$$

whose solution is

$$(5)\quad \omega = \{\omega_c', \omega_m\} = \frac{\omega_c}{2} \pm \sqrt{\left(\frac{\omega_c}{2}\right)^2 - \frac{\omega_z^2}{2}}$$

Substitution of the cyclotron frequency $\omega_c = eB/Mc$ into the left side of (3), and taking $\vec{r} = x + iy = r_o e^{-i\omega t}$, yields the equation of motion

$$(6)\quad \ddot{\vec{r}} + i\omega_c \dot{\vec{r}} = \omega_z^2 r/2 \quad ,$$

Its solution can be written

$$(7)\quad \vec{r} = \vec{r}_c e^{-i\omega_c' t} + \vec{r}_m e^{-i\omega_m t}$$

where ω_m is the drift or "magnetron" oscillation frequency. Thus, the motion of the charged particles in the x-y plane consists of a superposition of circular cyclotron orbits with larger magnetron orbits.

(*ii*) In the second version |9|, a radio-frequency electric field of amplitude V_o serves for ion containment, possibly superimposed by the dc field U, but with no magnetic field present. Then, the motion of charged particles in the trap potential (1) can be shown |10| to satisfy the Mathieu equation

$$(8)\quad \frac{d^2x_i}{d\tau^2} + (a_i - 2q_i\cos 2\tau)\, x_i = 0 \quad ,$$

where $x_i = x,y,z$; $a_i = a_x,a_y,a_z$, $\tau = \Omega t/2$, with the drive frequency Ω, and $a_x = a_y = -a_z/2 = 4eU/(MR_o^2\Omega^2)$, $q_x = q_y = -q_z/2 = 2eV_o/(MR_o^2\Omega^2)$.

Eq. (8) has stable solutions in certain ranges of a,q, (i.e., trapped particles move in closed orbits). In particular, for U = 0, stability requires M larger than a minimum particle mass. For Ω large compared with the orbital frequency, the motion, e.g. in z direction, can be described as the superposition

$$(9)\quad z(t) = \bar{z}(t) + \zeta(t)$$

of the orbital or "secular" motion $\bar{z}(t) = z_o\cos\omega_z t$ in a time-averaged quasi-potential, where $\omega_z = 2^{-1/2}eE_{oz}/zM\Omega$ is the secular frequency, $E_{oz} \propto z$ is the field component in z direction, and the driven or "micro" motion is

$$(10)\quad \zeta(t) = \zeta_o\cos\omega_z t\, \cos\Omega t \quad ,$$

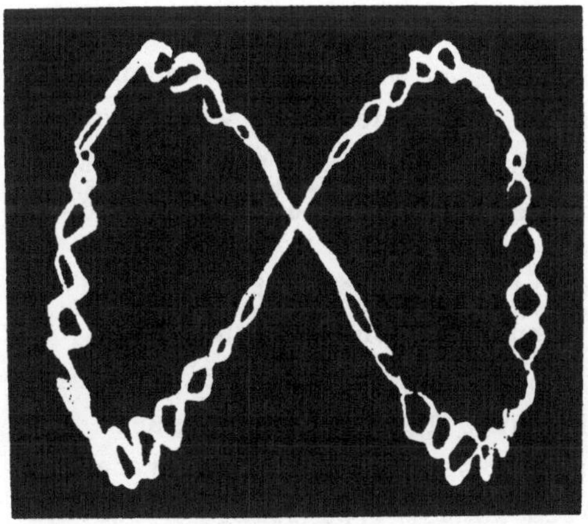

Fig. 2: Trace of aluminium dust particle in el.-dyn. trap (from Ref. 11).

with $\zeta_o = 2^{1/2}\omega_z z_o/\Omega$. This dynamics has been demonstrated long time ago using charged aluminium dust as trapped particles (see Fig. 2, |11|). With large enough Ω/ω_i, the micromotion is negligible. The secular oscillation can be reduced in amplitude, or "cooled" (see section 3), such that trapped particles approach the ideal of localization in free space.

2. Ionic Coherence

With any kind of interaction of light and atoms, the ultimate spectroscopic linewidth is determined either by the duration of the interaction, or by the lifetime of the excited coherence of states which oscillates at a well-defined phase. This is why systems capable of sustaining long-living coherence are of particular interest for applications in high-resolution spectroscopy and metrology. Dipole coherence, excited from the atomic (or ionic) ground state by the admixture of a short-living excited state, usually does not match those requirements (for exceptions, see e.g., Ref. |12|). However, the coherence of equal-parity states, say the Hertzian coherence of hyperfine states of the electronic ground level, can be remarkably long-lived, if collisions or other spurious interactions are avoided. This has been recently demonstrated in rather dramatic experiments: An ensemble of some 10^5 ^{171}Yb ions in an electrodynamic trap was excited and optically pumped, by a pulsed laser, out of the $S_{1/2}$, F = 1 hyperfine level. The coherence corresponding to the transition $\Delta F = 1$, $\Delta m_F = 0$ was excited by microwave radiation around 12.6 GHz |6|. This excitation was detected by optical double-resonance, i.e. more specifically by the increase of laser-excited resonance fluorescence. In the spectrum of fluorescence *vs.* microwave frequency, the narrowest observed resonance was 50 mHz wide. This value corresponds to the Q factor of the line being $2 \cdot 10^{11}$, and to the lifetime of the ground-state hyperfine coherence being on the order of 20 s.

With an optical double-resonance experiment on about twenty ^9Be ions in an electrodynamic trap |7|, a quadruple-resonance scheme was adopted to detect the 303-MHz field-independent spin-flip hyperfine transition in the electronic ground state. The double-resonance signal was recorded by the application of 2-sec rf pulses separated by 4 sec. It consisted of a well-resolved Ramsey-type fringe structure with the central fringe 75-mHz wide. In a similar experiment on $^{25}Mg^{+}$, the pulses for coherence formation and detection were even separated by 41 sec, demonstrating a remarkable persistence of the hyperfine coherence under the conditions of confinement in the electromagnetic trap.

Now that ultra-narrow resonance of trapped ions in the microwave frequency region are available, one might ask for analogous resonances at higher, in particular at optical frequencies. If those resonances can be preserved from undergoing critical shifts of frequency, they may be suitable for the high-precision frequency stabilization of laser oscillators, that is, for the representation of an optical standard frequency |13|.

Ionic species to be considered require a metastable state as part of a Λ-shaped three-level system including the ground state. The levels are linked by two E_1 transitions, which can be quasi-resonantly driven by two light fields (see Fig. 3). Ba^+ and Sr^+ are ionic species of this type.

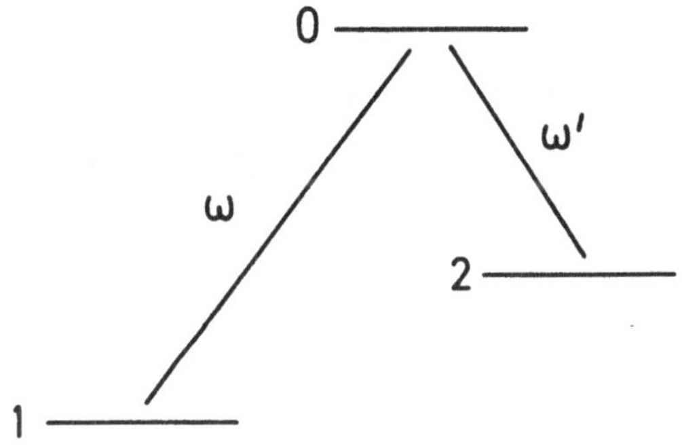

Fig. 3: Three-level system of Λ type.

In purely optical double-resonance experiments on trapped ions, again the excitation of resonance fluorescence, on one of the two coupled transitions, serves as monitor. In order to evaluate possible resonance frequencies, we restrict our consideration to the interaction of the trapped ions with one strong light field ("saturator"), whereas the other light field ("probe") is assumed weak. A complication arises from the presumed long life of the coherence excited in the oscillating ions. Their velocity component along the direction of light propagation periodically changes its sign such that alternate interactions with the saturator light, its frequency shifted up and down, may happen. This interaction processes from two classes: the first one is of probe absorption or emission type, identified by the exchange of one probe "photon" and an *even* number of saturation "photons", the second one is of Raman type with the exchange of one probe "photon" and an *odd* number of saturator "photons". In addition, slow gyrations of the ions' orbits establish a distribution of resonance frequencies corresponding to an inhomogeneously broadened line. Eventually these characteristics enable us to describe the ions moving in the trap and interacting with two travelling light waves - at least qualitatively - by a model of three-level absorbers with inhomogeneously broadened lines, interacting with two *standing* waves. A model including one standing saturator wave has been elaborated before and used with the computation of nonlinear absorption spectra |14|. A recent version |15| permits one to construct graphs which yield a synoptic outline of the resonances to occur, applicable even with a weak, detuned, standing probe wave. The condition imposed upon the ionic velocity component in the direction of light propagation, $v = \tilde{v}/k$, is

$$(11)\quad \Delta' - (k'/k)\tilde{v} = \begin{cases} q\tilde{v} & \text{for } q = 0, \pm 2, \ldots \\ \Delta + q\tilde{v} & \text{for } q = \pm 1, \pm 2, \ldots \end{cases}$$

When plotting the right and left-hand sides separately in a Doppler-type diagram - frequency detuning *vs.* axial velocity (Fig. 4) -, the intersections of the corresponding lines represent the actual resonance frequencies and the velocity classes which contribute to it. It can be easily shown |15| that in an alternative interpretation these lines represent the energy levels of the particles "dressed" by q saturator quanta,

$$< 0; q, 0 \,|H_0|\, 0; q, 0 > = q\tilde{v}$$

$$< 1; q, 0 \,|H_0|\, 1; q, 0 \quad = \Delta + q\tilde{v} \quad ,$$

or by one probe quantum,

$$< 2; 0, 1 \,|H_0|\, 2; 0, 1 > = \Delta' - (k'/k)\tilde{v} \quad ,$$

where the Hamiltonian H_0 does not account for the interaction of particles and light. If that interaction is included, avoided crossings of the levels of dressed ions evolve due to the diagonalization of the full Hamiltonian for velocity classes which can interact simultaneously via two of the multiphoton processes distinguished by q.

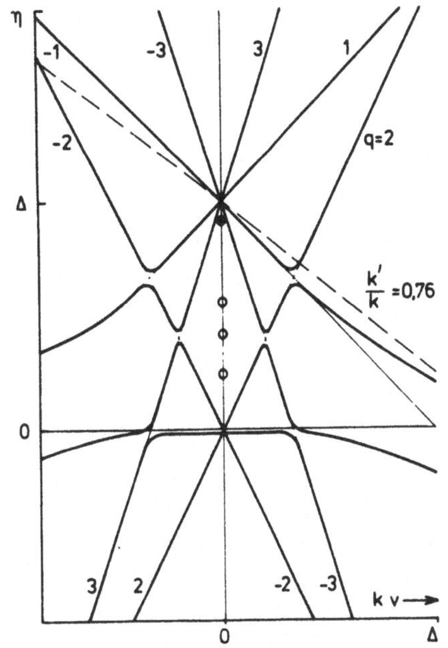

Fig. 4: Energy levels η of three-level particle plus q saturator (k) quanta ("dressed" levels, full lines) or one probe (k') quantum (dashed line) for k'/k of Ba^+ ($^2S_{1/2}$ - $^2P_{1/2}$ - $^2D_{3/2}$). After Ref. 15.

An actual spectrum is generated when the level corresponding to the probe light is scanned across the tuning range, and the overlap with the saturator levels is integrated for each position. The diagram shown in Fig. 4 is specified for Ba^+. Note that spectral resonances show up when the overlap of the probe level with saturator levels is maximum, i.e. if the probe level forms a tangent to the curved saturator level. Several resonances are predicted by this diagram, the most important of which is due to the interference of the fundamental Raman processes (q = -1) with single-photon probe emission (q = 0).

3. Cooling trapped ions

For a full evaluation of the capacity of trapped ions to develop narrow resonances it is indispensable to reduce the excursion of their oscillatory motion, i.e. to "cool" them, to a degree that the condition

$$2\pi r/\lambda \ll 1$$

holds. Here, r is the oscillation amplitude of the ions, and λ is the wavelength of the light to interact with them. Cryogenic cooling does not suffice for this purpose, and *optical* cooling |3,4| is the method of choice.

Let us restrict the discussion, for the moment, to only two internal states of the ions, S and P, which are involved in the corresponding resonance fluorescence. The moving ions must be described by two sets of harmonic oscillator levels (s. Fig. 5). Electronic transitions between the two sets obey the selection rule $\Delta n = 0$, i.e. the Franck-Condon principle holds. However, also sideband transitions with $\Delta n = \pm 1$ are weakly allowed. If the light interacts with the low-frequency sideband, ions of level S(n) will be excited to level P(n-1). Subsequent spontaneous emission will yield most ions in level S(n-1), with their oscillator energy reduced.

How is the fluorescence spectrum of the ions expected to vary in the course of cooling? Let us, for simplicity, assume that the excitation is performed by monochromatic light. Then the emission - also monochromatic in the weak-field limit for particles at rest |16| - is now characterized by phase modulation due to the periodic motion of frequency ω_s. The spectrum consists of lines at $\omega \pm m\omega_s$, equally spaced by ω_s, whose intensities vary as the squares of the m-th Besel function of the modulation index $2\pi r/\lambda$ (Fig. 6). At *large* oscillation

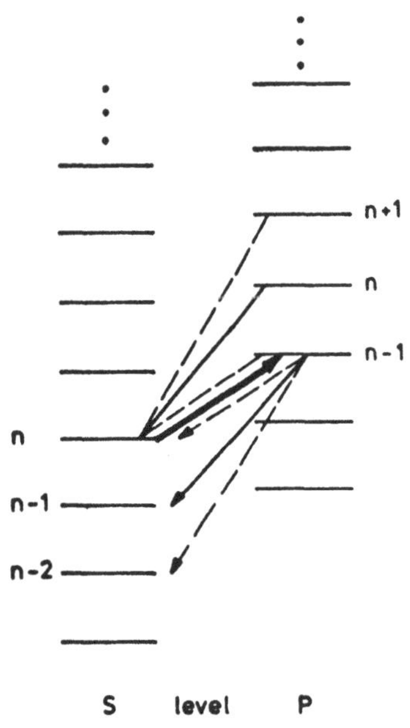

Fig. 5: Optical cooling of bound two-level particles. Full lines: transitions with $\Delta n = 0$, broken lines: sideband transitions. Heavy arrow: sideband excitation.

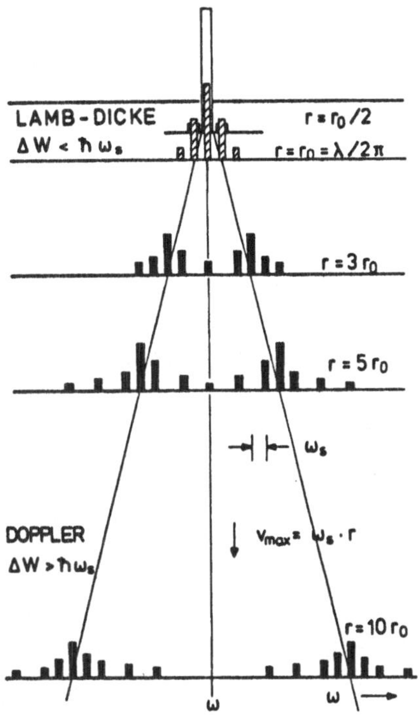

Fig. 6: Schematic fluorescence spectrum of two-level atom in harmonic potential under monochromatic excitation, *vs.* amplitude of oscillation (r).

amplitudes, they form clusters around frequency values which correspond to maximum speed of the ions in the direction of light, and opposite to it. The complete spectrum corresponds to an *inhomogeneously* broadened line: different sections of the spectral structure are generated, with an individual ion, at particular phase angles of its orbits, i.e. at different times and velocity values. The average over particular time intervals of idential phase in the orbit is equivalent to an ensemble average. This is the "Doppler" limit, when even an individual particle features a velocity *distribution*. In contrast, oscillatory motion at *small* amplitude makes the carrier frequency dominate the fluorescence spectrum. The small contributions of the nearest sidebands decrease with shrinking oscillation amplitude, but the sidebands do not shift in frequency. They cannot be attributed to individual orbital phases, and the complete spectrum (carrier plus two sidebands) corresponds to a homogeneously broadened line whose particular shape, i.e. its sideband contribution, is determined by the amplitude of the periodic motion. This is the "Lamb-Dicke" limit which is, of course, the situation most desirable for Doppler-free spectroscopic observation |3|.

Optical cooling or heating is the transfer of the photon momentum $\vec{p}_\nu$ to the interacting particle, and extraction of part of its kinetic energy W_ν from it:

$$\Delta\vec{p} = M\cdot\Delta\vec{v} = \vec{p}_\nu = \hbar\vec{k} \quad ,$$

$$\Delta W_{kin} = -\Delta W_\nu = \hbar\ (\omega_{abs} - \omega_{em}) = \hbar\vec{k}\vec{v} + 2\varepsilon \quad .$$

Here, $\vec{k}$ is the wave vector, and $\varepsilon = (\hbar\vec{k})^2/2M$ the recoil energy. Net cooling occurs when $\hbar\vec{k}\vec{v} < -2\varepsilon$ |17|. For velocity values well in the Doppler regime the recoil energy can be neglected. In a simplified model which considers the axial motion only,

$$\begin{aligned} \langle\Delta W_{min}\rangle &= \tfrac{1}{2}M\langle v_1^2 - v_2^2\rangle \\ &= M\langle\tfrac{1}{2}(v_1 + v_2)(v_1 - v_2)\rangle \\ &= (M/k)\cdot k\bar{v}\cdot\Delta v \end{aligned}$$

This relationship is visualized in Fig. 7 which shows, *vs.* the ionic velocity, the relevant electronic transition of the ions, excited by the light. The Doppler shift is indicated by the diagonal line. Here, $\langle\Delta W_{kin}\rangle$ is M/k times the shaded area. Macroscopic removal of thermal energy from the ions needs a *cyclic* process in the thermodynamic sense that can be sustained. In a gas, collisions would be required to change the velocity component v_2 along the direction of light propagation at random, such that a particle had a finite probability of arriving again at the component v_1, although at *reduced* kinetic energy

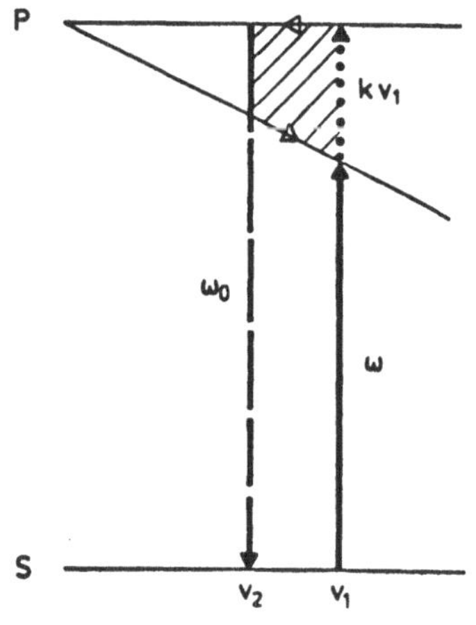

Fig. 7: Optical cooling of two-level atoms involving a cylic process.

$M\langle\vec{v}^2\rangle/2$. Fig. 7 also shows that the asymmetry of absorption and stimulated emission, and in particular the difference of their respective frequencies, are essential for energy transfer. No net cooling or heating is expected to occur with a purely stimulated process for the *two-level* interaction scheme.

Ions bound to a trap respond to the interaction with light in a way different from free atoms: the momentum of the "dressed" particle is *not* conserved, due to the presence of the potential well. Moreover, the ion will be found in a particular kinetic state *periodically*. Thus, a cyclic process can be closed without recurrence to collisions, which would be necessary when cooling a gas.

Manipulation of kinetic energy by light has been first demonstrated in the version cooling clouds of some 25 Ba^+ ions confined in an electrodynamic trap |3|, and simutaneously in another version with Mg^+ clouds in an electromagnetic trap |4|.

What kind of observable effects of optically cooling trapped ions can be predicted? First, in the excitation spectrum of resonance fluorescence, cooling is expected to dominate below resonance, heating above. Heating makes expand the ion cloud in the potential well. In combination with apertured detection, this effect will reduce the fluorescence signal. Second, the motional sidebands may be observed if they accompany a line of small enough width to resolve their spacings, when the ions are kept at a temperature well in the Lamb-Dicke regime. The separation of the cooling effects from variations of the fluorescence cross section poses a particular challenge to the experimentalist.

4. Spectroscopy of single cooled ions

Excitation spectra of a small ion cloud have been observed with Mg^+ in an electromagnetic trap |18|, and its equilibrium temperature has been derived from the residual Doppler width. In a magnesium ion, only the ground state and one resonance state are involved in the interaction, and the interpretation of the spectra is straightforward. Single-ion excitation spectra have been first recorded with Ba^+ |19|, and later with Mg^+ |20|. As we emphasize the possible characteristics of Raman coherence in the spectra, let us restrict ourselves to the barium experiments.

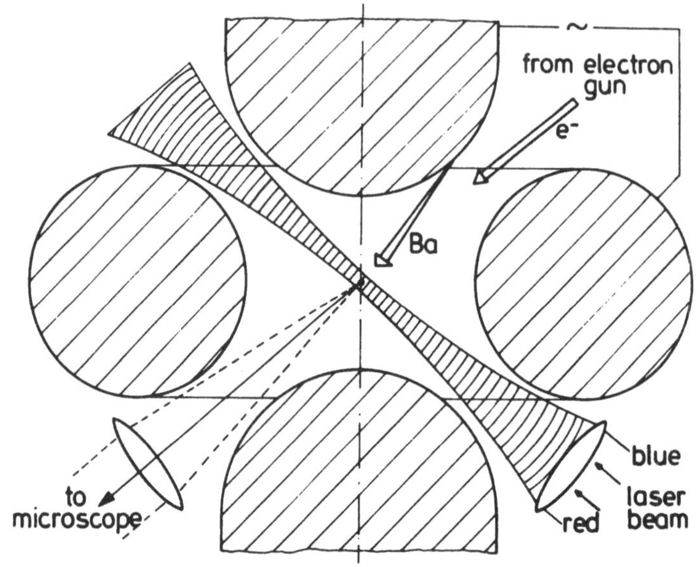

Fig. 8: Schematic cross section of miniature rf trap showing Ba^+ excitation and observation (from Ref. 13).

The resonance line at 493 nm connects the $^2S_{1/2}$ and $^2P_{1/2}$ states of Ba^+ (1 and 0, respectively, in Fig. 3). Upon excitation, the ions decay into the metastable $^2D_{3/2}$ level (2) with the branching ratio 1/3. Thus, steady generation of resonance fluorescence requires re-excitation from this level by a second light field at 650 nm.

The electrodynamic trap (Fig. 8) had 0.5-mm cap distance; the $2\pi\cdot 39.2$-MHz/0.8-kV drive generated a pseudo-potential well some 3 V deep. The background pressure was smaller than 10^{-10} mbar. The two coaxial cw dye laser beams were focuced to a 70-μm waist in the trap centre and crossed there with an atomic barium beam and an electron beam for impact ionization. Visual, photographic, and photo-electric detection of the resonance fluorescence, almost at right angles with the exciting light beams, was through a three-channel microscope. In the photoelectric version, a low-noise photo-multiplier and a photon counter were used. Removal of kinetic energy, and ion trapping in high vacuum, was achieved when the green laser was tuned some 250 MHz down from the centre frequency of the optical resonance. Cooling is indispensable for trapping in order to impose energy loss *in situ*, and to overcome some spurious heating by the hf drive.

Single barium ions have been cooled to less than 10 mK, as derived from the excess spot size of resonance fluorescence over the diffraction-limited image size of a point-like source |5|. Excitation spectra were recorded by scanning the red light across the $^2D_{3/2}$ - $^2P_{1/2}$ resonance |19|. It was noticed that also the interaction with the red light contributes to the thermal balance; when tuned below resonance centre by cooling, above by heating. Net heating again makes the orbit of the ion blow up and the fluorescence signal drop, as was readily noticed by visual observation. Bistable orbits may complicate the dynamics of the particle |21|. These peculiarities result in considerable signal fluctuations, However, a few spectral features have been reproducibly observed. They are demonstrated in the spectrum of Fig. 9:

1. At $\omega' - \omega_0 = (k'/k)(\omega - \omega_0)$, the fundamental Raman resonance, i.e. the interference of the q = 0 and q = -1 processes, marks the generation of coherence of the levels $^2S_{1/2}$ and $^2D_{3/2}$ (↑). Its width is due to the combined bandwidths of the two lasers (c. 3 MHz) and power broadening by the red light.
There have been hints to the appearance of higher-order resonances in some of the recorded spectra.

2. At centre tuning, a dramatic drop of the signal occurs due to the onset of heating by the red laser and the concomitant expansion of the ion orbit.

3. Surprisingly, at several hundred MHz *up*-tuning of the red laser, where considerable heating by the red light is expected to prevail, *contraction* of the orbit is observed by reappearance of the full signal. This phenomenon may either manifest orbit multi-stability |21|, or it seems to indicate the presence of an additional cooling mechanism.

5. Cooling by Raman coherence?

The formation of the electronic Raman coherence of the $^2S_{1/2}$ and $^2D_{3/2}$ levels in the barium ion has been readily proved by the appearance of the deep resonance in spectra like the one of Fig. 9. The Stokes and anti-Stokes transitions responsible for that forma-

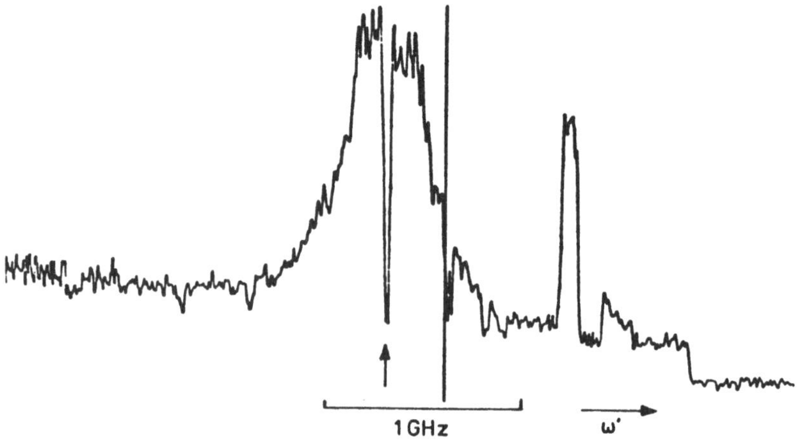

Fig. 9: Excitation spectrum of 493-nm resonance fluorescence of single Ba^+ *vs.* tuning of red laser ($\lambda' \cong 650$ nm). Raman resonance (↑) and decrease of signal with onset of heating by red light at centre frequency (vertical line). Cooling reappears at 600 MHz up-tuning.

tion do not involve the $^2P_{1/2}$ level. It seems that they are also accompanied by momentum transfer. For net heating or cooling to occur, an unbalance of the two processes would be required which could result from the difference of matrix elements, branching ratios, and respective light intensities. This unbalance also gives rise to optically pumping the ion into one of the levels - under the conditions of Fig. 9 into the ground state.

The transfer of mechanical energy may happen as follows: Resonant green light (ω) excites a low-frequency sideband of the Raman coherence by Stokes scattering. The emission may be stimulated, i.e. it may happen into the mode of the red light beam (ω') which is off-resonant in general (Fig. 10 right side). The non-zero energy transfer between mechanical oscillation and light is proportional to the shaded area. The readout of the coherence, e.g. by anti-Stokes scattering of red light (ω'), would undo this transfer. However, the readout may occur at a different position in the orbit, i.e. in a *different kinetic state* of the ion, since the Raman coherence is relatively persistent. Its average life-time τ,

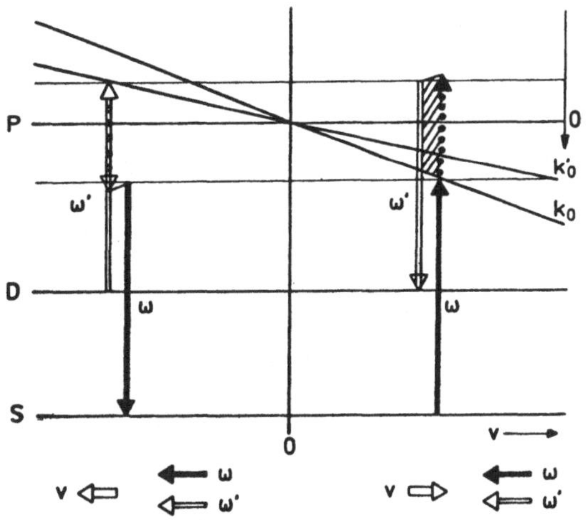

Fig. 10: Formation of Raman coherence in three-level ion with green light (ω) resonant (right), and readout at different v where red light (ω') is resonant (left).

determined by the bandwidths of the two lasers, and by their local intensity values, has to be comparable with the period T of the ion's oscillation in the trap. The most probable kinetic states for readout of the coherence are the states with resonantly enhanced interaction of ion and light fields: either the state at two-photon resonance, or the one where the resonant and off-resonant interactions with the two light fields of frequencies ω, ω' are exchanged as compared with coherence formation (Fig. 10, left side). This readout also yields net energy transfer. In the example of Fig. 10, the direction of the ionic momentum has even turned around during the lapse of time between coherence formation and readout. The orbital motion takes the ion back to a position where its axial velocity component equals the initial one. When T is taken as the secular period (1/T = 1.2 MHz), and the width of the Raman resonance from Fig. 9, τ is, however, about T/10, and one would expect weak effects only.

It is worth noting that the particular (down-) tuning of the green light identifies both electronic Stokes and anti-Stokes interactions at the respective locations in the ion's orbit as *anti-Stokes* transitions between levels of the harmonic oscillator, which are indispensable for a net transfer of energy |22|.

Cooling by a Raman transition would give access to much lower temp-

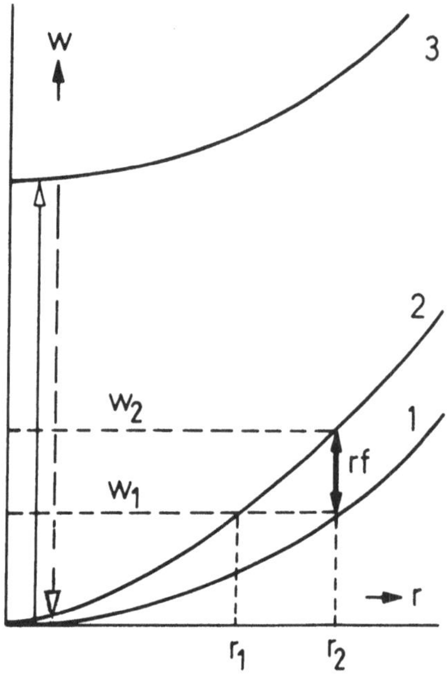

Fig. 11: Pritchard's scheme of cooling magnetically trapped Na atoms: The atoms are optically pumped to level 2 at *zero field* (centre of trap). Rf transitions at location r_2 equalize populations of 2 and 1 with v *unchanged* (Franck-Condon principle). Thus, total energy in 1 is $W_1 < W_2$. After Ref. 23.

perature of the ion |3,17|, since its natural line width, which determines the final temperature, is on the order of milli-Hz only. Cooling trapped particles by purely *stimulated* interactions would not undergo saturation and promise considerable efficiency. There is no analogue to this scheme for the interaction of two-level particles with a single light field.

Recently, a method of cooling magnetically trapped sodium atoms was proposed |23|, which is, in some respect, similar to the above scheme: Radio-frequency transitions are induced from state 2, populated by optical pumping, to a state 1 of equal parity - here another ground-state sublevel - at an off-centre location of the atoms at finite magnetic field (Fig. 11). Since the atoms keep their kinetic energy according to the Franck-Condon principle, their total energy will become smaller (W_1), and also their kinetic energy at trap centre. There, optical pumping - at zero magnetic field - takes the atoms back to state 2. By the orbital motion in the trap, the atoms shuttle between the regions of interaction and continue with the cyclic process of energy transfer.

6. Conclusions

The formation of Hertzian coherence of equal-parity states in trapped ions by the application of double-resonance interaction schemes has produced ultra-narrow resonances with quality factors on the order of 10^{11}. Optical coherences of equal-parity states may be responsible for the identification of finer details of the excitation spectrum of single trapped ions at well as for some novel contributions to their mechanical interaction with light. These effects seem to be undetectable in gases since collisions are present and the interaction time is limited.

*

Part of the described work has been supported by the Deutsche Forschungsgemeinschaft.

References

1, See, e.g. P.E. Toschek, *Tendances actuelles en physique atomique*, Les Houches, Session XXXVIII, G. Grynberg and R. Stora, eds., North Holland 1984, p. 381
2. R. Van Dyck, Jr., P. Ekstrom, and H. Dehmelt, Phys. Rev. Lett. *38*, 310 (1977)
3. W. Neuhauser, M. Hohenstatt, P.E. Toschek, and H.G. Dehmelt, Phys. Rev. Lett. *41*, 233 (1978)
4. D.J. Wineland, R.E. Drullinger, and F.L. Walls, Phys. Rev. Lett. *40*, 1639 (1878)
5. W. Neuhauser, M. Hohenstatt, P.E. Toschek, and H. Dehmelt, Phys. Rev. A*22*, 1137 (1980)
6. R. Blatt, H. Schnatz and G. Werth, Phys. Rev. Lett. *48*, 1601 (1982)
7. D.J. Wineland, J.J. Bollinger, and Wayne M. Itano, Phys. Rev. Lett. *50*, 628 (1983)
8. F.M. Penning, Physica *3*, 873 (1936)
9. W. Paul, O. Osberghaus, and E. Fischer, Forschungsber. d. Wirtsch.- u. Verkehrsmin. NRW. Nr. 415 (1958)
10. E. Fischer, Z. Phys. *156*, 1 (1959). - H.A. Schuessler, E.N. Fortson, and H.G. Dehmelt, Phys. Rev. *187*, 5 (1969)
11. R.F. Wuerker, H.M. Goldenberg, and R.V. Langmuir, J. Appl. Phys. *30*, 342 (1959)
12. J.C. Bergquist, R.L. Barger, and D.J. Glaze, *Laser Spectroscopy IV*, Rottach-Egern, H. Walther and K.W. Rothe, eds., Springer-Verlag, Heidelberg, 1979, p. 120
13. P.E. Toschek and W. Neuhauser, *Atomic Physics 7*, D. Kleppner and F.M. Pipkin, eds., Plenum, New York, 1981, p. 529

14. B.J. Feldman and M.S. Feld, Phys. Rev. *A5*, 899 (1972)
15. L. Roso, R. Corbalán, G. Orriols, R. Vilaseca, and E. Arimondo, Appl. Phys. *B31*, 115 (1983)
16. W. Heitler, *The quantum theory of radiation*, Clarendon, Oxford, 1954. - C. Cohen-Tannoudji, *Aux frontières de la spectroscopie laser*, Les Houches, Session XXXVII, R. Balian, S. Haroche, and S. Liberman, eds., North Holland, Amsterdam, 1977, p. 3
17. D.J. Wineland and W.M. Itano, Phys. Rev. *20*, 1521 (1979)
18. R.E. Drullinger, D.J. Wineland, and J.C. Bergquist, Appl. Phys. *22*, 365 (1980)
19. W. Neuhauser, M. Hohenstatt, P.E. Toschek, and H. Dehmelt, *Spectral Line Shapes*, B. Wende, ed., W. de Gruyter, Berlin 1981, p. 1045
20. D.J. Wineland and W.M. Itano, Phys. Lett. *82A*, 75 (1981)
21. J. Javanainen, this volume
22. A. Kastler, J. de Physique *11*, 255 (1950)
23. D.E. Pritchard, Phys. Rev. Lett. *51*, 1336 (1983)

LIGHT PRESSURE COOLING OF A TRAPPED THREE-LEVEL ION

Juha Javanainen §

University of Helsinki, Research Institute for Theoretical Physics
Siltavuorenpenger 20 C, 00170 Helsinki 17, Finland

1. Introduction

The present limitations on the resolution of optical spectroscopy, transit time broadening and second order Doppler shift, stem from the motion of the particles under investigation. It is thus particularly interesting to note that resonance-light pressure, possibly exerted by the very same laser beam as used to probe the atomic transition, may effectively control the motion of the atomic particles. Light pressure cooling of trapped ions [1,2] has recently been utilized to prepare an essentially immobile single ion subjected to continuous observation over a period of hours [3-5]. Neutral atoms in thermal beams have also been slowed down to zero velocity by applying a counter-running laser beam [6,7], and 'Bragg scattering' of an atomic beam from a standing wave pattern of light has been demonstrated [8]; see also [9]. Unlike charged ions, neutral atoms cannot be confined by electromagnetic fields that couple to the net charge of the particles, but light pressure traps for atoms have been suggested [10-12] and are being actively worked on.

The theorist's prototype of a light pressure problem consists of a two-level atom in a monochromatic travelling plane wave of light. Two main approaches to this case have developed. First, one may treat the center of mass as a quantized dynamical variable, and by eliminating adiabatically the internal degrees of freedom derive a diffusion equation for the Wigner function describing the center-of-mass motion of the atom [13,14]. Second, in the semiclassical approach the center of mass is treated classically, but the force becomes a quantum operator. One then extrapolates the classical expression of the force into the quantum regime and again derives a diffusion equation; the expectation value of the force operator provides the drift term, and a correlation function of the force supplies the diffusion [15,16]. Recently these two points of view have been unified [17]. More complicated configurations of the light field, notably the standing wave, have spurred a number of additional theoretical procedures; see e.g.

---

§Present address: Max-Planck-Institut für Quantenoptik, D-8046 Garching, Germany

[11,18] and references therein.

The perturbative approach with respect to the light intensity by Wineland and Itano [19-20] and the nonperturbative methods developed at the University of Helsinki ([21-24] are the most recent references) constitute nearly all that has been published about the detailed theory of light pressure cooling of a trapped ion. These theories consider a two-level ion, and employ a fully quantized representation of the center-of-mass motion. However, while for some ions the two-level picture is sufficient [2,4,5], some other ions might end up in a metastable state that must be emptied by another laser in order to allow the cooling to proceed [1,3]. Then an approach built around a three-level ion seems more appropriate - especially as quite dramatic effects have been seen experimentally that evidently depend on the presence of the third level [25]. For a free atom some treatments of light pressure exist that account for the level structure beyond the standard two-level model [18,26], but it turns out that these do not directly go over to the trapped ion.

In this paper we study the light pressure force on a trapped three-level ion in a travelling wave within the semiclassical approach, i.e. the center-of-mass motion is assumed classical. In section 2 we outline the basic theory of light pressure on a two-level atom from a partly new angle, which in section 3 leads immediately to a generalization to a three-level system. The predictions of the model, especially the existence of non-trivial limit cycle orbits of the ion in the trap, are exposed in section 4. The short discussion of section 5 concludes the paper.

2. Cooling of a trapped two-level ion

Imagine a closed two-level system whose upper level 2 decays at the rate Γ down to the ground state 1, in a laser beam characterized by the angular frequency Ω and wave number q. The ion is initially in the ground state, and moves with a velocity whose component in the direction of the laser beam is v. Next the ion absorbs one photon, and gets excited. However, in so doing the ion must also absorb the momentum of the photon, hence its new velocity is $v+\hbar q/M$. Suppose now that the subsequent de-excitation takes place through spontaneous emission. Since the angular distribution of spontaneous photons is invariant under inversion, in a spontaneous process the ion receives a recoil kick whose direction is random and average value zero. The net effect is that in a cycle of absorption and spontaneous emission the velocity of the ion changes by a random amount whose average is $\hbar q/M$.

The average change of the velocity is conveniently associated to the 'light pressure force'. Nevertheless, it ought to be borne in mind that the motion of the atom also inherits some stochastic or diffusive character from the underlying discrete photon processes [27].

In order to nail down the precise expression of the light pressure force we note that the same cycles of absorption and spontaneous emission also account for macroscopic absorption, i.e. loss of energy from the incoming laser field. Extrapolating the semiclassical expression for absorption to the limit of one ion we may express the rate of loss of energy in terms of the Rabi frequency $\kappa = dE/2\hbar$ and the off-diagonal density matrix element ρ_{21} as

$$\frac{dE}{dt} = -2\hbar\Omega\kappa \,\mathrm{Im}(\rho_{21}). \tag{1}$$

The rate of change of momentum, the force, is then

$$F = -\frac{1}{c}\frac{dE}{dt} = 2\hbar q\kappa \,\mathrm{Im}(\rho_{21}) . \tag{2}$$

The conventional two-level theory gives

$$F(v) = \hbar q \Gamma \frac{\kappa^2}{(\Delta+qv)^2+(\frac{\Gamma}{2})^2+2\kappa^2}, \tag{3}$$

where the detuning Δ is positive if the laser is tuned below the ion's resonance.

The force in (3) depends on the velocity through the Doppler shift qv. Another important point is that the force has an absolute maximum $\hbar q\Gamma/2$. This is because irreversible changes of the momentum only occur in spontaneous processes, whose rate is fixed by Γ. The factor 1/2 reflects the fact that at most half of the population may be in the excited level 2.

We now introduce the trapping potential that forces the ions to oscillate at the angular frequency ν in the direction of the beam z. If the linewidth of the transition, $\gamma = \Gamma/2$, is much larger than ν, the ion has time to adjust its state to the prevailing instantaneous velocity, and the expression of the force (3) still applies. The equations of motion of the ion are

$$\dot{z} = v, \tag{4a}$$

$$\dot{v} = -\nu^2 z + \frac{F(v)}{M} . \tag{4b}$$

If the laser is tuned below resonance, the ion predominantly absorbs photons when it moves towards the laser. The light pressure then counteracts the oscillatory motion, and one expects the ion to slow down. Indeed, the analysis of (4) shows that for $\Delta > 0$ the point

$$z_0 = \frac{F(0)}{M\nu^2} \approx 0, \quad v_0 = 0 \tag{5}$$

is a stable attractor in the phase space of the ion, and all orbits tend asymptotically ($t \to \infty$) to (z_0, v_0). Equations (4) predict that, irrespective of the initial conditions, the ion is eventually stopped at the point z_0. Actually this will not happen, as the random diffusion associated with light pressure smears the orbit of the ion. A detailed analysis [19,20,21-24] shows that the Doppler width of the diffusion-limited velocity distribution is of the order

$$q\,\delta v \simeq \left[\frac{\hbar q^2}{2M}\max(\Delta,\gamma,\kappa)\right]^{\frac{1}{2}} . \qquad (6)$$

Because the recoil energy in frequency units appearing inside the square root is typically in the 100 kHz regime and the linewidth might be 10 MHz, this expression shows that the Doppler width is ordinarily much narrower than the frequency scales characterizing the resonance of the ion. From the point of view of spectroscopy the ion may be regarded as staying fixed at the point z_0.

3. Force on a three-level ion

With the aim of generalizing the considerations of the previous section we first study the

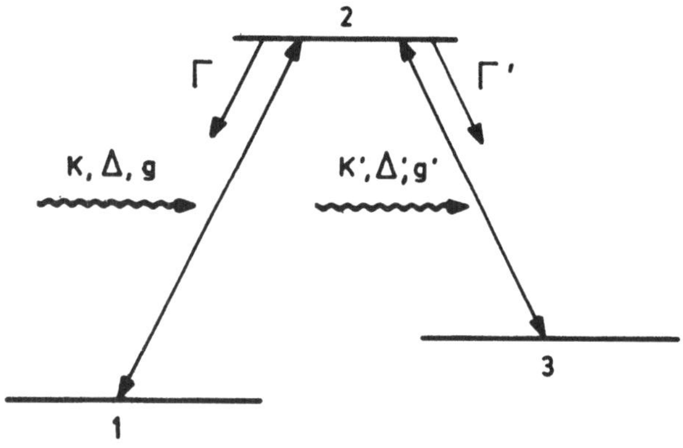

Figure 1. The three-level Λ configuration.

light pressure force on a three-level ion with the Λ configuration of states (Fig.1). The transition 1-2 is driven by a laser characterized by the Rabi frequency κ, detuning Δ and wave number q; the corresponding parameters for the second co-propagating light field that excites the transition 3-2 are denoted with primes. The excited level now decays spontaneously back to either 1 or 3 with the rates Γ and Γ', respectively. These rates fix completely all spontaneous decay terms in the equations of motion for the density matrix ρ_{ij}, i,j=1,2,3. For instance,

$$\dot{\rho}_{22}=-(\Gamma+\Gamma')\rho_{22}+\dots,\ \dot{\rho}_{21}=-\frac{\Gamma+\Gamma'}{2}\rho_{21}+\dots\ . \tag{7}$$

With these ingredients one can write down the equation of motion for the density matrix ρ_{ij}, see e.g. [28-30] and references therein. Furthermore, for the present closed three-level system it pays to eliminate the conserved total population $\rho_{11}+\rho_{22}+\rho_{33}=1$ from the equations. One arrives at an equation of the form

$$\frac{\partial\rho}{\partial t}=A\rho+vB\rho+\lambda\ . \tag{8}$$

Here ρ is an eight-component vector consisting of suitably chosen linear combinations of ρ_{ij}, and A and B are 8×8 matrices. We have split the matrix in (8) into two parts in order to display explicitly the dependence of the equations on the velocity of the ion v. The remaining vector λ originates from the elimination of the conserved total population.

We do not dwell on the details of the theory of the three-level system any longer, but merely indicate some reasons why it shows a much richer physical structure than the two-level atom. First, in general optical pumping between levels 1 and 3 takes place. Second, there are two population resonances, which are located at $\Delta=0$ and $\Delta'=0$ when the velocity is v=0. Third, when again v=0, the Λ system exhibits the non-absorbing two-photon resonance at $\Delta=\Delta'$. Mathematically this emerges via the two-photon coherence ρ_{13} between the extreme levels of the system. Since in our model neither the level 1 nor the level 3 decays, this coherence lives very long and is damped because of parasitic reasons like the linewidths of the lasers. The two-photon resonance may then be extremely sharp.

Suppose now that we deal with a three-level ion that oscillates in the trapping potential at the angular frequency ν, and we want to calculate the density matrix at a given time t when the ion has the coordinates (z,v). In this case the long-lived two-photon coherence may store the memory of the past velocities, i.e. the Doppler shifts, for much longer times than the period of oscillations $2\pi\nu^{-1}$, and in contrast to the two-level system of section 2 one usually cannot assume that the state of the ion is determined by its 'present velocity' v alone. Similarly, the optical pumping may also be slow and store information about the history of the ion. We have to invoke the past velocities of the ion,

$$v(t')=v\cos\nu(t-t')+\nu z\sin\nu(t-t'),\ t'\leq t\ , \tag{9}$$

and integrate the equation (8) with the time dependent velocity v(t'),

$$\frac{\partial}{\partial t'}\rho(t')=A\,\rho(t')+v(t')\,B\,\rho(t')+\lambda\ , \tag{10}$$

from the distant past $t'=-\infty$ to the present time $t'=t$. The result gives the density matrix elements ρ_{ij} (z,v) for an ion that is at (z,v) at the time of the inspection t.

Equation (10) constitutes a coupled set of linear differential equations with sinusoidally oscillating coefficients. The solution can be found by expanding ρ as a Fourier series

$$\rho(t') = \sum_{n=-\infty}^{\infty} \rho_n \exp\{-in[\nu(t-t')+\phi]\} \tag{11a}$$

with

$$v = u \cos \phi \ , \ \nu z = u \sin \phi. \tag{11b}$$

Then (10) gives a three-term recursion relation for the coefficients ρ_n . Much in the line of Refs. 24 we have written a program that solves the three-term recursion relation numerically with the aid of matrix continued fractions.

Having at our disposal the density matrix $\rho_{ij}(z,v)$, we finally obtain the light pressure force on an ion at z,v:

$$F(z,v) = 2\hbar[\ q\kappa\ \mathrm{Im}\ \rho_{21}(z,v) + q'\kappa'\ \mathrm{Im}\ \rho_{23}(z,v)\ ] \ . \tag{12}$$

This is the precise counterpart of (2), and states that the forces acting on the dipoles 2-1 and 2-3 are simply added. All complications of the three-level dynamics are hidden in the procedure needed to obtain the dipoles. While a microscopic justification of (12) is still missing, we note that the ideology behind the result seems to agree both with the corresponding treatment of a trapped two-level ion [24] and with a recent three-level theory for a free atom [26].

4. New features of the three-level system

Although the number of variable parameters for the cooling of a trapped three-level ion is very large and so far we have only explored just a few aspects of the problem, some unexpected new concepts have already emerged. We highlight here two of them: existence of stable oscillations of the ion with a nonzero amplitude, and transitions between motional states that resemble phase transitions.

To investigate the nature of the orbits of the ion in the phase space (z,v) we first note that the light pressure force is in general small compared to the trapping forces. Consequently, during one cycle of the mechanical oscillations the orbit of the ion remains almost unperturbed and may be characterized by its velocity amplitude u. The mechanical energy gained by the ion during one cycle in the phase space may then be expressed as the integral of power over one unperturbed cycle. For convenience we divide this energy by the radius of the orbit u, and study instead the quantity

$$\Delta E(u) = \frac{1}{u}\int_0^{2\pi\nu^{-1}} dt\; v(t)F(z(t),v(t)) \; . \qquad (13)$$

If $\Delta E(u)$ is positive (negative), the velocity amplitude of the orbit gradually tends to increase (decrease); if $\Delta E=0$, the orbit is stationary. Stationary orbits may further be classified as stable and unstable according to whether $d\Delta E/du$ is negative or positive. In the former case the light pressure restores the stationary orbit after it has been disturbed slightly, whereas in the latter case the deviation from the stationary orbit tends to grow with time.

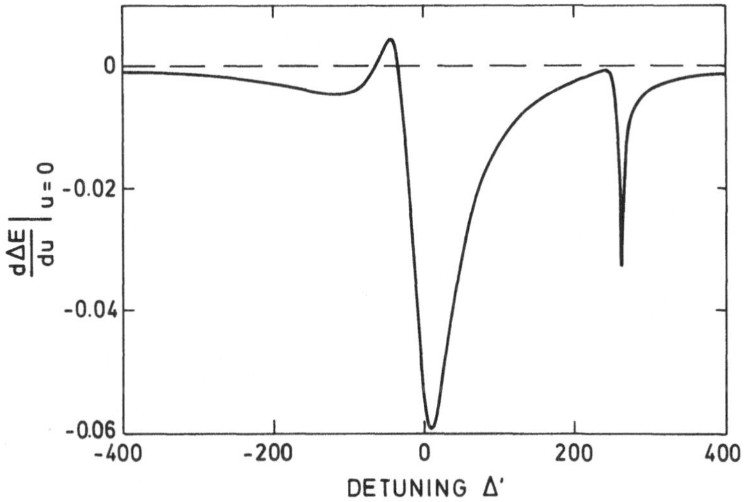

Figure 2. The quantity $d\Delta E(u)/du$ at u=0 is plotted as a function of the detuning Δ'. Positive values indicate that the state of the ion with zero velocity is unstable.

The orbit with u=0 representing a non-moving ion is always stationary. In order to check if it is necessarily stable we plot in Fig.2 $d(\Delta E)/du$ at u=0 as a function of the detuning of the second laser Δ'. The detuning of the first laser is $\Delta=250$, and the Rabi frequencies are $\kappa=60$, $\kappa'=10$. Here and below we use the ionic parameters $\Gamma=16$, $\Gamma'=5$, $q=2.03$ and $q'=1.54$. If the units of frequency, velocity and wave number are 2π MHz, ms^{-1} and 2π MHz s m^{-1}, then these parameters correspond to Ba^+ used in some of the experiments [1,3,25]. The mechanical oscillation frequency is $\nu=1$, and to avoid singularities we choose the decay rate of the two-photon coherence $\gamma_{13}=0.5$. It can be seen that when Δ' is between -60 and -30, the orbit with u=0 is unstable.

To learn what happens in this particular regime of Δ', we fix $\Delta'=-50$, and plot $\Delta E(u)$ as a function of the velocity amplitude u in Fig.3. At the origin the curve starts off with a positive

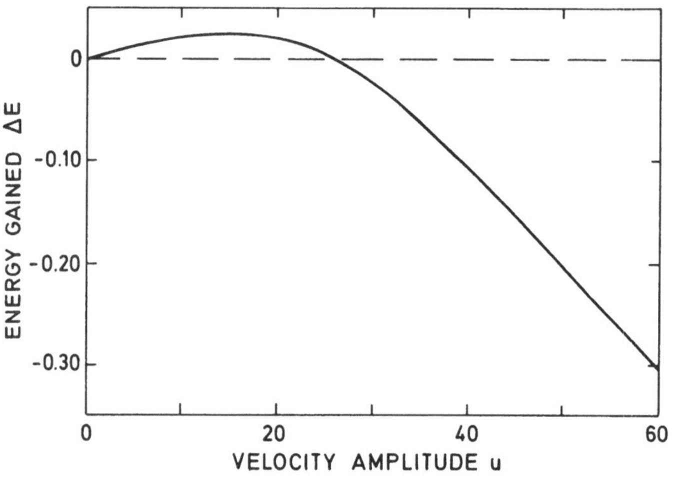

Figure 3. The mechanical energy gained by the ion during one cycle ΔE(u) is plotted as a function of the velocity amplitude u. The value u=0 corresponds to an unstable orbit, the second zero of ΔE determines a stable orbit with a nonzero velocity amplitude.

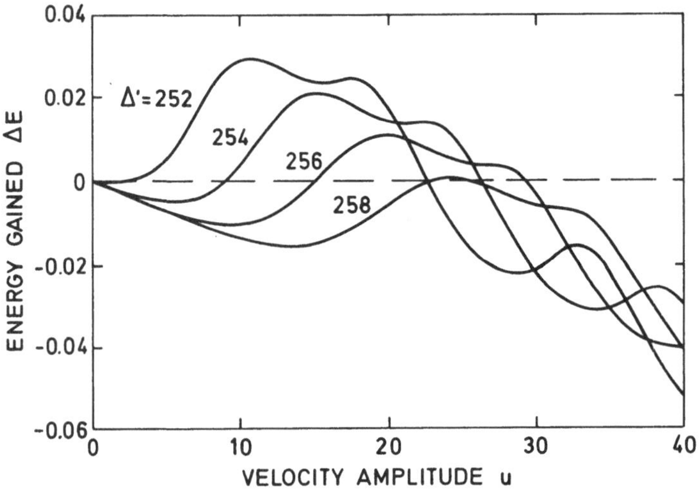

Figure 4. The energy gain ΔE(u) as a function of the velocity amplitude u for four different detunings Δ'=258,256,254 and 252 chosen in the neighbourhood of the two-photon resonance Δ'=Δ=250. Two stable orbits may coexist, and as Δ' is varied a first-order phase transition takes place.

slope as is appropriate for an unstable orbit u=0. But the curve ΔE(u) has another zero at u≈25, and this u gives a stable orbit. This orbit corresponds to a nontrivial stable limit cycle in the phase space.

A moment's reflection on the shape of the curve in Fig.3 shows that the transition from the orbit with u=0 to an orbit with u≠0 is continuous as a function of the detuning Δ', and therefore resembles a second-order phase transition.

In Fig.4 we have plotted the energy gained during one cycle, $\Delta E(u)$, as a function of the velocity amplitude u for four different detunings Δ'=258,256,254 and 252. The detuning of the first laser is Δ=250, so we are scanning the neighbourhood of the two-photon resonance $\Delta=\Delta'$. Other parameters are ν=2, κ=15 and κ'=30. The curves show that two stable orbits may coexist. Furthermore, when Δ' is decreased, the orbit u=0 becomes unstable and the ion suddenly switches to a new orbit with u≈20. This is reminiscent of a first-order phase transition. Like first-order phase transitions, also this one may exhibit metastable states, transitions between them induced by random perturbations such as diffusion, and hysteresis.

5. Discussion

In this paper we have outlined a numerical method of calculating the light pressure force on a trapped three-level ion, and studied the combined influence of the trapping force and the light pressure on the phase space orbits of the ion. Our main result is that in addition to, or as a substitute of, the usual zero-velocity state of the ion there may appear stable stationary states that correspond to oscillations of the ion with a nonzero amplitude.

The physical origin of the results is at present not fully understood. There are indications that the nontrivial stationary orbit in Fig.3 might be connected to optical pumping between the levels 1 and 3, whereas those in Fig.4 would follow from the memory of past velocities conveyed in the two-photon coherence ρ_{13}. However, a quantitative study of these aspects remains to be done.

This work was triggered by experimental results obtained by Peter Toschek and his collaborators [25]. So far a comparison with experiments has not been attempted, but judged by the frequency scales of the experiments [25] it seems that the intensities may have been much higher than in the present theoretical study. This is another point we are presently pursuing.

Investigations into laser cooling of a trapped three-level ion gain additional momentum from the perspective of applying the narrow two-photon resonance as a frequency standard. To this end it is necessary to understand thoroughly the dynamics of the cooling, and methods of monitoring the cooled state also have to be devised. The field will no doubt provide a fruitful forum for joint effort of experimenters and theoreticians alike.

References

1. W. Neuhauser, M. Hohenstatt, P. Toschek and H.G. Dehmelt, Phys. Rev. Lett. **41**, 321 (1978)
2. D.J. Wineland, R.E. Drullinger and F.L. Walls, Phys. Rev. Lett. **40**, 1639 (1978)

3. W.Neuhauser, M. Hohenstatt, P. Toschek and H.G. Dehmelt, Phys. Rev. **A 22**, 1137 (1980)

4. D.J. Wineland and W.M. Itano, Phys. Lett. **82A**, 75 (1981)

5. W. Nagourney, G. Janik and H. Dehmelt, Proc. Natl. Acad. Sci. USA **80**, 643 (1983)

6. J.V. Prodan, W.D. Phillips and H. Metcalf, Phys. Rev. Lett. **49**, 1149 (1982)

7. V.O. Balykin, V.S. Letokhov and A.I. Sidorov, Optics Comm. **49**, 248 (1984)

8. P.E. Moskowitz, P.L. Gould, S.R. Atlas and D.E. Pritchard, Phys. Rev. Lett. **51**, 370 (1983)

9. V.A. Grinchuk, A.P. Kazantsev, E.F. Kuzin, M.L. Nagaeva, G.A. Ryabenko, G.I. Surdutovich and V.P. Yakovlev, Sov. Phys. JETP **59**, 56 (1984)

10. See the articles in 'Laser-Cooled and Trapped Atoms', ed. W.D. Phillips; Progress in Quantum Electronics **8**, Numbers 3/4 (1984)

11. J. Dalibard, S. Reynaud and C. Cohen-Tannoudji, J. Phys. B **17**, 4577 (1984)

12. A. Ashkin, Optics Lett. **9**, 454 (1984)

13. V.G. Minogin, Sov. Phys. JETP **52**, 1032 (1980)

14. R.J. Cook, Phys.Rev. **A 22**, 1078 (1980)

15. R.J. Cook, Phys. Rev. Lett. **44**, 976 (1980)

16. J.P. Gordon and A. Ashkin, Phys. Rev. **A 21**, 1606 (1980)

17. J. Dalibard and C. Cohen-Tannoudji, to appear in J. Phys. B

18. C. Tanguy, S. Reynaud and C. Cohen-Tannoudji, J. Phys. B **17**, 4623 (1984)

19. D.J. Wineland and W.M. Itano, Phys. Rev. **A 20**, 1521 (1979)

20. W.M. Itano and D.J. Wineland, Phys.Rev. **A 25**, 35 (1982)

21. M. Lindberg, J. Phys. B **17**, 2129 (1984)

22. M. Lindberg in 'Proceedings of the Finnish-Soviet Seminar on Theoretical Problems in Quantum Electronics', ed. V.P. Chebotayev and S. Stenholm, to appear

23. J. Javanainen, M. Lindberg and S. Stenholm, J. Opt. Soc. Am. **B 1**, 111 (1984); M. Lindberg and S. Stenholm, J. Phys. B **17**, 3375 (1984)

24. J. Javanainen, J. Phys. B **14**, 2519 (1981); **14**, 4191 (1981); and to appear in J. Phys. B

25. P. Toschek's talk in this Seminar

26. V.G. Minogin and Yu.V. Rozhdetsvensky, Appl. Phys. **B 34**, 161 (1984)

27. R.J. Cook, Phys. Rev. **A 23**, 1243 (1981)

28. S. Stenholm, 'Foundations of Laser Spectroscopy', Wiley, New York, 1984

29. C. Feuillade and P. Berman, Phys. Rev. **A 29**, 1236 (1984)

30. M. Kaivola, P. Thorsen and O. Poulsen, to be published

NONLINEAR AND COHERENT PROPERTIES OF LASER RADIATION PRESSURE ON ATOMS

V.G. Minogin

Institute of Spectroscopy, USSR Academy of Sciences, 142092 Troitzk Moscow Region, USSR

1. Introduction

Laser radiation pressure is a universal mechanism which provides an effective control over the motion of neutral atoms. Basic physical reason for such a status of laser radiation pressure is its selective dependence on atomic velocity due to the linear Doppler effect. Moving atom absorbs efficiently light photons only when the detuning between the frequency ω of light and the atomic transition frequency ω_0 is compensated for by an appropriate Doppler shift $\vec{k}\vec{v}$. This means that the atom is subjected to the action of the light pressure force only at a definite value of its velocity.

Along with this basic property of laser radiation pressure, the use of laser radiation for mechanical action on atoms has some important advantages. First of all, due to high spectral brightness and monochromaticity of laser radiation the laser light at resonance with atomic transition creates such a high value of force on atom that cannot be achieved using conventional thermal light sources. For example, CW laser radiation of moderate intensity (0.1 W/cm^2), that is sufficient to saturate the allowed dipole atomic transition, creates a light pressure force which produces an acceleration of atom four to five orders higher than the gravitational acceleration. Second, the control over the laser radiation frequency and polarization allows effective atom-laser light interaction during long periods of time. The classical illustration of one of such possibilities is optical orientation of atoms by circularly-polarized radiation that provides long-time circulation of atom between two quantum states. Finally, high directivity of laser beams make it possible to create force fields of different spatial configurations.

Nowadays laser radiation pressure is successfully used in experiments on atomic velocity control [1] . The selective dependence of resonant light pressure on atomic velocity is especially important in experiments on velocity monochromatization and deceleration of atomic beams [2-9] and collimation of atomic beams [10,11] .

The dependence of light pressure force on atomic velocity is always a principal point in applications of laser radiation pressure. This dependence is relatively simple only in the case of interaction of a two-level atom with a travelling monochromatic laser wave where the light pressure force is a Lorentzian function of atomic velocity. In more complicated cases, when the light field is a superposition of several laser beams or the atom is considered to have many levels, the dependence of the light pressure force on atomic velocity usually differs from the Lorentzian function due to the nonlinear interaction of laser beams with an atomic transition and atomic level coherence.

When the light field consists of several waves, an atom can take part in nonlinear processes absorbing photons from some waves and emitting them to others. Each of these processes can occur due to the Doppler effect at a definite atomic velocity. Therefore, nonlinear processes deform the dependence of the light pressure force on atomic velocity. Similarly, different atomic levels become coherent only at definite atomic velocities because the probabilities of atomic transitions between separate quantum states depend on the Doppler shifts of exciting waves. Accordingly, the atomic level coherence induced by light field modifies the velocity dependence of the light pressure force, too.

This paper gives a short review of nonlinear and coherent properties of laser radiation pressure on atoms and describes the related experiments. To illustrate some typical cases, the discussion is concerned with two- and three-level atoms and light fields consisting of one or two light waves.

2. Conditions of applicability of the concept of resonance radiation pressure

Assume that the light field exciting the dipole transitions of an atom consists of waves whose wave vector difference is comparable with the wave vectors themselves. Let the rate of spontaneous relaxation of an atom from excited states be 2γ and $R = \hbar^2 k^2/2M$ be the recoil energy corresponding to the absorption (emission) of a photon with the wave vector $\vec{k}(|\vec{k}| = \omega_{\vec{k}}/c)$. Atomic transition frequencies are assumed to be close to the light field frequencies $\omega_{\alpha\beta} \simeq \omega_{\vec{k}}$.

Under above assumptions the concept of resonant radiation pressure has a physical meaning when two conditions are satisfied

$$t \gg \gamma^{-1}, \tag{1}$$

$$\hbar\gamma \gg R. \tag{2}$$

These conditions mean that the translational motion of an atom in a resonant light field can be considered as classical. Condition (1) on the atom-light field interaction time τ is necessary for the classical atomic momentum to exceed the photon momenta $\hbar k_i$ being a measure of quantum fluctuations. The condition (2) means that the atomic momentum change under the light pressure force $\Delta p \simeq M\gamma/k$ always exceeds the photon momenta $\hbar k_i$.

When conditions (1) and (2) are satisfied, the translational motion of an atom in a light field is described by the infinite Fokker-Planck equation for the atomic distribution function $w(\vec{r},\vec{p},t)$, where $\vec{r}$ and $\vec{p}$ are classical coordinate and momentum of an atom [12-16]

$$\frac{\partial w}{\partial t} + \vec{v}\frac{\partial w}{\partial \vec{r}} = -\frac{\partial}{\partial \vec{p}}(\vec{F}w) + \sum_{i,j=x,y,z} \frac{\partial^2}{\partial p_i \partial p_j}(D_{ij} w) + \dots \tag{3}$$

The coefficients of Eq.(3) determine the light pressure force $\vec{F}$, the momentum diffusion tensor D_{ij} and other quantum-statistical average quantities. They are the functions of atomic velocity (momentum).

The velocity dependence of the light pressure force $\vec{F}$ is of particular interest for practical applications. This is due to the fact that the light pressure force induces a directional drift of atoms and it is thus responsible for the basic effects of the light pressure. Having in mind the fundamental role of the light pressure force we discuss below the contributions of nonlinear and coherent processes by analizing the velocity dependence of the light pressure force. Also, we consider some applications of resonant light pressure for velocity-selective control over atomic motion.

3. Two-level atoms in the travelling wave field

Let a two-level atom interact with a plane travelling wave

$$\vec{E} = \frac{1}{2}\vec{e}E_0 e^{ikz-i\omega t} + c.c. \tag{4}$$

Assume that the lower level of the atom is the ground state and the upper level decays to the ground level with the spontaneous emission rate 2γ.

In the field (4) a two-level atom is acted by the light pressure force [17]

$$\vec{F} = \hbar\vec{k}\gamma \frac{G}{1+G+(\Omega-kv_z)^2/\gamma^2}, \tag{5}$$

having a Lorentzian dependence on the velocity projection v_z. Here $\Omega = \omega-\omega_0$ is the detuning between the wave frequency and the atomic transition frequency ω_0, $G = 1/2(dE_0/\hbar\gamma)^2$ is the saturation parameter, d is the matrix element of atomic transition dipole moment.

The Lorentzian dependence of the force (5) on the velocity projection is due to the linear Doppler effect. The dependence of the force on the parameter G is caused by the atomic transition saturation. When $G \to \infty$, the force (5) tends to a quantity $\hbar k\gamma$ given by the product of the photon momentum $\hbar k$ by the spontaneous decay rate 2γ and the relative population $n_2 = 1/2$ of the upper level of atomic transition.

The nonlinear dependence of the force (5) on velocity has recently found wide applications for velocity-selective deceleration of atomic beams and formation of monovelocity atomic beams (longitudinal cooling of atomic beams). Qualitatively, the longitudinal cooling of atomic beams may be understood as following (Fig.1). The light pressure force acting on the atoms in atomic beam interacting with resonant counter-propagating radiation decreases the velocities of resonant atoms. The slowing down of atoms which are at reso-

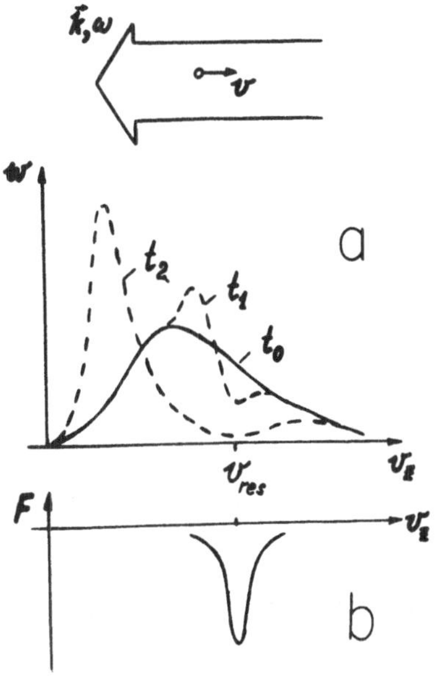

Fig.1. The velocity monochromatization of atomic beam decelerated by laser wave. The initial velocity distribution is shown by solid curve. The dashed lines show the velocity distribution for times t_1 and t_2 ($t_o < t_1 < t_2$)

nance with radiation gives rise to a dip in the velocity distribution centered at the velocity $v_{res} = (\omega_0 - \omega)/k$ (Fig.1a). The formation of the velocity dip is followed by the formation of a peak of decelerated atoms centered at a velocity smaller than v_{res}. For a long interaction time, almost all the atoms are decelerated and, since the velocity dependence of the light pressure force F (Fig.1b) has a nonlinear belled shape, the initial wide velocity distribution always converts to a narrow monovelocity distribution. The longitudinal cooling of atomic beam was studied in experiments where beams of sodium atoms were decelerated by CW dye laser radiation [2-9]. Below we describe the experiments [4] in which effects of velocity monochromatization were first observed and the experimental velocity distribution was compared with the distribution calculated from the Fokker-Planck equation.

Figure 2 schematically shows the experimental scheme for velocity distribution monochromatization of Na atomic beams decelerated by laser radiation. A narrow atomic beam (2) coming from an oven (1) was irradiated by a light beam (3) tuned to resonance with the transition $3S_{1/2} - 3P_{3/2}$ of sodium atom. Since the ground state of Na atom is split in two hyperfine structure sublevels, the laser radiation had two frequencies to make the atom-radiation interaction cyclic. One frequency was in resonance with the transition

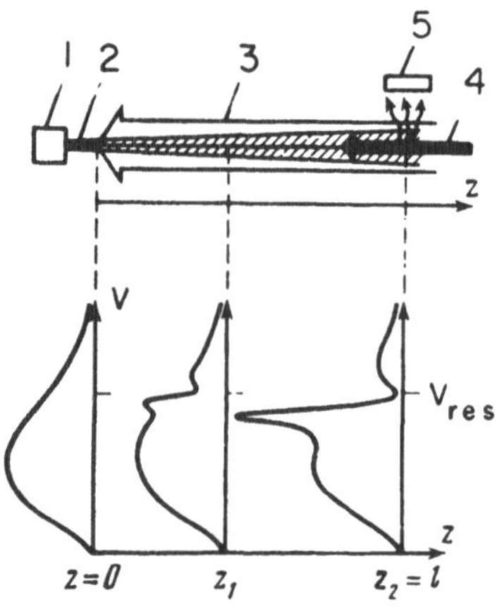

Fig.2. The scheme of an experiment on longitudinal cooling of atomic beams and velocity distribution deformation for different distances from the point of the atoms entering the light beam.

$3S_{1/2}(F=1)$ - $3P_{3/2}$, the second one with the transition $3S_{1/2}(F=2)$-$3P_{3/2}$. The difference of these two laser frequencies was equal to the frequency of the hyperfine structure interval between the sublevels $3S_{1/2}(F=1)$ and $3S_{1/2}(F=2)$ which is 1772 MHz. The velocity distribution of atoms along the beam axis was registered by observing the fluorescence signal excited by a probe laser beam (4). The probe beam had one frequency. Its frequency could be tuned within the atomic beam absorption line. In order that the strong two-frequency laser beam should not deform the fluorescence signal excited by the probe beam, the two-frequency laser beam was periodically

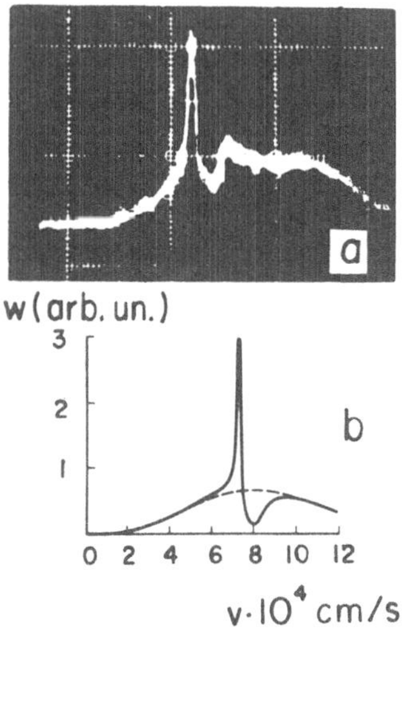

Fig.3. Experimental (a) and theoretical (b) profiles of velocity distribution of a Na atomic beam decelerated by laser radiation.

interrupted by a mechanical chopper. The fluorescence signal was recorded by a detector (5) only at time intervals when the strong beam was switched off by the chopper.

One of the experimental dependences of fluorescence intensity on probe laser beam frequency is shown in Fig.3. The curve presents the longitudinal velocity distribution of atoms in a beam caused by the nonlinear deceleration of the atoms by counter-propagating laser radiation. When the dependence $w = w(v)$ was measured, the laser beam was tuned to resonance with the atoms at the penk of initial thermal velocity distribution.

Since it was the light pressure force that made a basic contribution to the velocity distribution deformation in the experiment [4], the experimental results proved to be similar to the results of calculations done on the basis of the Liouville equation (Fig.3b).

In the first experiments [4] the ratio of the initial velocity distribution width to the width of a narrow peak of monochromated atoms was $\Delta v_{in}/\Delta v_{fin} = 19$. Such a degree of monochromating was consistent with a decrease in the temperature of relative atomic motion from its initial value $T_{in} = 573$ K to $T_{fin} = 1.5$ K. In further experiments [5-7] the beams of Na atoms were produced with the absolute temperature T = 0.07 K.

4. Two-level atoms in the field of counter-propagating waves

If a two-level atom interacts with the field of two travelling waves (4) propagating in the directions $\pm z$, the light pressure force F is rather a complicated function of atomic velocity (Fig.4).

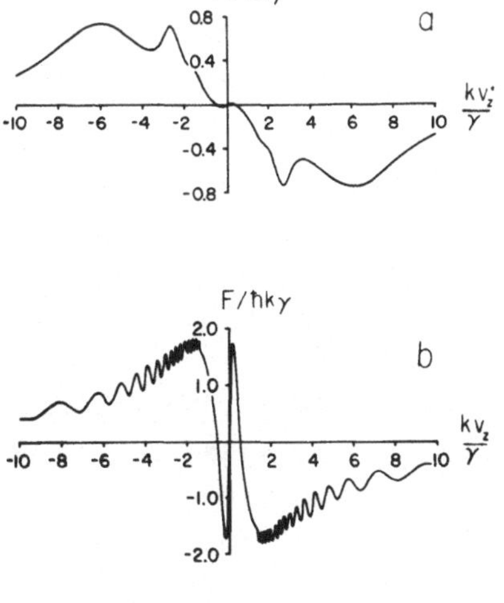

Fig.4. The dependence of the light pressure force on the ptojection of atomic velocity onto the z axis in the case of interaction of a two-level atom with two plane waves propagating in the $\pm z$ directions. $\Omega = -3\gamma$, G=10(a), G=1000(b).

For small values of the saturation parameter G, when two counter-running waves saturate the atomic transition independently, effective atom-field interaction can take place at two velocity values

$$\pm k v_z \simeq \omega - \omega_0 = \Omega \; . \tag{6}$$

Therefore, in the case of weak saturation of atomic transition the light pressure force contains two resonances. For the reasons set forth below they can be called first-order resonances.

As the saturation parameter increases, the force acquires higher-order resonances due to nonlinear interaction of the atoms with the counter-propagating waves. The simplest of them are second-order resonances (Fig.5a). These resonances arise when the absorption of photons from one travelling wave occurs simultaneously with

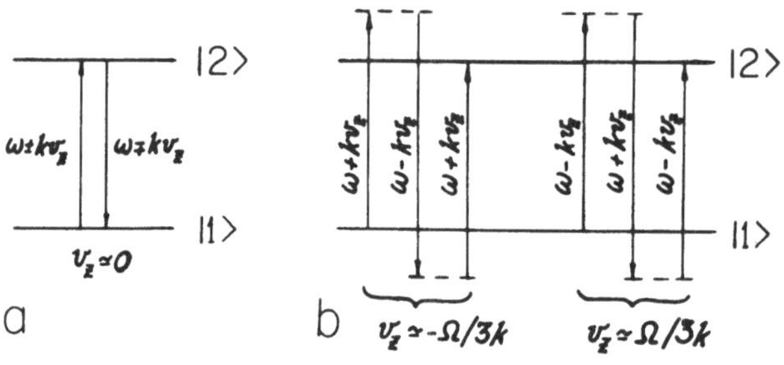

Fig.5. Second- and third-order resonant processes.

photon emission into the other travelling wave. Since the frequency of one wave in a rest frame of an atom is $\omega + kv_z$ and the frequency of the other wave is $\omega - kv_z$, in accordance with the energy conservation law

$$(\omega \pm k v_z) - (\omega \mp k v_z) = 0 \tag{7}$$

the second-order resonances are centred at the zeroth velocity (v=0). Next are third-order resonances (Fig.5b). They arise due to nonlinear interaction where the photon absorption from two counter-propagating waves comes about simultaneously with photon emission into one of the waves. According to the energy conservation law written in a rest atomic frame

$$(\omega \pm kv_z) - (\omega \mp kv_z) + (\omega \pm kv_z) = 0 \tag{8a}$$

the third-order resonances are centred near the velocities

$$\pm kv_z \simeq (\omega - \omega_0)/3. \tag{8b}$$

Fourth-order resonances centred near $v_z \simeq 0$ and fifth-order resonances centred near the velocities

$$\pm kv_z \simeq (\omega - \omega_0)/5. \tag{9}$$

The examples considered show that odd- and even-order resonances are caused by nonlinear processes which either change the internal atomic state or leave it unchanged [18]. The processes in which the absorption (emission) of n+1 photons from one wave is followed by the emission (absorption) of n photons into the other wave are responsible for the change of the internal atomic state and result in 2n+1 -order resonances centred near the velocities

$$\pm kv_z \simeq \Omega/(2n+1).$$

The processes in which the absorption of n photons from one wave is followed by the emission of n photons into another wave are responsible for 2n -order resonances localized near the zeroth velocity.

The specific dependence of the light pressure force on the atomic velocity in the case of counter-propagating waves has been recently used for two-dimensional transverse cooling of atomic beams [11]. The scheme of transverse cooling realized for Na atomic beams is shown in Fig.6. In this scheme an atomic beam (2) coming out of

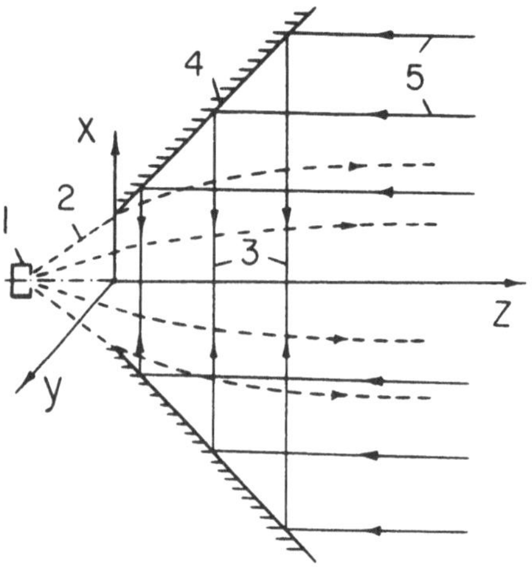

Fig.6. The scheme of atomic beam transverse cooling and atomic beam collimation.

the source (1) was irradiated by an axisymmetric field (3) the frequency ω of which was red-shifted about the atomic transition frequency. The axisymmetric field was formed by the laser beam (5) reflecting from the inner surface of a hollow metal cone (4).

The axisymmetric field (3) can be presented at any point inside the cone as a sum of counter-propagating waves. Since the field frequency ω is assumed to be smaller than the atomic transition frequency ω_0 and the saturation parameter G is not very large, the light

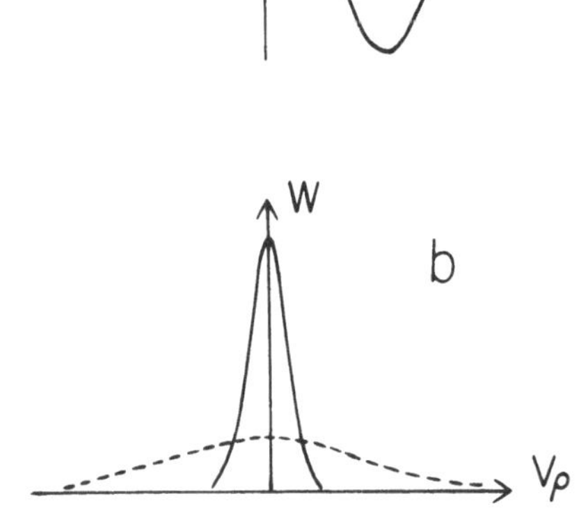

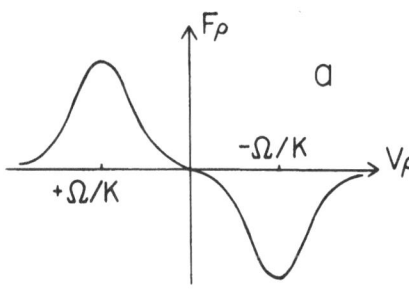

Fig.7. a)The light pressure force F_ρ in the axisymmetric field as a function of radial atomic velocity v_ρ .
b)Narrowing of transverse velocity distribution of an atomic beam under the action of the force F_ρ .

pressure force at any point inside the cone is directed towards the radial velocity $\vec{v}_\rho$ (Fig.7a). Accordingly, the light pressure force narrows the velocity distribution transverse to the beam propagation axis, that is, it cools the atomic beam (Fig.7b). The transverse cooling of atomic beam is followed by the collimation of the beam, that is, by a decrease in its divergence as it has been pointed out firstly by Hansch and Schawlow [19].

The basic parameter determining the transverse cooling of atomic beam is effective temperature that characterizes the width of atomic velocity distribution transverse to the beam propagation axis. For the detuning Ω this temperature is [10]

$$T_{tr}^{0} = \frac{1}{4k_B}\hbar\gamma\left(|\Omega|/\gamma + \gamma/|\Omega|\right). \tag{10}$$

For $\Omega = -\gamma$ the temperature has a minimum value

$$T_{min} = \hbar\gamma/2k_B \tag{11}$$

which ranges from 10^{-4} to 10^{-3}K.

In [11] a two-frequency field of dye laser radiation was used to investigate experimentally the transverse cooling of Na atomic beams. The transverse temperature of the atomic beam was reduced to $3.5 \cdot 10^{-3}$K. The cross-section area of the atomic beam was decreased by one half due to collimation.

5. Three-level atoms in the field of two travelling waves

When multilevel atoms interact with a resonant radiation field, the effects of atomic level coherence are of great importance along with nonlinear effects. Some manifestations of atomic level coherence

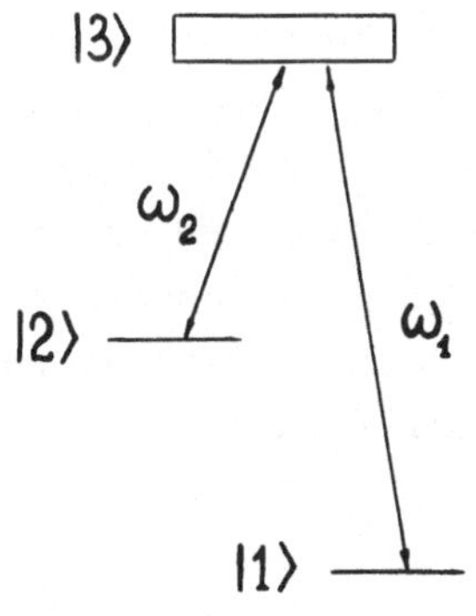

Fig.8.

can be understood in the case of a three-level atom.

Let a three-level atom have two ground states. The upper state $|3\rangle$ is assumed to decay into ground states with the spontaneous decay rates $2\gamma_1$ and $2\gamma_2$ (Fig.8). Assume also that the atom is excited by two waves. Wave 1 with the frequency ω_1 is resonant to the transition $|1\rangle - |3\rangle$ and wave 2 with the frequency ω_2 is resonant to the transition $|2\rangle - |3\rangle$. Each of the waves is assumed to be given by the relation (4). The matrix elements of the dipole transitions $|1\rangle - |3\rangle$ and $|2\rangle - |3\rangle$ are considered to be equal.

When the waves propagate in the same direction, for example in the z-direction, the dependence of the light pressure force on the velocity projection v_z is resonant like in the case of interaction of a two-level atom with one travelling wave (Fig.9). The case of a three-level atom is however characterized by a sharp dependence

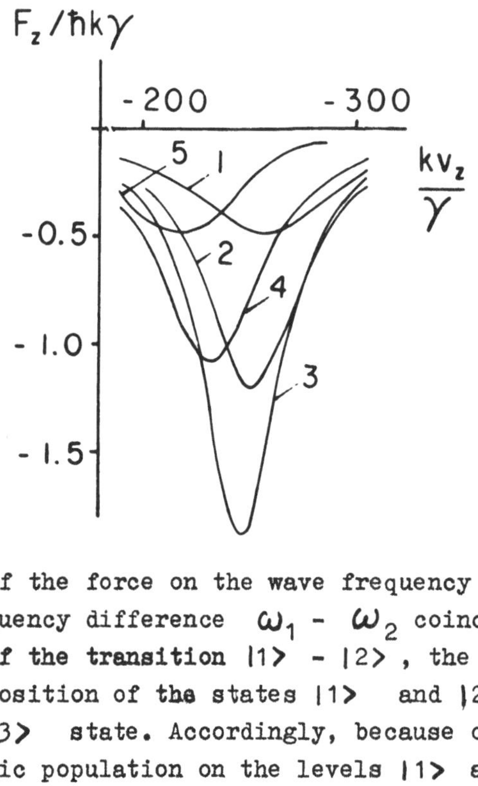

Fig.9. The dependence of the light pressure force on the velocity projection v_z for three-level atom excited by plane waves propagating in the -z direction for $\gamma_1 = \gamma_2 = \gamma$, $G=10^3$, $(\omega_1 - \omega_{31})/\gamma = -270$. The numbers 1,2,...,5 near the curves refer to the detuning $(\omega_2 - \omega_{32})/\gamma = -250, -240, -230, -200, -170$.

of the force on the wave frequency difference [20]. When the frequency difference $\omega_1 - \omega_2$ coincides with the frequency ω_{21} of the transition $|1\rangle - |2\rangle$, the atom comes to a coherent superposition of the states $|1\rangle$ and $|2\rangle$ and is not excited to the $|3\rangle$ state. Accordingly, because of the coherent trapping of atomic population on the levels $|1\rangle$ and $|2\rangle$ the light pressure force becomes zero. The atom is effectively excited only when the wave frequency difference reaches the value of natural transition line width

$$\omega_1 - \omega_2 - \omega_{12} \simeq \gamma(1+G)^{1/2}. \qquad (12)$$

With further increase of detuning the light pressure force drops on account of optical pumping of the atom to one of the low-lying levels.

In the case of counter-propagating waves the light pressure force becomes zero for a definite value of v_z. This can be explained by the fact that the coherent superposition of the states $|1\rangle$ and $|2\rangle$ takes place when the wave frequency difference in a rest atomic frame coincides with the frequency ω_{21}. Let wave 1 propagate

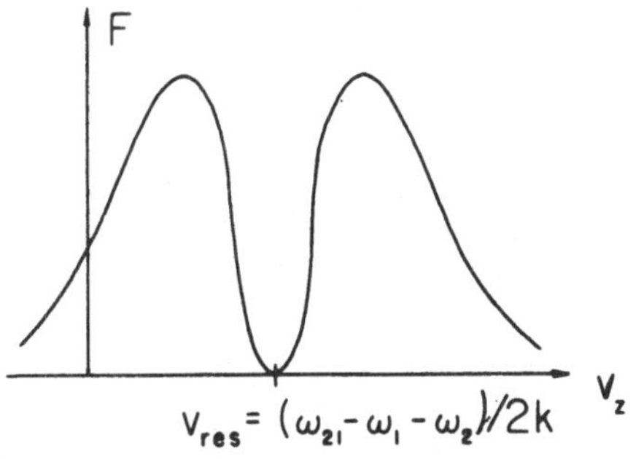

Fig.10. The dependence of the light pressure force on the velocity projection v_z for a three-level atom excited by two waves propagating in the $\pm z$ directions for $2\gamma_2 = \gamma_1 = \gamma$

in the -z direction and wave 2 in the +z direction. Then in a rest atomic frame the frequency of the wave 1 will be $\omega_1 + kv_z$, and the frequency of the wave 2 will be $\omega_2 - kv_z$. Accordingly, at the condition

$$(\omega_1 + kv_z) - (\omega_2 - kv_z) \simeq \omega_{21} \qquad (13)$$

the light pressure force becomes zero (Fig.10).

6. Conclusion

The above considered examples thus show that the value and the velocity dependence of the light pressure force are determined by nonlinear processes of atom-light field interaction and atomic level coherence. Because of this, a directed control over nonlinear and coherent processes allows light pressure to be applied for effective action on atomic motion in different physical situations.

References

1. Letokhov V.S., Minogin V.G.- Phys. Rep. 73, 1 (1981)
2. Balykin V.I., Letokhov V.S., Mishin V.I.- Pis'ma Zh. Eksp. Teor. Fiz. 29, 614 (1979); Zh. Eksp. Teor. Fiz. 78, 1376 (1980)
3. Balykin V.I., Letokhov V.S., Minogin V.G.- Zh. Eksp. Teor. Fiz. 80, 1779 (1981)
4. Andreev S.V., Balykin V.I., Letokhov V.S., Minogin V.G.- Pis'ma Zh. Eksp. Teor. Fiz. 34, 463 (1981); Zh. Eksp. Teor. Fiz. 82, 1429 (1982)
5. Phillips W.D., Metcalf H.- Phys. Rev. Lett. 48, 596 (1982)
6. Prodan J.V., Phillips W.D., Metcalf H.- Phys. Rev. Lett. 49, 1149 (1982)
7. Phillips W.D., Prodan J.V., Metcalf H.J.- In: Laser-Cooled and Trapped Atoms, Ed. by W.D. Phillips, NBS Special Publication, Washington, 1983, p. 1
8. Blatt R., Ertmer W., Hall J.L.- Ibid., p. 142
9. Balykin V.I., Letokhov V.S., Sidorov A.I.- Opt. Comm. 49, 248 (1984); Zh. Eksp. Teor. Fiz. 86, 2019 (1984)
10. Balykin V.I., Letokhov V.S., Minogin V.G., Zueva T.V.- Appl. Phys.B 35, 149 (1984)
11. Balykin V.I., Letokhov V.S., Sidorov A.I.- Pis'ma Zh. Eksp. Teor. Fiz. 40, 251 (1984)
12. Baklanov E.V., Dubetzkii B.Ya.- Opt. Spektr. 41, 3 (1976)
13. Minogin V.G.- Zh. Eksp. Teor. Fiz. 79, 2044 (1980)
14. Gordon J.P., Ashkin A.- Phys. Rev.A 21, 1606 (1980)
15. Cook R.J.- Phys. Rev.A 22, 1078 (1980)
16. Stenholm S.- Phys. Rev.A 27, 2513 (1983)
17. Ashkin A.- Phys. Rev. Lett. 24, 156 (1970); 25, 1321 (1970)
18. Minogin V.G., Serimaa O.T.- Opt. Comm. 30, 373 (1979)
19. Hänsch T.W., Schawlow A.L.- Opt. Comm. 13, 68 (1975)
20. Minogin V.G., Rozhdestvensky Yu.V.- Appl. Phys.B 34, 161 (1984)

CONTRIBUTED PAPERS

(Abstracts)

PART I: Collisions in Laser Fields

MULTIPHOTON TRANSITIONS IN THE COULOMB CONTINUOUS SPECTRUM

Alfred Maquet and Valérie Véniard

Laboratoire de Chimie Physique[a]
Université Pierre et Marie Curie
11, Rue Pierre et Marie Curie
F. 75231 PARIS Cedex 05- FRANCE

We report an **exact** perturbative calculation of two-photon free-free transition amplitudes in the Coulomb continuous spectrum. The corresponding cross sections are associated to an experimental situation in which, while scattered in the field of a nucleus, incoming electrons absorb (or emit) two photons of an external single-mode laser field. Both the non-relativistic and dipole approximations are assumed. Within this context the term **exact** means that we have used **exact Coulomb wave functions** for describing the incoming and outgoing electron states. The infinite summation over the complete hydrogenic spectrum, contained into the second-order amplitude, is implicitly performed by using a compact representation of the **Coulomb Green's function** [1]. Our results represent, in some sense, the two-photon generalization of Sommerfeld's celebrated formulae for the bremsstrahlung in a Coulomb potential [2].

We have carried out the analytical calculation as far as possible, in order to get a closed form expression of the amplitude. This has the advantage of facilitating the study of the behaviour of the cross sections in various **limiting cases** of physical interest. The general form of the amplitude depends in a complicated way on the physical parameters of the problem, namely the laser frequency ω and intensity I, the magnitudes of the initial and final electron energies E_i and E_f and the respective orientations of their momenta $\vec{p}_i$ and $\vec{p}_f$ with respect to the laser polarization $\vec{s}$. It simplifies considerably however, in three limiting cases (atomic units are used throughout):

i). If $\omega \ll E_i, E_f$, corresponding to the **soft-photon** approximation, the cross section goes over smoothly to the low intensity ($I \ll 1$) limit of the Kroll and Watson result for the two- photon free-free transitions [3]. An interesting point is that this latter result was derived for the specific case of a short range potential: our calculation shows that, to lowest order in $\omega/E_{i,f}$, the result holds also for the infinite range Coulomb field [4]. As expected, one observes also that the Rutherford elastic scattering cross section factors out: this represents an extension to the two-photon case of the so-called Low theorem related to the soft-photon limit of the bremsstrahlung cross section [5].

ii). If, in addition, one allows E_i, E_f to become large, one obtain a simple closed form expression, corresponding to the (lowest order non vanishing) **Born approximation** of the cross section.

iii). Finally, the opposite limit, $\omega \gg E_i$,(6), also leads to simplifications. Then the amplitudes become independent of the energy of the incoming electron and of the scattering angle, the overall angular dependence of the cross section being contained in the scalar products $(\vec{s}.\vec{p}_i)$ and $(\vec{s}.\vec{p}_f)$.

We have checked that our numerical results go over smoothly into the analytical results for the three limiting cases considered. Similar checks have been also performed in the simpler one-photon case. By comparing our results for the one- and two-photon processes with the predictions of Kroll and Watson's theory we have been able to discuss the transition between the two approaches.

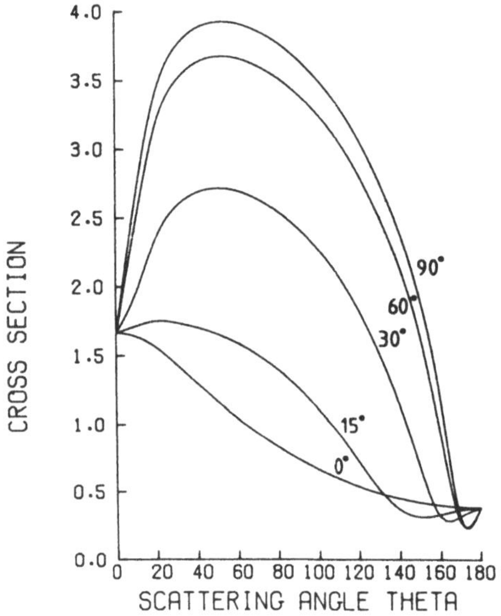

The angular behaviour of the cross section is displayed in the figure, where the variations of $\log(d\sigma/d\Omega)$ are plotted in terms of the polar angle θ of the final momentum $\vec{p}_f$ for various values of φ. We have illustrated here the case of two-photon absorption for the special geometry $\vec{p}_i.\vec{s} = 0$, the polar axis Oz being set along $\vec{p}_i$ and Oy along the laser polarization $\vec{s}$. The energies of the incoming electron and of the laser photon are 1. Ry. and .5 Ry. respectively.

Acknowledgements. This work is the result of a fruitful collaboration with M. Gavrila (FOM, Amsterdam).

References: (a) Laboratoire Associé au CNRS.

(1) J. Schwinger, J. Math. Phys. 5, 1606,(1964).

(2) A. Sommerfeld, *Atombau und Spektrallinien,* (Vieweg, Braunschweig, 1939).

(3) N. M. Kroll and K. M. Watson, Phys. Rev. A8, 804 (1973).

(4) A. Maquet, in *Resonances- Models and Phenomena,* edited by S. Albeverio, L. S. Ferreira and L. Streit (Springer, Berlin, 1984) p. 257.

(5) F. E. Low, Phys. Rev. 110, 974 (1958).

(6) M. Gavrila and J. Z. Kaminski, Phys. Rev. Lett. 51, 613 (1984).

GAUGE PROBLEMS IN LASER FIELDS

M. Zarcone[+], G. Ferrante[+], C. Leone[°] and M. Zukowski[^]

[+] Istituto di Fisica, Via Archirafi 36, 90123 Palermo, Italy

[°] Istituto di Fisica, Facoltà di Ingegneria, Parco d'Orleans, 90128 Palermo, Italy

[^] Institute of Theoretical Physics and Astrophysics, University of Gdansk, Poland.

In recent times, the problems concerning the choice of e.m. gauge to describe the interaction of radiation with matter have received a lot of attention, mostly in connection with optical transitions. At the same time little or no attention at all has been paid to the same problems within nonperturbative methods of radiation matter interaction as in the case for field assisted atomic collisions. Gauge aspects need to be considered anew in the context of field-assisted collisions, first, because it is not clear in advance if the esperience gained in the optical transition investigations is adequate for nonperturbative treatments; second, because there is by now enough evidence that in the nonperturbative treatments gauge consistency appears to be more relevant than in the perturbative ones. Schematically, the reasons to address this subject may be summarized in two points: (i) to learn which gauge in a given problem is able to provide better results in approximate treatments of radiation-matter interaction within collision processes; and (ii) to gain experience on how to guarantee in these treatments the maximum of gauge consistency avoiding mere artefacts. In fact, for instance, in collisions involving particles with internal structure, the total wavefunction of the collision event must account simultaneously for free, relative and bound motion. Usually the problem is never solved exactly. Resorting to approximate derivations advantage is generally taken of any feature able to yield simplifications of the problem under consideration . One of these is to solve separately the Schrödinger equations for the free and the bound motions choosing the e.m. gauges such as to give the results in the simplest form and/or in the most straightforward way within a given accuracy. A result of the above procedure may be that parts of the total wavefunctions of the collision process are solved in different gauges and to a different degree of accuracy. This possibility suggests that care must be taken to have gauge consistent total wavefunctions, and in order that different approximations are mutually consistent.

We have considered, first, the case of Radiation and Electric field gauges in the context of potential scattering in the presence of a laser field taken in dipole approximation. As expected on general grounds, it is readly shown that the results in the two gauges are the same provided no approximations are made as far as the particle-field interaction is concerned. When the interaction is instead treated not exactly, as it occurs for instance in the so called low frequency approximation, new and interesting differencies are found. Developing the full Green function in powers of the particle-field interaction we have: a) In the R-gauge, if we take only the first three terms of the expansion and consider only terms proportional to ω we obtain the low frequency Kroll and Watson result fot the T-matrix; b) the same result is obtained in the E-gauge taking only the first two terms of the Green function expansion. The E-gauge appears then more straigthforward in getting a given approximation than the R-gauge. In this context, we have started a study of rearrangement collisions in the presence of a laser field. Again we have paid due attention to an accurate gauge-consistent construction of the total wavefunctions. In this respect, rearrangement collisions present some novel features, which are absent in other kind of atomic collisions. For instance, the A^2 term does not cancel as generally happens when there is no rearrangement; even more interesting is the presence inside the spatial matrix element of a Bessel function containing the lasers parameters.

ELECTRON STATES IN A CONSTANT MAGNETIC FIELD AND THE ZERO FIELD LIMIT IN POTENTIAL SCATTERING

J. Bergou and S. Varró
Central Research Institute for Physics, H-1525 Budapest

In the recent times much effort has been devoted to the investigation of induced scattering processes in the presence of strong laser fields and/or magnetic fields. These precesses are usually described within the framework of perturbation theory in Furry picture. This means that the effects of the strong fields are taken exactly into account while the other agents are considered as small perturbations which cause transitions between the dressed states of the scattered particle. One inconsistency of this approach is that the transition probabilities do not reduce to their zero-field form in the zero magnetic field limit. This shortcoming of the theory can be traced back to be existing already on the level of the electron states in a magnetic field used as in and out states of the scattering. We, therefore, reconsider these states.

First let us make a few remarks on the gauge invariance in a homogeneous constant magnetic field. It is clear that the vector potentials $\underline{A}(\theta) \equiv B(-y \sin^2\theta,\ x \cos^2\theta,\ 0)$ parametrised by θ always give us the same magnetic induction $\underline{B}=(0,0,B)$. This means that the relation $\underline{A}' = \underline{A}(\theta')$ generates a one parameter set of gauge transformations. Applying the gauge transformation on the wavefunctions it turns out that there is a special class of states which are gauge independent in the sense that they do not depend on θ. These states coincide with the wavefunctions of the electron if $\theta=\pi/4$. Thus, they are solutions to the Schrödinger equation with the symmetric gauge vector potential, and they are represented by the Johnson and Lippmann-type wavefunctions.[1)]

It is well-known that the coordinates x_o and y_o of the gyration center in the plane perpendicular to the direction of the magnetic field are gauge invariant constants of motion. One of the main points in our consideration is that if we characterize the degenerate states belonging to a fixed energy by one of the following combinations: x_o(or y_o), $x_o^2+y_o^2$, and x_o+iy_o; independently from the particular choice of gauge we are using, we can generate Landau[2], Johnson and Lippmann and helical coherent states. In order to illustrate how this method works let us specify the gauge as $\underline{A}=B(-y,0,0)$. In this case the Hamiltonian does not depend on x. The usual Landau solutions are

$$F_{np_x} \equiv \langle x|p_x\rangle \langle y-\frac{p_x}{M\omega_c}|n\rangle , \qquad n=0,1,2,\ldots , \tag{1}$$

where ω_c is the cyclotron frequency and M is the electron's mass. The continuous parameter p_x represents the degeneracy. The eigenstates of the squared radial position of the center of gyration, $x_o^2+y_o^2$, are

$$G_{sp_y} \equiv \langle y|p_y\rangle \langle x-\frac{p_y}{M\omega_c}|s\rangle , \qquad s=0,1,2,\ldots , \tag{2}$$

where now p_y is a continuous degeneracy parameter. It is clear that any (normalised) superposition

$$F_{nf} \equiv \int F_{np_x} f(p_x)\,dp_x \tag{3}$$

is an eigenstate with the same energy, and the superposition

$$G_{sg} \equiv \int G_{sp_y} g(p_y)\,dp_y \tag{4}$$

is eigenstate with the same radial position of the center of gyration. If we require that the functions G_{sg} and F_{nf} be identical we get an equation for the functions $f(p_x)$ and $g(p_y)$ which can easily be solved. This simple procedure gives us the simultaneous eigenfunctions of both the energy and the radial position, which turn out to be just the Johnson-Lippmann states $|n,s\rangle$.

By using the above method we can construct the simultaneous eigenfunctions of the energy and the complex position x_o+iy_o. These eigenfunctions $|n,\alpha\rangle$[3] are coherent superpositions of the states $|n,s\rangle$.

If we keep the energy fixed and take the zero limit of the magnetic field then from the Johnson-Lippmann states, for example, we can derive cylindrical free electron states with definite angular momentum. From the coherent state $|n,\alpha\rangle$ in this limiting case we obtain a plane wave the wave vector of which is determined by the real and the imaginary part of α. However, if we take into account the correct normalisation of these states then we must notice that all of them vanishes in the mentioned limit. If we wish to reproduce normalisable free electron states in the zero magnetic field limit we must use normalised superposition of $|n,s\rangle$ or $|n,\alpha\rangle$[4] states with different energies (n). The need for that can easily be justified by considering the fact that no free electron energy eigenstate can be normalised to unity.

1) M.H. Johnson and B.A. Lippmann, Phys. Rev. 76, 828 (1949)
2) L.D. Landau, Z.f. Physik 64, 629 (1930)
3) A. Janussis, Z.f. Physik 190, 129 (1966)
4) I.A. Malkin and V.I. Manko, Sov. Phys. - JETP 28, 527 (1969), see also S. Varró, J. Phys. A17, 1631 (1984)

LASER-ASSISTED IONIZATION ON $He(2\,^1, 2\,^3S) + He(1^1S)$ COLLISION SYSTEM

P.PRADEL, P.MONCHICOURT, D.DUBREUIL, J.HEUZE,
J.J.LAUCAGNE AND G.SPIESS

Centre d'Etudes Nucléaires de Saclay, Service de Physique des Atomes et des Surfaces, 91191 Gif-sur-Yvette Cedex, France

Laser-assisted collisions can be defined [1] as processes where photon absorption is made possible only in the course of a collision between two atoms, whereas it is not possible when these two atoms are far apart.

The observed bound-free assisted process is :

$He^*(2^1S, 2^3S) + He(1^1S) + \hbar\omega \longrightarrow He^+(1^2S) + He(1^1S) + e^-$ (1)

The photon energy $\hbar\omega = 3.49$ eV is chosen so as not to be resonant with any atomic transition. The $He(1^1S)$ atom acts only as a perturber which shifts the He* levels and allows photon absorption. Figure 1 shows the potential curves of He^*_2 and He_2 systems [2, 3] together with the field-dressed curves for one photon absorption. Ionization becomes possible for approximately $R \lesssim 2Å$.

Our working energies allow field-free collisional ionization via a diabatic channel [4] :

$He^*(2^1S, 2^3S) + He(1^1S) \longrightarrow He^+(1^2S) + He(1^1S) + e^-$ (2)

The value of the cross-section σ_d of reaction (2) permits a determination of the cross-section σ_a for the reaction (1). The experimental work is performed for 2 kinetic energies 50 eV and 35 eV in the center-of-mass system.

The He* beam comes from a charge-exchange reaction between an He^+ beam colliding on a cesium target [5]. Charged particles emerging with He* atoms from the Cs cell are removed out off the beam by a transverse electric field. This neutral beam is then incident on the He gas target cell. He^+ resulting from reactions (1) or (2) are charge-analysed by a transverse electric field and then are incident on a detector. In the interaction chamber, the He* beam is crossed with an orthogonal laser light beam (3rd harmonic of a Nd:YAG laser, $\lambda = 355$ nm,

$I \sim 8 \times 10^{6}$ Wcm $^{-2}$. The ions correlated with the laser shot (via the process (1)) are superimposed on the background diabatic signal from reaction (2). Fig.2 shows the histograms of the data. For E_{CM} =50 eV we see a laser correlated ion peak which is equal to the diabatic signal. In the same way for E_{CM} =35 eV the assisted ion peak is 2.5 times larger than the diabatic one. σ_d increases with increasing energy by approximately a factor 1.7; on the contrary σ_a decreases by a factor ~ 0.5 showing that the colliding system exhibits a completely different kinematics when illuminated by the laser light. Competing reactions such as i) two photon-ionization ii) photoexcitation of He* followed by collisional ionization and iii) collisional excitation of He* followed by photoionization have been investigated and cannot account for the measured additional ion contribution.

REFERENCES

/1/ L.I. Gudzenko and S.I.Yakovlenko, Sov. Phys.JETP 35, 877 (1972).
/2/ R.P. Saxon, K.T. Gillen and B. Liu, Phys. Rev. A15, 543 (1977).
/3/ S.L. Guberman and W.A. Goddard III, Phys. Rev. A12, 1203 (1975).
/4/ K.T. Gillen, J.R. Peterson and R.E. Olson, Phys. Rev. A15, 527 (1977).
/5/ P. Pradel and J.J. Laucagne, J. Phys.(Paris) 44, 1263 (1983).

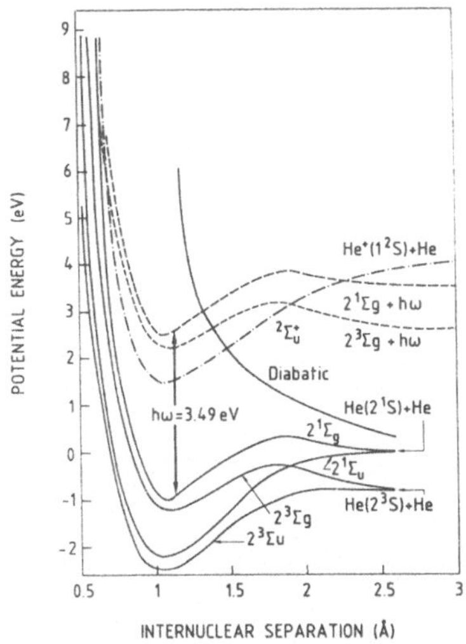

Figure 1 : Relevant potential curves of He^*_2 and He^+_2. Broken lines represent the field dressed potential curves.

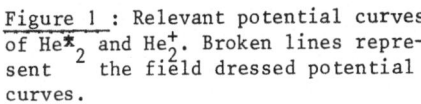

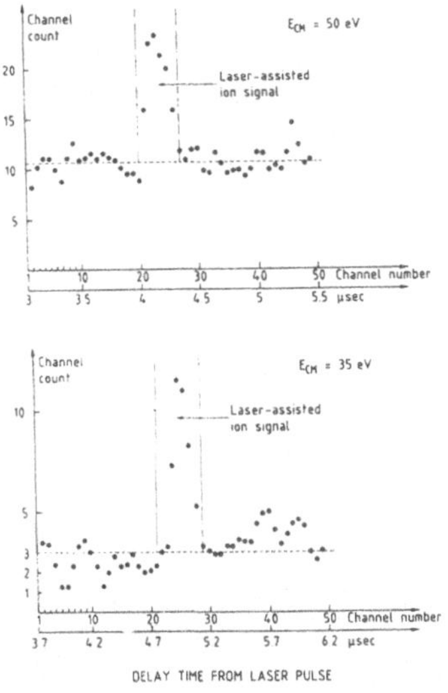

Figure 2 : Diabatic and assisted signal histogram for E_{CM}=50eV and E_{CM}=35eV. The horizontal broken lines represent the averaged value of the pure diabatic signal.

PART II: Multiphoton Ionization

VARIATION OF k INDEX IN ATI PROCESSES

Z. Deng and J.H. Eberly*
Department of Physics and Astronomy
University of Rochester
Rochester, NY 14627, USA

Abstract: We present a model for above threshold ionization (ATI) phenomena in which the continuum-continuum matrix elements decrease with increasing energy.

Above-threshold ionization (ATI) is a term that refers to photon absorption, usually multiphoton absorption, by an electron that is already energetically above its ionization threshold[1]. This phenomenon was first reported by Agostini, et al.[1], and the most recent observations have been reported by Lompre et al.[1], and Kruit, et al.[1].

Perhaps the most unexpected aspect of ATI is the possibility that transitions among continuum electron states above threshold can be saturated. This is the interpretation we have given[2] for the constant value of the multiphoton index $k = \partial \log$ (ionized population)/$\partial \log I$ reported by Kruit, et al.[1].

In this note we extend our earlier work and consider the variation of the k index with laser intensity I. First we recall the features of our model (see also Białynicka-Birula[3]). The Hilbert space of the electron has a "ladder" of states which are sequentially connected by the usual dipole interaction $exE(t)$. We restrict our attention to an atom with a discrete bound state $|0\rangle$, and an infinite sequence of continua. The continuum-continuum matrix element is denoted $V_{mm'}$, and the mth density of states by P_m/π. We assume here that V_{mm+1} is independent of the mth continuum energy ω_m, but still depends on m. We have shown[2] that certain continued fractions determine the character of ATI processes in our model, and that saturation is governed by a dimensionless parameter Z:

$$Z_{mm+1} = P_m P_{m+1} \mid V_{mm+1} \mid^2 ,$$

which is linearly proportional to laser intensity I.

In the figure we show the dependance on I of the index k for the first five ATI electron peaks, assuming that $Z_{mm+1} = S^2 Z_{m-1m}$. The factor S introduces variation of the matrix elements from one continuum to the next. In the two graphs we show S = 1 and S = 0.56. The former value represents no variation in V's at all, of course, and the first graph shows that in this case k drops quite

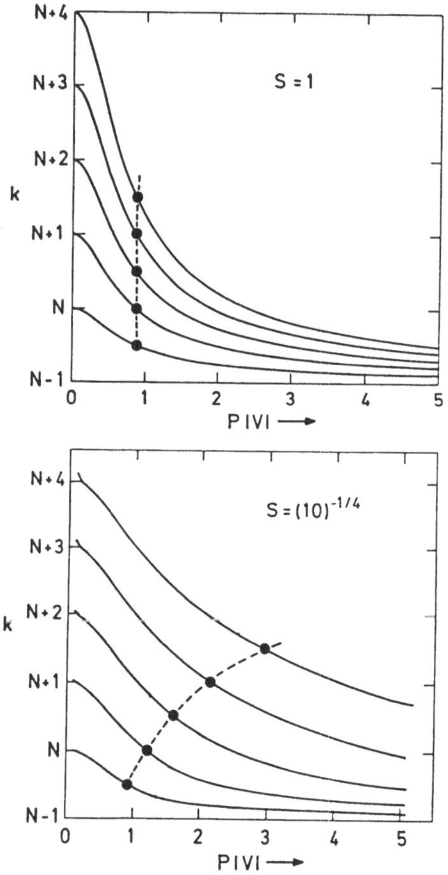

<u>Figure Caption:</u> The k index for the first five ATI peaks as a function of laser field strength. At low fields each peak shows the perturbative value: k = N-1+m for the mth peak. For high fields each k approaches N-1, the fully saturated value. The cases discussedin the text behave differently. The dotted lines connect points halfway between full saturation and perturbative limits. We label field strength by PV = $\sqrt{Z_{12}}$.

rapidly to saturation from the perturbative value for all photopeaks. The second graph shows the effect of variation in V's. The difference due to matrix-element variation is emphasized by the dashed lines, which connect the points where each curve has dropped halfway to its saturated value.

Acknowledgements: JHE acknowledges receipt of an Alexander von Humboldt senior fellowship, and the kind hospitality of the Max-Planck-Institut für Quantenoptik. This work was partially supported by the US Air Force Office of Scientific Research.

References and Footnotes:

*Temporary Address: Max-Planck-Institut für Quantenoptik
D 8046 Garching, West Germany

1. Agostini P., Fabre G., Petite G., and Rahman N.K. (1979) Phys. Rev. Letts. 42, 1127; Lompre L.A., L'Huiller A., Mainfray G. and Fan J.Y. (1984), J.Phys.B 18, L817 - 822; and Kruit P., Kimman J., Muller H.G., van der Wiel M.J. (1983) Phys.Rev. A 28 248-255

2. Deng Z. and Eberly J.H. (1984), Phys.Rev.Letts. 53 1810-1814; and Deng Z. and Eberly J.H. (1985), J.Opt.Soc.Am. B 2 (in press)

3. Białynicka-Birula Z. (1984), J.Phys.B. 17 3091-3101; and Edwards M., Pan L. and Armstrong, Jr., L. (1984) J.Phys.B L515-520

MULTIPHOTON IONIZATION OF ATOMS IN A STRONG FIELD. A NON-PERTURBATIVE METHOD

Michèle CRANCE Jocelyne SINZELLE

Laboratoire Aimé Cotton, CNRS II, bât. 505 91405 ORSAY FRANCE

When an atom with ionisation potential E is irradiated by a field of frequency w and intensity I, ionisation occurs after absorption of, at least, n photons, n being the first integer larger than $E/\hbar w$. In weak field, the energy of ejected electrons is $n\hbar w - E$. In fact, more than n photons may be absorbed and the electron energy spectrum consists of several peaks at energies $(n+k)\hbar w - E$; k is a positive integer. Perturbation theory applied at minimum non vanishing order predicts that the peaks intensity varies as I^{n+k}. As far as the first peak is dominant, ionisation probability still varies as I^n. When secondary peaks develop, deviations from power laws are expected and perturbative treatment at first non vanishing order is no longer valid. Strong field effects have been observed on electron energy spectra obtained in multiphoton ionisation of Xenon (1). When the field intensity is increased, saturation is observed on the peaks of electron energy spectrum while ionisation probability still obeys a power law with the index predicted by perturbation theory.

It is not realistic to undertake a calculation of multiphoton absorption probabilities beyond the minimum non vanishing order. A large number of terms is involved, they partially cancel each other and numerical problems become rapidly untractable (2). This work is an attempt at understanding the puzzling problem set by the results obtained in Xenon. It is too difficult to calculate multiphoton ionisation in Xenon, so we have chosen a simpler case to investigate the validity limit for a perturbative treatment of multiphoton ionisation towards several continua : photoionisation of Lithium ground state.

We propose a non perturbative method to study multiphoton ionisation of alkali atoms in strong field. The system atom plus field is described in the dressed atom picture. The atom is represented by a model potential (3). Radial part of the wavefunction for the outershell electron is expanded on a finite basis of complex square integrable functions (4). In the dipole approximation, the Hamiltonian matrix is written on a basis of tensorial products of atomic states and field states in number representation. Eigenstates of the Hamiltonian can be followed by continuity, as a function of the field intensity. The eigenstate $\hat{g}$ deduced from the initial state g , has a complex energy $\hat{E}$. The ionisation probability of g is $(2/\hbar)\mathrm{Im}(\hat{E})$. Contributions to various peaks of the electron energy spectrum can be calculated from the components of $\hat{g}$ on each multiplicity of the dressed atom.

As a function of frequency and intensity, we have calculated the probability P_k to eject an electron after absorption of k photons and the ionisation probability P(I). Up to an intensity of 10^{15} W cm^{-2}, for frequencies between 0.2 and 0.5au, ionisation probability hardly departs from a linear law. Figure I shows the variation of P(I)/I as a function of intensity for three frequencies. In the same range of intensity, saturation of the P_k's is observed: Figure II shows the example of frequency 0.3 a.u. . P_1 saturates first when the intensity exceeds 10^{13} W cm^{-2} . No saturation of P(I) is observed before 10^{15} W cm^{-2}, but P_2 and P_3 exhibit saturation already. These results are in qualitative agreement with experimental observations on Xenon. The depletion of first peaks in the electron energy spectrum is compensated by the appearance of additional peaks.

In calculating strong field effects we extend the basis (number of multiplicities and number of l-values)until the perturbed energy becomes stable with respect to the size of the basis. The basis obtained contains the states which would be involved in a perturbative treatment and thus indicates at which order perturbation expansion should be taken. For an intensity of 5 10^{15} W cm^{-2} the basis consists of ten multiplicities with eight l-values. On this example, we have shown that the ionisation probability may remain approximately equal to its weak field value for an intensity too high for a perturbative treatment at first non vanishing order to be valid. In other words, this example shows that observation of a power law with an index corresponding to the number of photons absorbed does not implies the validity of perturbation theory applied at first non vanishing order.

Figure I

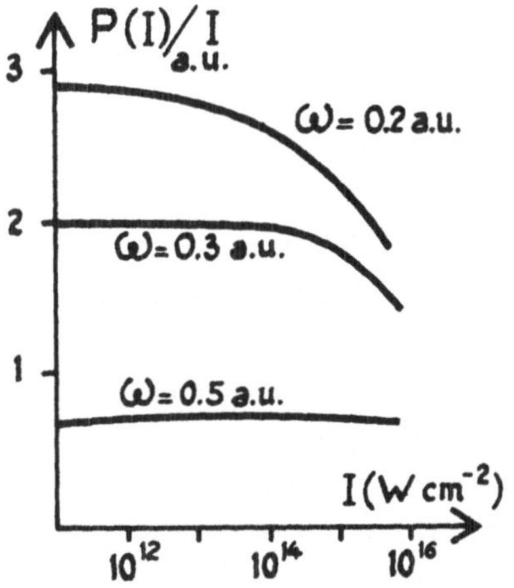

Figure II

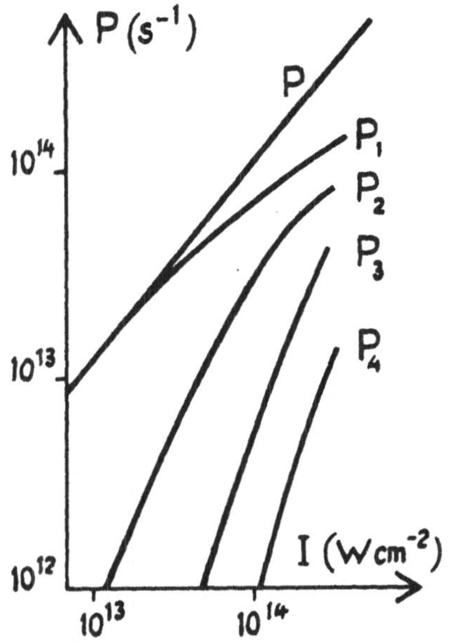

We acknowledge with thanks the support of the "Conseil Scientifique du Centre de Calcul Vectoriel pour la Recherche".

(1) P.Kruit,J.Kimman,H.G.Muller,M.J.Van der Wiel Phys.Rev. A28 248 (1983)
(2) M.Aymar,M.Crance J.Phys. B14 3585 (1981)
(3) M.Aymar,M.Crance,M.Klapisch J.Physique 31 C4 141 (1970)
(4) S.I.Chu,W.P.Reinhardt Phys.Rev.Lett. 39 1195 (1977)

TRANSITION MATRIX METHOD FOR MULTIPHOTON IONIZATION PROCESSES

Anthony F. Starace
Department of Physics and Astronomy
The University of Nebraska
Lincoln, NE 68588-0111
U.S.A.

and

Peter Zoller
Institute for Theoretical Physics
University of Innsbruck
A - 6020 Innsbruck
Austria

A transition matrix theory is developed to treat effects of electron correlations on two-photon ionization transitions in closed shell atoms and ions in the Random Phase Approximation (RPA). The theory extends the treatment of Chang and Fano [1] for single-photon ionization of closed shell atoms. The electromagnetic field interaction is treated in second order perturbation theory and electron correlations of the RPA type are included to infinite order. Ground and excited intermediate states are represented by a sum of configurations having a pair of virtually excited electrons in addition to the ground state or singly excited configurations. It is found that only one-particle functions, representing certain projections of excited two-particle wavefunctions, need to be calculated for the intermediate state in order to describe electron correlation in the RPA. The transition matrix equations for the unknown single particle functions in the intermediate and final states are derived using the graphical method of Starace and Shahabi [2]. The summations over intermediate states, including the continuum, is represented by the solution of an inhomogeneous set of equations for the effective intermediate state by the well-known Dalgarno-Lewis method [3]. Solutions of the equations allow one to obtain non-resonant two-photon ionization cross sections, photoelectron angular distributions, etc.

Two photon ionization of argon, i.e.

$$\text{Ar } 3p^6 + 2\gamma \longrightarrow \text{Ar } 3p^5(^2P) + e^-(\ell = 1 \text{ and } 3). \tag{1}$$

will serve to illustrate the theory. We choose the following configurations for the final, intermediate, and initial states respectively:

$$|f\rangle \equiv |3p^5\ \psi_\ell^f\ {}^1L\rangle \tag{2a}$$

$$|\lambda\rangle \equiv |3p^5\ \psi_d^\lambda\ {}^1P\rangle + |3p^5\ \phi_d^\lambda\ \phi_\ell^\lambda\ {}^1P\rangle \tag{2b}$$

$$|i\rangle \equiv |3p^6\ {}^1S\rangle + \sum_{L'S'} C(L'S')|3p^4(L'S')\ \phi_d^i\ \phi_d^i\ {}^1S\rangle \tag{2c}$$

The initial state correlation functions, ϕ_d^i, and coefficients, C(L'S'), can be calculated using the multiconfiguration Hartree-Fock code of Froese-Fischer [4]. The use of an average function, ϕ_d^i, instead of functions dependent on L'S', has been found to be a good approximation [5].

The dipole matrix element for process (1) is calculated from $\langle f|D|\lambda\rangle$, where D is the electric dipole operator and $|\lambda\rangle$ satisfies [5], [6]:

$$(E_i + \omega - H)\,|\lambda\rangle = D|i\rangle, \tag{3}$$

where E_i is the initial state energy, ω is the photon energy, and H is the system Hamiltonian. The matrix element $\langle f|D|\lambda\rangle$ depends on the unknown functions ψ_ℓ^f (ℓ = 1 or 3), ψ_d^λ, and $\phi_d^\lambda\langle\phi_\ell^\lambda|\psi_\ell^f\rangle$. In the simple representation of states in (2), ψ_ℓ^f is a HF continuum wave function and ψ_d^λ and $\phi_d^\lambda\langle\phi_\ell^\lambda|\psi_\ell^f\rangle$ are obtained from a pair of uncoupled transition matrix [1] equations derived by integrating the commutator relation,

$$[H,|f\rangle\langle\lambda|] = \omega|f\rangle\langle\lambda| + |f\rangle\langle i|D, \tag{4}$$

over the N-1 coordinates of the non-interacting electrons.

REFERENCES

[1] T. N. Chang and U. Fano, Phys. Rev. A 13, 263 (1976); Phys. Rev. A 13, 282 (1976).
[2] A. F. Starace and S. Shahabi, Phys. Rev. A 25, 2135 (1982).
[3] A. Dalgarno and J. T. Lewis, Proc. Roy. Soc. A 233, 70 (1955).
[4] C. Froese-Fischer, Comput. Phys. Commun. 4, 107 (1972).
[5] J. R. Swanson and L. Armstrong, Jr., Phys. Rev. A 15, 661 (1977).
[6] T. N. Chang and R. T. Poe, J. Phys. B 9, L311 (1976).

ONE- AND TWO-PHOTON DETACHMENT OF NEGATIVE HYDROGEN IONS: A HYPERSPHERICAL APPROACH.

Michael G. J. Fink and Peter Zoller
Institut fuer Theoretische Physik
Universitaet Innsbruck
Sillgasse 8
6020 Innsbruck

The investigation of electron correlation effects is presently one of the main lines of research in multiphoton ionisation (MPI) (Chin and Lambropoulos 1983). Experimental work has so far been confined to heavy alkaline earth atoms and rare gases (see also the various contributions to this conference). Enormous difficulties, however, are encountered in a theoretical description of these processes in properly accounting for the complicated atomic structure together with the high order of nonlinearity. At the present stage of our understanding, a promising line of investigation is to study electron correlation effects in low order MPI of light systems. Negative ions seem to be ideal candidates in this context: they are characterized by strong electron correlation effects of the outer shell electrons in the absence of a long range Coulomb potential and can be ionized by the absorption of a few photons of presently available laser wavelenghts. In particular, the negative hydrogen ion, which is the simplest three body system, can serve as a model system to develop and understand various approximation schemes. A second aspect which makes H^- an interesting object to be studied is the applicability of the hyperspherical adiabatic formulation (Fano 1983). From the point of view of multiphoton absorption this approach is appealing because it accounts for most of the electron correlation while having the simple structure of a one channel formulation.

The adiabatic hyperspherical approximation $\Psi(R,\omega) \approx R^{-5/2} F_\mu(R)\,\Phi_\mu(R;\omega)$ is based on a parametrisation of the two electron configuration space by a Hyperradius R and five angle coordinates ω. This formulation is now well known (see the review article by Fano, 1983). In lowest order perturbation theory (N=1,2) in the light field with frequency the one and two photon detachment cross sections are given by

$$\sigma^{(N)}_{f\leftarrow i} = 2\pi(2\pi\alpha)^N \omega_L^N \left| T^{(N)}_{f\leftarrow i} \right|^2 \qquad (a.u.) \;, \tag{1}$$

where the $T^{(N)}_{f\leftarrow i}$ are the transition matrix elements

$$T^{(1)}_{f\leftarrow i} = \langle f | D_q | i \rangle \qquad \text{and} \qquad T^{(2)}_{f\leftarrow i} = \langle f | D_q | \lambda \rangle \tag{2}$$

of the spherical q - component of the dipole operator in length form between the ground state $|i\rangle$ and the final state $|f\rangle$ normalized to the energy of the outgoing electron. Following Dalgarno and Lewis the second order matrix element is evaluated by replacing the infinite summation in the second order matrix element by the laser induced virtual state $|\lambda\rangle$ which can be found by solving the inhomogeneous equation

$$(E_i + \omega_L - H_{at} + i\varepsilon)\,|\lambda\rangle = D_q\,|i\rangle \tag{3}$$

The single photon dipole transition matrix element for He was calculated by Miller and Starace in adiabatic approximation (1979). Generalizing the adiabatic approximation for the two photon case, we assume that the laser induced virtual state $|\lambda\rangle$ is in the lowest $^1P^o$ channnel

$$\lambda(R,\omega) \approx R^{-5/2} F_\lambda(R)\, \Phi_{1P^o}(R,\omega) \tag{4}$$

Consequently, we have for the radial function of the intermediate state the inhomogeneous equation

$$\left(E_i + \omega_L + \frac{1}{2}\frac{d^2}{dR^2} - U^{ad}_{1P^o}(R) + i\varepsilon\right) F_\lambda(R) = R(\Phi_{1P^o}|d_q|\Phi_{1S})\, F_i(R) \tag{5}$$

where $(\Phi_{1P^o}, d_q, \Phi_{1S})$ is a slowly varying dipole matrix element between the channels λ and i. It is remarkable that eqs. (5) together with (2b) have the simple structure of a one - channel problem .

Figure 1 shows our results for one photon detachment in adiabatic approximation (solid line) and compares it with the results by Daskhan and Ghosh (1983) and the references cited there. The overall agreement depends on the reference and is between 3% and 10% and best near threshold. The maximum of the cross section is overestimated while with increasing energy our cross section becomes too small. In Figure

2 we plot the two photon detachment cross section in the photon energy region 0.35 eV < $2\hbar\omega$ < 0.7 eV, where one photon is insufficient to ionize the system. We assumed circularly polarized light so that the final state has $^1D^e$ symmetry. Near the two photon threshold the cross section starts proportional to $E^{L+1/2}$ (L=2) with decreasing slope for larger energies. Like in the one photon case we expect our results to be most accurate near threshold. In this energy region the intermediate state wave function $F_\lambda(R)$ is still well localized near intermediate values of R; in addition the continuum phase shifts are best in this energy region.

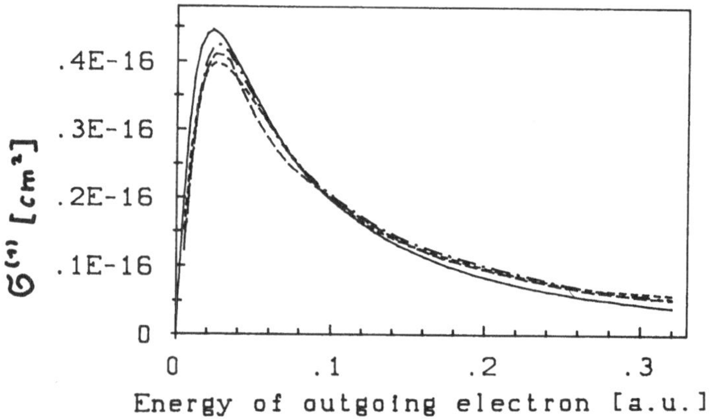

Fig. 1 One photon detachment cross section of H^-

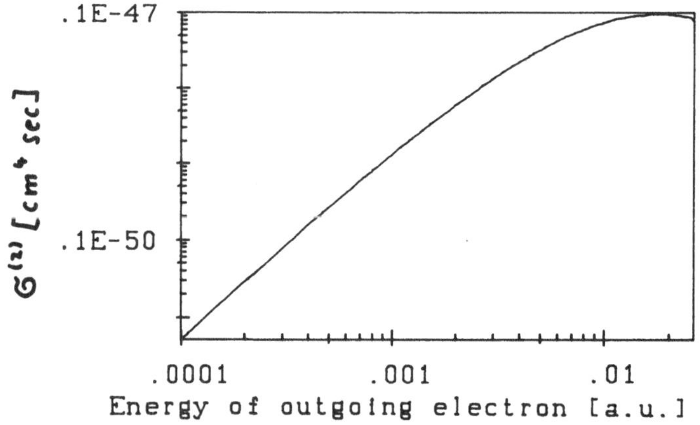

Fig. 2 Two photon detachment cross section of H^-

Literature:

S.L. Chin and P. Lambropoulos, Multiphoton Ionisation of Atoms, Academic Press (1984)
M. Daskhan and A.S. Ghosh, Phys. Rev. A 28, 2767 (1983)
U. Fano, Rep. Prog. Phys. 46, 97 (1983)

PART III: Laser Spectroscopy

NEW ASPECTS OF THE RADIATION COUPLING OF TWO BOUND STATES WITH A PREDISSOCIATING (AUTOIONIZING) RESONANCE

A.Lami and N.K.Rahman
Istituto di Chimica Quantistica ed Energetica Molecolare, CNR
and
Dipartimento di Chimica e Chimica Industriale, Università di Pisa
Via Risorgimento, 35 - I-56100 Pisa, Italy

We report here some results of a theoretical study regarding the double resonance Λ-configuration in which two bound states are made to interact via the continuum containing a resonance /1,2/ (Fig. 1). It has been known for some time

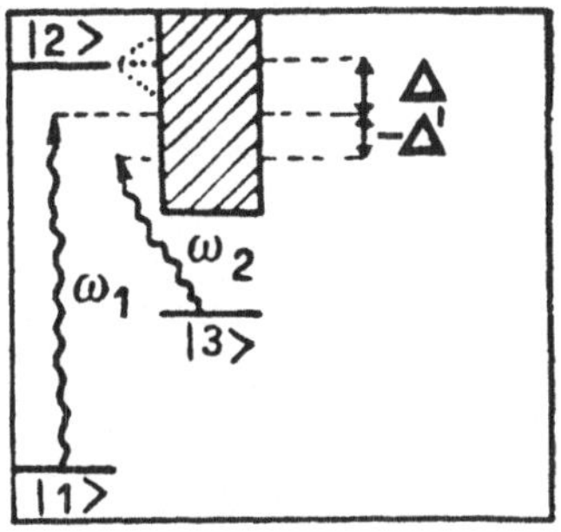

Fig. 1

that by a suitable choice of parameters, (laser detunings and field strengths) one can induce a photon-dressed non-decaying state in the continuum in which part of the population may remain trapped /3,2,1/. As a consequence, the atom or the molecule is prevented from being totally ionized or dissociated. Since the scheme in Fig.1 is essentially a 3-level configuration, it is natural to ask if it is at all possible to choose detunings and field strengths, such that two non-decaying states are generated. In the following, we show that this can be done, and that this leads to some interesting possibilities such as quantum beating of trapped population.

We proceed by examining the poles of the resolvent of the system in the parameter space and constructing the curves on which at least a single stable state resides. A typical locus in the plane of the two detunings (keeping the field strengths fixed) is shown in fig 2a. Varying the field strengths, the form of this three-branched curve remains invariant. Such a curve pinpoints exhaustively the location of stable states in the Δ, Δ' plane. The singular point P at which two branches intersect is where the stable states coexist.

Finally, we explore the time dependence of the system at the singular point P. In fig. 2b the populations of the levels |1> and |3> as well as the total ionization (or dissociation) are plotted as a function of time. They all show the effects of quantum beats between the two dressed bound states. Especially noteworthy is the permanent oscillation in the ionization yield.

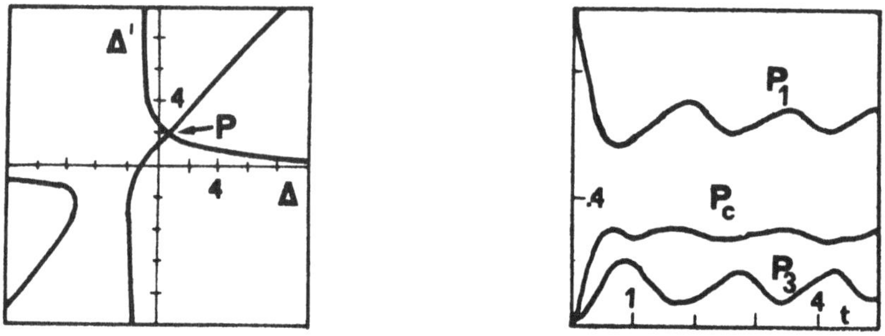

a b

Fig. 2

2a) The locus of stable states in the Δ, Δ' plane (see Fig. 1). The field strengths have been chosen so that the widths of |1> and |3> are respectively .4 and .6 in units of γ_2 (the natural linewidth of the resonance). The generalized Fano parameters (see ref. 1) are $q_1 = 2.$, $q_3 = 4.$, $q_{13} = 1.$
2b) Time dependence of the populations of states |1>, |3> and of the continuum (the time is in units $1/\gamma_2$) at the singular point P (Fig. 2a) at which there are two stable states.

References

1. A.Lami and N.K.Rahman, Phys.Rev.A 26, 3360 (1982); paper in "Collisions and half collisions with lasers", Eds. N.K.Rahman and C.Guidotti (Harwood, Chur-New York, 1984).
2. Z.Deng and J.H.Eberly, J.Opt.Soc.Am. B 1, 102 (1984).
3. P.E.Coleman, P.L.Knight and K.Burnett, Optics Comm. 42, 171 (1982).

LASER INDUCED RESONANCES IN THE MPI-SPECTRUM
OF SODIUM ATOMS

D.Feldmann,G.Otto,D.Petring and K.H.Welge
Fakultät für Physik,Universität Bielefeld
D 4800 Bielefeld 1

The multiphoton ionization (MPI) spectrum of sodium atoms in a beam has been measured in the presence of a second strong laser field. Intensities of both lasers up to about 10^{10} Wcm^{-2} have been applied. One laser (ω_1) was tuned between 547 nm and 580 nm,the second one (ω_2) had a fixed wavelength of 532 nm. Na^+-ions have been detected in a TOF mass spectrometer.Figure 1 shows the ionization spectrum obtained when both lasers are linearly polarized parallel to each other.

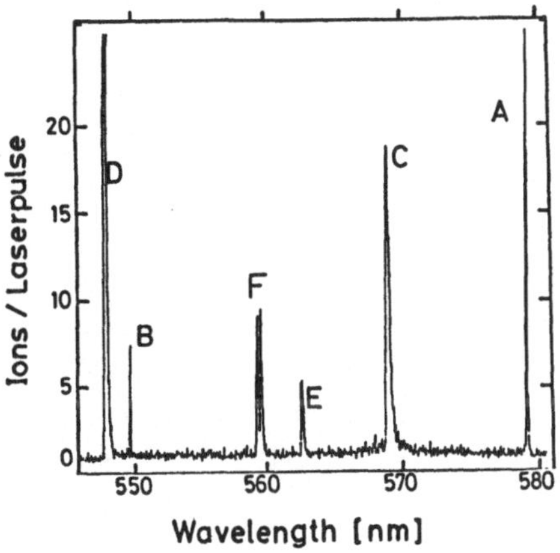

Figure 1 : Na^+-ion yield versus wavelength of laser 1.

The small background signal is mainly caused by $3\omega_2$-three-photon ionization.The peaks labeled A,B,C and D correspond to two-photon resonant three-photon ionization: A: $2\omega_1 = E(4^2D)$, B: $2\omega_1 = E(6^2S)$, C: $\omega_1+\omega_2 = E(6^2S)$ and D: $\omega_1+\omega_2 = E(5^2D)$,where E( ) denotes the excitation energy of the atomic state in brackets,and ω_1,ω_2 the energy of a photon from the respective laser.this type of processes has often been observed in MPI experiments.
The additional peaks E and F can be explained as resonantly enhanced five-photon ionization processes and will be discussed in more detail:
Peak E is observed at a photon energy ω_1 which which fullfills the

energy equation $3\omega_1 = E(4^2D) + \omega_2$ .Two pathways of 4^2D-excitation shown in figure 2 can be associated with this equation.

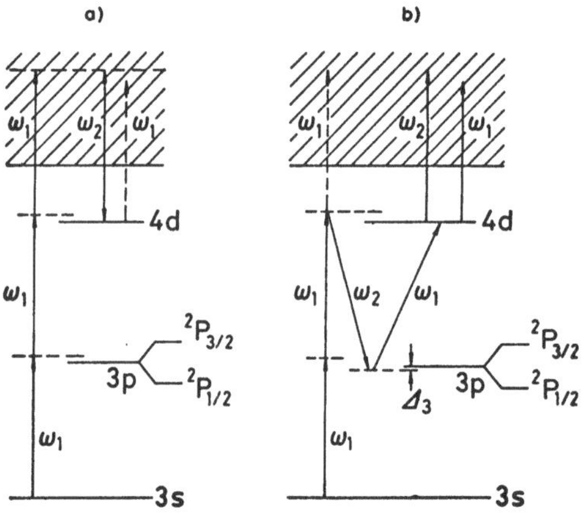

Figure 2 : Two pathways by which the 4D-state can be coupled into the ionization process: a) via the $3\omega_1$-ionization continuum and b) via a Raman-type $(2\omega_1-\omega_2+\omega_1)$ excitation.

They differ with respect to the sequence of absorption and stimulated emission of photons. a)The 4^2D-state can be coupled with ω_2 into an otherwise unstructured $3\omega_1$-ionization continuum,creating an "autoionizing like" resonance.This type of process has recently gained much interest because it offers the possibility to change an ionization continuum by photon-dressed excited states and it is closely related to stimulated ion-electron recombination.Calculations have been performed by several groups (1) and "autoionizing like" states have been used to explain the results of experiments on frequency tripling and induced polarization anisotropy (2) in sodium vapour.
b) The second pathway is a Raman-type $(2\omega_1-\omega_2+\omega_1)$-four-photon excitation of the 4^2D-state with subsequent one-photon ionization.
For the following reasons we favour this process b) to explain our results: Two additional "autoinizing like" resonances are to be expected within the experimental wavelength range.They can be characterized by the equations $3\omega_1 = E(5^2S) + \omega_2$ and $3\omega_2 = E(6^2D) + \omega_1$ and are expected near 577 nm and 555 nm respectively.We do not observe peaks at these wavelengths and from the experimental detection limit we can conclude that they must be at least one order of magnitude weaker than the process correlated with peak E . We see no obvious reason why the second process involving the 6^2D-state should be so

much weaker. - On the other hand the different strengths of the processes can simply be explained for the Raman-type excitation b). The detunings from the 3^2P-state after three photons : $\Delta_3 = (2\omega_{1,2} - \omega_{2,1}) - E(3^2P)$ are different which implies different near-resonant enhancement. Δ_3 is $\sim$3300 cm^{-1} and $\sim$2600 cm^{-1} for the processes not observed and only $\sim$ -200 cm^{-1} for the observed one.

Peak "E" consists of two peaks with a wavelength separation of 0.3 nm and they are observed at photon energies which obey the energy equations: $2\omega_1 = E(3^2P_{1/2}) + \omega_2$ and $2\omega_1 = E(3^2P_{3/2}) + \omega_2$. Again two models might be considered to explain this process. (a) Analog to the "autoionizing like" resonance it might be interpreted as a coupling of the 3^2P-state with ω_2 into the manyfold of "virtuell" states scanned by $2\omega_1$ in the $3\omega_1$-ionization process,- or (b) as a Raman-type $(2\omega_1-\omega_2)$-excitation of the 3^2P-state with subsequent two-photon ionization. Although this differentiation is somewhat arbitrary one can expect that in process (a) at least two photons ω_2 are involved in the resonantly enhanced five-photon $(2\omega_1-\omega_2+\omega_2+\omega_{1,2})$-ionization, whereas in (b) only one ω_2-photon is necessary when the ionization from the 3^2P-state proceeds by $2\omega_1$-absorption. From the linear ω_2-intensity dependance we conclude that peak "F" is mainly caused by process (b) : $(2\omega_1-\omega_2+2\omega_1)$-ionization.-Preliminary results of experiments in which the electron energy has been analyzed are in agreement with the interpretations for both peaks "E" and "F".

Conclusions: 1. Our results show that at intensities of about 10^{10} Wcm^{-2} five-photon processes which are resonantly enhanced can be stronger than nonresonant three-photon ionization processes.
2. We find no evidence for "autoionizing-like" resonances in the MPI-spectrum .

References:

(1) L.Armstrong,B.L.Beers and S.Feneuill(1975)Phys.Rev.A12,1903, Yu.I.Geller and A.K.Popov (1980) Sov.Phys.JETP 51 ,255, P.E.Coleman,P.L.Knight and K.Burnett (1982) Opt.Comm.42,171, and several others.
(2) S.S.Dimov,L.I.Pavlov,K.V.Stamenov,Yu.I.Heller and A.K.Popov (1983) Appl.Phys. B30,35, S.S.Dimov,L.I.Pavlov,K.V.Stamenov and G.B.Altshuller (1983) Opt.Quant.Electr. 15,305 .

EFFECTIVE GAS IONIZATION WITH SIMULTANEOUS IRRADIATION OF PULSED CO_2 AND EXCIMER LASERS

Jun Sasaki*, Sanichiro Yoshida, Yasuhiko Arai, Kimiya Tateishi, M.P.Lei and Taro Uchiyama

Department of Electrical Engineering, Faculty of Science and Technology Keio University, Yokohama, Japan

*) present address: Institut f. Experimentalphysik, Universität Innsbruck, Austria

Here we present the recent results of our experiments, in which it is confirmed that the preionization which is induced by excimer laser relaxes the requirements for optical gas breakdown which is induced by IR (Infra-Red) laser. Pulsed CO_2 laser (10.6 µm) is adopted as a IR light source.

Generally, the dominant gas ionization processes by lasers greatly depend on the wavelength region of the applied laser (ref. 1). Multi-photon ionization process may be dominant in UV laser induced ionization, on the other hand the IR laser may enhance the number of electrons through a cascade ionization process, in which primary electrons play an important role.

Experiments and Results

The different experiments are mentioned below. (1) Simultaneous irradiation of the focus region of a pulsed CO_2 laser beam by spark UV light. The probability of optical gas breakdown occurrence as a function of all laser shots is measured for the two cases of UV light on and off. This experiment is performed to confirm the effect of UV light irradiation. (2) Simultaneous irradiation by KrF laser (248 nm) and pulsed CO_2 laser. The optical gas breakdown occurrence probability and created electron number density are measured. Here, the induced current methode is used to measure the electron number density.

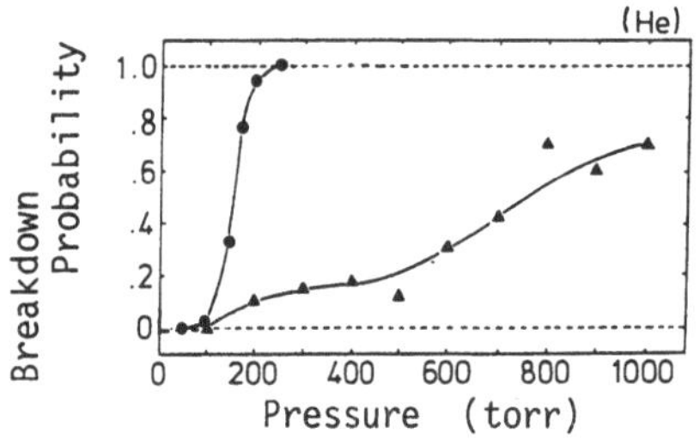

Fig. 1.: Optical Breakdown occurence probability as a function of pressure. (●) UV flash on and (▲) UV flash off. CO_2 laser 2J per pulse.

Ineach experiment, an apparent increase of the optical gas breakdown occurrence probability is observed with the simultaneous irradiation of UV light on the focus region of a CO_2 laser beam. Fig. 1 shows a result of experiment (1). The effect of UV irradiation is evident here. Fig. 2 shows the measured electron number density produced by irradiation of the KrF laser alone. Fig. 3 shows the measured number density times optical breakdown occurence probability, with the simultaneous irradiation of KrF and CO_2 lasers.

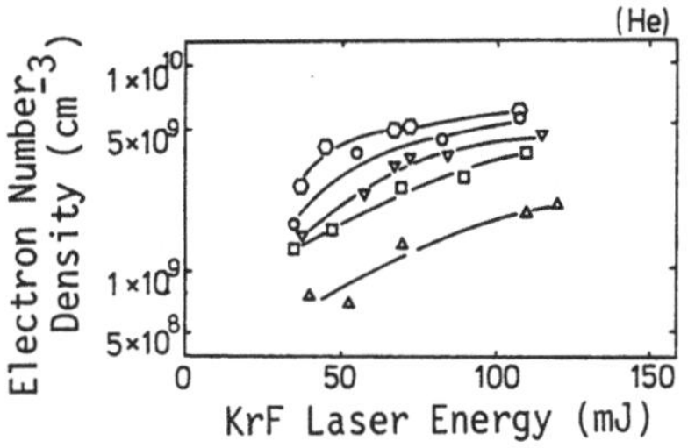

Fig. 2.: Electron number density created by irradiation of KrF laser alone, as a function of laser energy. Pulse width is about 20nsec (FWHM), (△) 100 torr, (□) 200 torr, (▽) 300torr, (O) 400 torr, (⚬) 500 torr, Helium

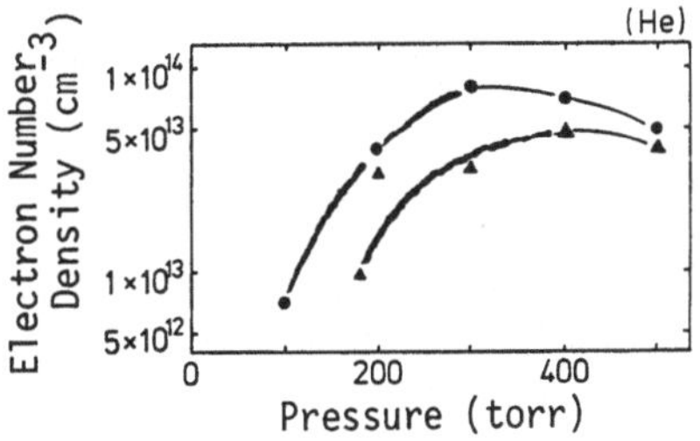

Fig. 3.: Created electron number density times optical breakdown occurrence probability, as a function of the pressure. (▲) only CO_2 laser is irradiated (●) KrF and CO_2 are simultaneously irradiated. CO_2 2J per pulse, KrF 40mJ per pulse.

These results tell us that a number of primary electrons on the order of 10^9 cm^{-3}, which are provided by the KrF excimer laser whose intensity is on the order 10^8 W/cm^2, is sufficient to relax the optical breakdown occurrence conditions.

Our final goal is to relax the threshold of laser power requirements for long distance gas breakdown control. In a third experiment, it is observed that triggering of flashover between non-uniform gap electrodes where DC voltage is applied, can be achieved only when optical breakdown occurs between the electrodes.

We conclude that the simultaneous irradiation by excimer laser and CO_2 laser should be also effective for breakdown control.

References

/1/ C. Grey Morgan, J. Phys. (D), No.6, pp. 720-729, (1973); George Bekefi, Principles of Laser Plasmas (Wiley, New York, 1976), Chap. 11

PART IV: Laser Cooling and Trapping

LASER MANIPULATION OF ATOMIC BEAM VELOCITIES: DEMONSTRATION OF STOPPED ATOMS AND VELOCITY REVERSAL

W.Ertmer*, R.Blatt**, J.L.Hall and M.Zhu

Joint Institute for Laboratory Astrophysics
University of Colorado and National Bureau of Standards
Boulder, Colorado 80309, USA

The deceleration of atoms in an atomic beam by resonant radiation of a counterpropagating laser beam has received considerable attention for several years now[1]. The basic scheme of most of these experiments consists of an atomic beam and a counterpropagating laser beam. The experiments differ mainly in the experimental solution of two fundamental problems accompaning these methods: when the atoms are decelerated by the successive absorption of photon momentum, their Doppler shift changes very fast and thus the atoms run out of resonance after the accumulated shift of a few homogeneous line widths. On the other hand, the stopping process needs very many photons (e.g.20 000 of them, if Na atoms shall be stopped to zero velocity from an initial velocity of about 600 m/sec, using the Na-D-line) and thus optical pumping even on a verv minute scale has to be avoided or counteracted.

In our experiment[2], performed with a Na atomic beam and a laser beam tuned to the transition 3s $^2S_{1/2}(F=2) \rightarrow 3p\ ^2P_{3/2}(F=3)$ (σ^+-light), we maintained the resonance condition during the deceleration process by an adiabatic sweep of the laser frequency corresponding to the change in the Doppler shift. Optical pumping to the F=1 hfs state of the ground level was compensated by the application of a second laser frequency, which was synchroneously swept with the main deceleration frequency. Both procedures were achieved by using very efficient phase modulation of the laser light in two electrooptic modulators. The first produced the second frequency as one of the two sidebands, shifted by 1.77 GHz (equivalent to the hfs splitting of the ground state) from the carrier. For each of these two frequencies the second modulator produced sidebands, varying its frequencies in time in such a way, that they stayed in resonance with the decelerated atoms. With a new modulator design it was possible to transfer 1/3 of the incoming laser beam into each first sideband, leaving about 1/10 of the initial laser power in the carrier and the rest in higher sidebands. For driving the modu-

---

* Now at: Institut für Angewandte Physik der Universität Bonn, Bonn, West Germany

**Now at: I.Institut für Experimentalphysik der Universität Hamburg, Hamburg, West Germany

lation an rf power of a few watts was necessary. During each frequency sweep the cooling sideband of the carrier frequency (= incoming laser frequency) was tuned in the beginning for resonance with atoms of a chosen velocity (e.g. 600 m/sec as the most probable velocity of the atoms in the beam) and then swept within 1 msec over a range of 1 GHz corresponding to a Doppler tuning of a velocity intervall of about 600 m/sec. Thus most of the atoms of the beam belonging to this velocity intervall from 0 to 600 m/sec could be decelerated to a very narrow velocity intervall. The velocity distribution of the decelerated atoms was measured with a separate probe laser beam from an independent laser system crossing the atomic beam at an angle of 56° from perpendicular incidence. From the frequency width of the fluorescence signals from this probe beam we can conclude that the width of the velocity intervall of the decelerated atoms is about 8 m/sec in longitudinal direction and about 5 m/sec in transverse direction. With a fixed tuning intervall of 1 GHz, the final velocity of the decelerated atoms could be varied, depending on the start frequency, over a wide range without major losses in intensity. In this way it was not only possible to stop the atoms completely, but also to drive them back to the oven with a velocity of about -100 m/sec, when starting the deceleration process for atoms with a velocity of 500 m/sec.

In an improved version of our experiment we produced the laser frequency needed for counteracting the optical pumping not by electrooptical modulation but by using a two-frequency laser. With this arrangement we could reconfirm our earlier results.

References:

(1) See, e.g.: Proceedings of the Workshop on Spectroscopic Application of Slow Atomic Beams, held at NBS, Gaithersburg (Maryland) USA, April 1983, edited by W.D.Phillips, NBS Special Publication 653 (1983);
V.O.Balykin, V.S.Letokhov and A.I.Sidorov: Opt.Comm.49,248 (1984);
J.Prodan, W.D.Phillips, I.So, H.Metcalf and J.Dalibard: Phys.Rev. Lett., accepted for publication

(2) W.Ertmer, R.Blatt, J.L.Hall and M.Zhu: Phys.Rev.Lett., accepted for publication

DYNAMICS OF THE LASER-COOLING OF A TRAPPED ION

Markus Lindberg

Institut für Theoretische Physik der Universität Frankfurt,
Robert-Mayer-Str. 8, D-6000 Frankfurt-Main, Fed. Rep. Germany.

Abstract:

It has been experimentally shown by Neuhauser et al. [1] that it is possible to confine a single ion to a small spatial region, cool it and study it for a considerable long time.

The system is modelled by a two level system coupled to a harmonic potential well interacting with a strong semiclassical laser field mode [2] . The cooling dynamics are studied from the point of view of breaking the degeneracy in the mechanical energy states. When the momentum of the laser field is taken to be zero only the internal states are able to relax to a steady state. The initial mechanical state reaches a steady state when $\hbar k$ is non zero and the different states are able to couple. The time scale compared to the internal relaxation times is characterized by a parameter $\eta^2 = \frac{\hbar k^2}{2M\nu}$, which is the ration of the recoil energy to the trap energy quantum.

The calculations are made by constructing projectors to the degenerate eigenstates of the unperturbed time evolution operator corresponding to the internal steady state. The complementary states follows adiabatically the evolution of the steady state. Going to second order in η we obtain coupling between the degenerate states. A set of rate equations is obtained. The solution gives a time independent steady state with exponentially distributed population on the vibrational levels. All coherences between the states which would lead to oscillations in the trap are shown to decay to zero.

[1] Neuhauser W, Hohenstatt M, Toschek P and Dehmelt HG
1978 Phys. Rev.Lett. 41, 233-6

[2] Javanainen J and Stenholm S
1980 Appl.Phys. 21, 35-45

[3] Javanainen J, Lindberg M and Stenholm S
1984 J.Opt.Soc.Am. B1, 111-5.

LIST OF PARTICIPANTS

G. Alber	JILA, Boulder, USA
N. Andersen	Univ. of Copenhagen, Denmark
F.R. Aussenegg	Univ. Graz, Austria
J. Bergou	Central Research Institute for Physics, Budapest, Hungary
R. Blatt	Univ. Hamburg, FRG
W.E. Cooke	Univ. of Southern California, USA and FU Berlin, FRG
M. Crance	Laboratoire Aime Cotton, Orsay, France
J.H. Eberly	Univ. of Rochester, USA and MPI für Quantenoptik, Garching, FRG
F. Ehlotzky	Univ. Innsbruck, Austria
F.H.M. Faisal	Univ. Bielefeld, FRG
D. Feldmann	Univ. Bielefeld, FRG
G. Ferrante	Univ. of Palermo, Italy
M. Fink	Univ. Innsbruck, Austria
W. Freysinger	Univ. Innsbruck, Austria
M. Gravrila	FOM, Amsterdam, The Netherlands
K. Harth	Univ. Kaiserslautern, FRG
H. Helm	SRI International, Menlo Park, USA
W. Henle	Univ. Innsbruck, Austria
G. Holzmüller	Univ. Innsbruck, Austria
J. Javanainen	Univ. Helsinki, Finland and MPI für Quantenoptik, Garching, FRG
C.J. Joachain	Univ. Libre de Bruxelles, Belgium
M. Kaivola	Univ. of Aarhus, Denmark
H. Klar	Univ. Freiburg, FRG
A. Lami	Univ. of Pisa, Italy
M. Lindberg	Univ. Frankfurt am Main, FRG and Univ. of Helsinki, Finland
L.A. Lompre	CEN Saclay, France
A. Maquet	Univ. Pierre et Marie Curie, Paris, France
M. Marte	Univ. Innsbruck, Austria
E. Pelikan	Univ. Freiburg, FRG
S. Penselin	Univ. Bonn, FRG
P. Pradel	CEN Saclay, France
G. Rempe	MPI für Quantenoptik, Garching, FRG

C.K. Rhodes	Univ. of Illinois, Chicago, USA
R. Roussel	CEN Saclay, France
J. Sasaki	Keio University, Japan and Univ. Innsbruck, Austria
T. Sauter	Univ. Hamburg, FRG
A.F. Starace	Univ. of Nebraska, Lincoln, USA
P. Stehle	Univ. of Pittsburg, USA and Univ. Innsbruck, Austria
P.E. Toschek	Univ. Hamburg, FRG
C.R. Vidal	MPI für Extraterrestrische Physik, Garching, FRG
K. Voigtländer	Univ. Ulm, FRG
H.D. Vollmer	Univ. Ulm, FRG
K.H. Welge	Univ. Bielefeld, FRG
M. Zarcone	Univ. Palermo, Italy
P. Zoller	Univ. Innsbruck, Austria

Lecture Notes in Physics

Vol. 195: Trends and Applications of Pure Mathematics to Mechanics. Proceedings, 1983. Edited by P. G. Ciarlet and M. Roseau. V, 422 pages. 1984.

Vol. 196: WOPPLOT 83. Parallel Processing: Logic, Organization and Technology. Proceedings, 1983. Edited by J. Becker and I. Eisele. V, 189 pages. 1984.

Vol. 197: Quarks and Nuclear Structure. Proceedings, 1983. Edited by K. Bleuler. VIII, 414 pages. 1984.

Vol. 198: Recent Progress in Many-Body Theories. Proceedings, 1983. Edited by H. Kümmel and M. L. Ristig. IX, 422 pages. 1984.

Vol. 199: Recent Developments in Nonequilibrium Thermodynamics. Proceedings, 1983. Edited by J. Casas-Vázquez, D. Jou and G. Lebon. XIII, 485 pages. 1984.

Vol. 200: H.D. Zeh, Die Physik der Zeitrichtung. V, 86 Seiten. 1984.

Vol. 201: Group Theoretical Methods in Physics. Proceedings, 1983. Edited by G. Denardo, G. Ghirardi and T. Weber. XXXVII, 518 pages. 1984.

Vol. 202: Asymptotic Behavior of Mass and Spacetime Geometry. Proceedings, 1983. Edited by F. J. Flaherty. VI, 213 pages. 1984.

Vol. 203: C. Marchioro, M. Pulvirenti, Vortex Methods in Two-Dimensional Fluid Dynamics. III, 137 pages. 1984.

Vol. 204: Y. Waseda, Novel Application of Anomalous (Resonance) X-Ray Scattering for Structural Characterization of Disordered Materials. VI, 183 pages. 1984.

Vol. 205: Solutions of Einstein's Equations: Techniques and Results. Proceedings, 1983. Edited by C. Hoenselaers and W. Dietz. VI, 439 pages. 1984.

Vol. 206: Static Critical Phenomena in Inhomogeneous Systems. Edited by A. Pękalski and J. Sznajd. Proceedings, 1984. VIII, 358 pages. 1984.

Vol. 207: S. W. Koch, Dynamics of First-Order Phase Transitions in Equilibrium and Nonequilibrium Systems. III, 148 pages. 1984.

Vol. 208: Supersymmetry and Supergravity/Nonperturbative QCD. Proceedings, 1984. Edited by P. Roy and V. Singh. V, 389 pages. 1984.

Vol. 209: Mathematical and Computational Methods in Nuclear Physics. Proceedings, 1983. Edited by J. S. Dehesa, J. M. G. Gomez and A. Polls. V, 276 pages. 1984.

Vol. 210: Cellular Structures in Instabilities. Proceedings, 1983. Edited by J. E. Wesfreid and S. Zaleski. VI, 389 pages. 1984.

Vol. 211: Resonances – Models and Phenomena. Proceedings, 1984. Edited by S. Albeverio, L. S. Ferreira and L. Streit. VI, 369 pages. 1984.

Vol. 212: Gravitation, Geometry and Relativistic Physics. Proceedings, 1984. Edited by Laboratoire "Gravitation et Cosmologie Relativistes", Université Pierre et Marie Curie et C.N.R.S., Institut Henri Poincaré, Paris. VI, 336 pages. 1984.

Vol. 213: Forward Electron Ejection in Ion Collisions. Proceedings, 1984. Edited by K.O. Groeneveld, W. Meckbach and I.A. Sellin. VII, 165 pages. 1984.

Vol. 214: H. Moraal, Classical, Discrete Spin Models. VII, 251 pages. 1984.

Vol. 215: Computing in Accelerator Design and Operation. Proceedings, 1983. Edited by W. Busse and R. Zelazny. XII, 574 pages. 1984.

Vol. 216: Applications of Field Theory to Statistical Mechanics. Proceedings, 1984. Edited by L. Garrido. VIII, 352 pages. 1985.

Vol. 217: Charge Density Waves in Solids. Proceedings, 1984. Edited by Gy. Hutiray and J. Sólyom. XIV, 541 pages. 1985.

Vol. 218: Ninth International Conference on Numerical Methods in Fluid Dynamics. Edited by Soubbaramayer and J. P. Boujot. X, 612 pages. 1985.

Vol. 219: Fusion Reactions Below the Coulomb Barrier. Proceedings, 1984. Edited by S.G. Steadman. VII, 351 pages. 1985.

Vol. 220: W. Dittrich, M. Reuter, Effective Lagrangians in Quantum Electrodynamics. V, 244 pages. 1985.

Vol. 221: Quark Matter '84. Proceedings, 1984. Edited by K. Kajantie. VI, 305 pages. 1985.

Vol. 222: A. García, P. Kielanowski, The Beta Decay of Hyperons. Edited by A. Bohm. VIII, 173 pages. 1985.

Vol. 223: H. Saller, Vereinheitlichte Feldtheorien der Elementarteilchen. IX, 157 Seiten. 1985.

Vol. 224: Supernovae as Distance Indicators. Proceedings, 1984. Edited by N. Bartel. VI, 226 pages. 1985.

Vol. 225: B. Müller, The Physics of the Quark-Gluon Plasma. VII, 142 pages. 1985.

Vol. 226: Non-Linear Equations in Classical and Quantum Field Theory. Proceedings, 1983/84. Edited by N. Sanchez. VII, 400 pages. 1985.

Vol. 227: J.-P. Eckmann, P. Wittwer, Computer Methods and Borel Summability Applied to Feigenbaum's Equation. XIV, 297 pages. 1985.

Vol. 228: Thermodynamics and Constitutive Equations. Proceedings, 1982. Edited by G. Grioli. V, 257 pages. 1985.

Vol. 229: Fundamentals of Laser Interactions. Proceedings, 1985. Edited by F. Ehlotzky. IX, 314 pages. 1985.